Cutting-Edge Strategies in Drug Delivery and Immunotherapy for Autoimmune Disorders

Young-Ok Son • Mrinmoy Ghosh
Editors

Cutting-Edge Strategies in Drug Delivery and Immunotherapy for Autoimmune Disorders

Springer

Editors
Young-Ok Son
Department of Animal Biotechnology, Faculty of Biotechnology, College of Applied Life Sciences
Jeju National University
Jeju, Korea (Republic of)

Mrinmoy Ghosh
Department of Animal Biotechnology, Faculty of Biotechnology, College of Applied Life Sciences
Jeju National University
Jeju, Korea (Republic of)

ISBN 978-981-95-0519-7 ISBN 978-981-95-0520-3 (eBook)
https://doi.org/10.1007/978-981-95-0520-3

This Springer imprint is published by the registered company Springer Nature Singapore Pte Ltd.
The registered company address is: 152 Beach Road, #21-01/04 Gateway East, Singapore 189721, Singapore

About This Book

This book is a specialized and timely compilation dedicated to advancing our understanding of autoimmune diseases, with a focused exploration of Rheumatoid Arthritis (RA), Type 1 Diabetes (T1D), and Multiple Sclerosis (MS), three of the most challenging and impactful autoimmune disorders. It brings together original research, case reports, and comprehensive reviews that dissect the multifactorial nature of autoimmunity through genetic, epigenetic, molecular, and environmental lenses. Structured to guide readers from foundational concepts to cutting-edge therapeutic innovations, this volume begins with a global overview of autoimmune diseases as a rising public health crisis. It then investigates the genetic predisposition and hereditary patterns that underpin disease susceptibility, alongside environmental triggers and their role in immune system dysregulation. The book also explores the broad spectrum of autoimmune manifestations from systemic to organ specific and dissects the molecular and cellular mechanisms driving these conditions. A particular strength of this book lies in its forward-looking chapters on therapeutic innovation. Readers will gain insights into current challenges and future directions in treatment, including computational modeling, the application of novel compounds, advanced drug delivery systems, and targeted next-generation nanotechnologies.

This book will serve as a valuable resource for researchers, clinicians, biomedical scientists, and graduate students seeking to deepen their knowledge of autoimmune diseases. It provides not only a solid scientific foundation but also highlights translational strategies that may inform clinical decision-making and inspire future research directions. By fostering interdisciplinary dialogue and showcasing emerging technologies, this volume aims to accelerate the development of personalized and more effective therapies ultimately contributing to better patient outcomes and improved quality of life for those afflicted by autoimmune conditions.

Contents

Chapter 1
Introduction to Autoimmune Diseases: A Global Health Challenge

Md. Meraj Ansari, Godagama Gamaarachchige Dinesh Suminda, Mrinmoy Ghosh, and Young-Ok Son

1.1 Introduction

A class of diseases known as autoimmune diseases occurs when the body's own cells and tissues are mistakenly attacked by the immune system (Li et al. 2024a; Song et al. 2024). Inflammation, tissue damage, and disturbances in body processes may result from this. There are more than 80 varieties of autoimmune diseases, each of which affects a distinct body component. The most prevalent are T1D, MS, and RA. Depending on the particular autoimmune disease, symptoms can vary greatly; however, frequent symptoms include fever, weight fluctuations, muscular aches, joint pain and swelling, exhaustion, and skin rashes. Although the precise causes of autoimmune diseases are unknown, a number of factors, including hormones, environmental factors, and genetics, are thought to play a role in their development. An increased risk may result from a family history of autoimmune diseases. Women are more likely to have several autoimmune diseases that may indicate hormonal

Md. M. Ansari · Y.-O. Son (✉)
Department of Animal Biotechnology, Faculty of Biotechnology, College of Applied Life Sciences, Jeju National University, Jeju, Korea (Republic of)

Interdisciplinary Graduate Program in Advanced Convergence Technology and Science, Jeju National University, Jeju-si, Republic of Korea
e-mail: ansarimm@jejunu.ac.kr; sounagi@jejunu.ac.kr

G. G. D. Suminda
Interdisciplinary Graduate Program in Advanced Convergence Technology and Science, Jeju National University, Jeju-si, Republic of Korea
e-mail: godagama@jejunu.ac.kr

M. Ghosh
Department of Animal Biotechnology, Faculty of Biotechnology, College of Applied Life Sciences, Jeju National University, Jeju, Korea (Republic of)

Y.-O. Son, M. Ghosh (eds.), *Cutting-Edge Strategies in Drug Delivery and Immunotherapy for Autoimmune Disorders*,
https://doi.org/10.1007/978-981-95-0520-3_1

components. Chronic inflammatory autoimmune diseases known as RA have a major effect on global health (Bhamble et al. 2024; Jori et al. 2024; Kumar et al. 2023). Approximately 18 million patients have RA worldwide. Women are more likely to have RA than men; approximately 70% of RA patients are female, and people over 55 are more likely to have RA. Perhaps because of greater average age, environmental pollutants, and lifestyle choices, RA is more common in developed nations (Finckh et al. 2022; Tobón et al. 2010). Chronic joint pain, stiffness, and edema caused by RA can significantly impair everyday activities and lower the quality of life (Ali et al. 2024; Vyawahare et al. 2023).

RA can lead to complications in other body systems, including the heart, lungs, and nervous system (Kim and Suh 2020; Misra 2024; Wu et al. 2022). The cost of managing RA, including medical treatment, rehabilitation, and loss of productivity, places a significant economic burden on individuals and healthcare systems. Although RA-related deaths have decreased over the past three decades, the number of cases continues to rise. In 2020, RA was responsible for approximately 3.6 million DALYs, with a higher burden in females.

1.1.1 Increasing Prevalence

The prevalence of RA is projected to continue to increase, with an estimated 31.7 million people living with RA by 2050. Addressing the global impact of RA requires improved diagnostic tools, effective treatment, and increased public awareness.

MS is a significant global health issue with a substantial impact on individuals and healthcare systems worldwide (Qian et al. 2023). There are more than 1.8 million MS patients with MS worldwide. It is more common in young adults, especially in those aged 20–40 years, and is more common in women. High-income areas like North America, Western Europe, and Australasia are more likely to have MS. Its prevalence is lower in regions such as sub-Saharan Africa. MS can lead to significant disability, affecting cognitive, emotional, motor, sensory, and visual function. Symptoms include vision problems, difficulty walking, numbness, and fatigue. The cost of managing MS, including medical treatment, rehabilitation, and loss of productivity, places a significant economic burden on individuals and the healthcare system (Qian et al. 2023). In 2019, approximately 22,439 deaths due to MS were reported worldwide (Qian et al. 2023). According to Disability-Adjusted Life Years (DALYs), which measure the number of years lost as a result of illness, disability, or early death, MS accounted for approximately 1.15 million DALYs in 2016. Over the past few decades, there has been an increase in the incidence, mortality, and DALYs associated with MS. This trend is expected to continue, particularly in areas with high sociodemographic indexes (SDI) (Qian et al. 2023). Better diagnostic methods, efficient therapies, and improved public awareness are all necessary to address the worldwide effects of MS.

T1D is a long-term autoimmune disease with major effects on global health (Addissouky et al. 2024; Popoviciu et al. 2023). There are approximately nine million people with T1D worldwide. Most of these people live in wealthy nations. Although it can happen at any age, T1D usually shows up during childhood or adolescence. If left untreated, T1D can cause serious side effects, such as renal failure, nerve damage, cardiovascular illness, and eyesight loss (Eid et al. 2023; Julián et al. 2024; Kahanovitz et al. 2017; Kropp et al. 2023; Wu et al. 2025). Diabetes has killed more than two million people worldwide in 2021. These figures are specifically influenced by T1D because of its complexities. Insulin, monitoring supplies, and medical services can be expensive for managing T1D. People and healthcare systems are severely impacted financially by this, particularly in low- and middle-income nations, where access to care is restricted. It is anticipated that the prevalence of T1D will continue to increase owing to environmental triggers and genetic predisposition. There are glaring disparities in T1D care around the world, with many people in low- and middle-income nations not having access to necessary drugs and treatment (Gregory et al. 2022). Better diagnostic technologies, greater public knowledge, and easier access to insulin and diabetes care are necessary to address the global effects of T1D. Complications can be avoided and symptoms can be managed with early diagnosis and prompt action. Because autoimmune disease have a wide range of symptoms, their diagnosis can be difficult. Complete medical history, radiological studies, and blood tests are frequently required. With the use of pharmaceuticals, including immunosuppressants, corticosteroids, nonsteroidal antiinflammatory drugs (NSAIDs), and biologics, treatment usually focuses on controlling symptoms and lowering inflammation. Balanced diet, consistent exercise, and stress reduction are examples of lifestyle modifications that can be extremely important in the management of autoimmune disease.

1.2 Prevalence of Autoimmune Diseases

Although the prevalence of autoimmune disease varies worldwide, a sizable fraction of the population is affected by them. In the United States, autoimmune disease affect approximately 3% of the population, translating to approximately ten million people. It is estimated that autoimmune disease impact 5–10% of the global population, with millions of people affected worldwide. RA is a common autoimmune disease that affects individuals worldwide (World Health Organization 2023). As of 2019, there were 18 million RA patients globally (World Health Organization 2023). Women make up about 70% of patients with RA, and 55% of them are over 55 years of age (World Health Organization 2023). Industrialized nations have a higher frequency, most likely because of lifestyle risk factors, environmental pollutants, and demographics. MS is a prevalent autoimmune disease with a significant global impact. According to the National Multiple Sclerosis Society, as of 2023, there were 2.9 million individuals with MS worldwide. High-income areas like North America, Western Europe, and Australasia have higher rates of MS (Qian et al. 2023). The prevalence is lower in regions such as sub-Saharan Africa; however, data collection in these areas may be less comprehensive. MS is approximately three times more common in females than in males. MS typically manifests in young adults, particularly those aged between 20 and 40 years.

The number of persons with MS has grown over time, rising from 2.3 million in 2013 to 2.8 million in 2020 and 2.9 million in 2023, according to the MS International Federation. Longer life expectancy and better diagnosis could be contributing factors to this increase. The central nervous system is affected by multiple sclerosis (MS), a chronic illness that causes a variety of symptoms such as fatigue, numbness, vision problems, and trouble walking. Early diagnosis and prompt action are essential for managing illness and enhancing quality of life. Type 1 diabetes (T1D) is a chronic autoimmune disease that affects millions of individuals globally. Globally, there are approximately nine million people with T1D. T1D prevalence varies according to geography, with high-income nations having higher rates. According to estimates, the prevalence of T1D varies between 0.8 and 4.6 cases per 1000 people worldwide, with the majority of cases occurring between 1 and 1.5 cases per 1000. Although it can happen at any age, T1D usually shows up during childhood or adolescence. Although T1D affects both sexes, some studies have indicated that men are slightly more likely to have it. Owing to a combination of environmental causes and genetic susceptibility, the incidence and prevalence of T1D have been increasing over time. Geographical location, ethnicity, diagnostic criteria, age, sex, and birth season all affect the occurrence of T1D. Insulin, monitoring devices, and medical services can all be expensive when controlling T1D, particularly in low- and middle-income nations, where access to treatment is restricted. Better diagnostic technologies, greater public knowledge, and easier access to insulin and diabetes care are necessary to address the global effects of type 1 diabetes. Complications can be avoided and symptoms can be managed with early diagnosis and prompt action. The frequency and prevalence of autoimmune disease have steadily increased over the past few decades, especially in Western nations. Women are more likely to suffer from many autoimmune disease than men. High-income areas, such as North America, Western Europe, and Asia, have higher rates of autoimmune disease (Guan et al. 2024; Li et al. 2023; Luo 2024). Although data gathering may be less thorough in locations such as sub-Saharan Africa, the prevalence is lower. Because autoimmune disease are chronic, require long-term care, and have high medical expenses, they represent a substantial public health burden.

1.3 Pathophysiology of Autoimmune Diseases

Complex interactions between genetic, immunological, and environmental factors are involved in the pathogenesis of autoimmune disease (Colafrancesco et al. 2013; Pisetsky 2023; Yasmeen et al. 2024). Autoimmune disease are more likely to develop in individuals with certain genotypes. Human leukocyte antigen (HLA) genes are closely linked to a number of autoimmune disease. The majority of autoimmune disease are polygenic, which means that several genes, each with a minor impact, increase risk. The immune system unintentionally attacks body tissues during autoimmune disease. Both the innate and adaptive immune systems may be involved in this process. Many autoimmune disease are characterized by the formation of autoantibodies or antibodies that target self-antigens (Elkon and Casali 2008; Ma et al. 2022). T-cell and B-cell dysregulation are key factors in the etiology of autoimmune disease (Wang and Zheng 2013; Wang et al. 2024; Xiao et al. 2022). In genetically predisposed individuals, environmental factors such as infections, stress, and exposure to specific chemicals can cause autoimmune disease. The term "exposome" describes the entirety of environmental exposures that occur from the moment of conception onward and can have an impact on the onset of autoimmune disease. A trigger (such as infection) that activates the immune system frequently initiates the disease process. Chronic inflammation and tissue damage result from persistent immunological activation. The afflicted tissues and organs gradually sustain damage because of continuous immunological reactions.

1.3.1 Pathophysiology of RA

Genetic, immunological, and environmental variables intricately interact with RA pathophysiology (Guo et al. 2018; Zamanpoor 2019). Human leukocyte antigen (HLA) genes increase the risk of RA (Medhasi and Chantratita 2022; van Drongelen and Holoshitz 2017). The risk is influenced by several genes, each of which has a minor impact. In RA, inflammation and tissue damage result from the immune system inadvertently attacking the body's own tissues. One characteristic feature of RA is the generation of autoantibodies or antibodies that target self-antigens. T cell and B cell dysregulation are key factors in the pathophysiology of RA. In people who are genetically predisposed, environmental factors, such as infections, stress, and exposure to specific chemicals, can cause RA to develop. The development of RA may be influenced by environmental exposure from conception onward. A trigger (such as infection) that activates the immune system frequently initiates the disease process. Chronic inflammation and tissue damage result from persistent immunological activation. The afflicted tissues and organs gradually sustain damage because of continuous immunological reactions. The synovium or joint lining thickens and becomes inflamed in RA, resulting in the formation of a tissue known as the pannus. The contiguous bone and cartilage were invaded and eroded by this pannus. Proteolytic enzymes and cytokines cause cartilage to lose its water content, resilience, and integrity. TNF-α, IL-1, and IL-6 are important cytokines that promote inflammation and joint degradation in RA (Kondo et al. 2021; McInnes et al. 2016) (Fig. 1.1). MMPs, or matrix metalloproteinases, facilitate the decomposition of cartilage and other joint tissues (Ali et al. 2024).

1.3.2 Pathophysiology of MS

Multiple sclerosis (MS) is caused by a complex interaction of environmental, genetic, and immunological factors (Miljković and Spasojević 2013; Sriwastava et al. 2024; Wu and Alvarez 2011). The hallmark of multiple sclerosis (MS) is an autoimmune reaction in which the immune system unintentionally targets the myelin sheath, which protects nerve fibers in the central nervous system (CNS)

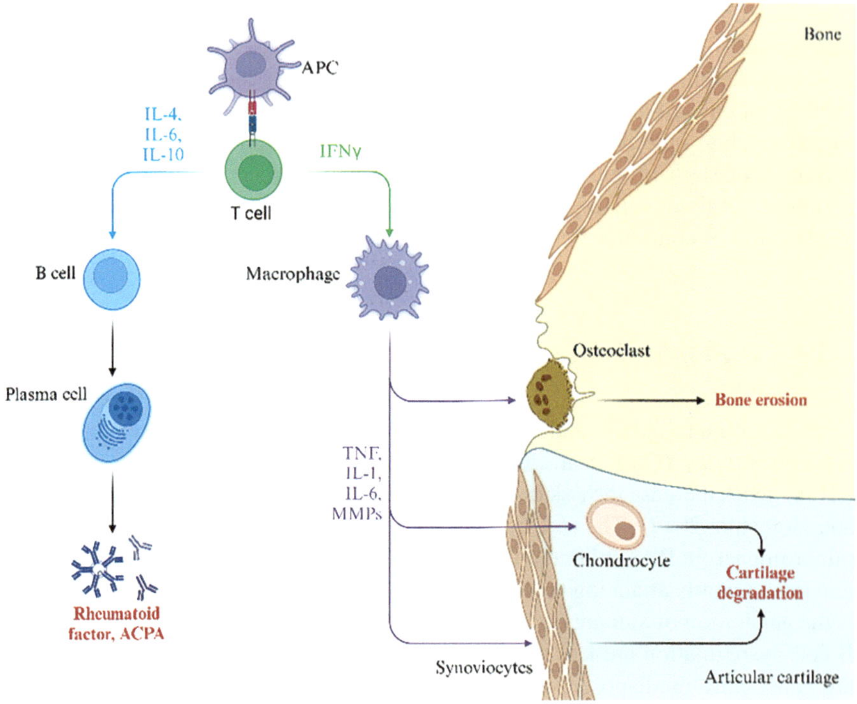

Fig. 1.1 Schematic representation of RA pathophysiology. (Image created with Biorender.com)

(Barkhane et al. 2022; Shah et al. 2023). Inflammation caused by the immune system's assault damages the myelin coating and nerve fibers underneath. T cells and B cells are examples of activated immune cells that enter the central nervous system and fuel inflammation (Fig. 1.2). The risk of multiple sclerosis is increased by certain genetic factors, such as the presence of particular human leukocyte antigen (HLA) genes. There may be a hereditary component of MS, as having a family member with the condition increases the risk. Reduced sun exposure can lower vitamin D levels, which are associated with a higher risk of multiple sclerosis (Grant et al. 2024; Hartley et al. 2015). In genetically predisposed individuals, some infections, especially those caused by specific viruses, may cause autoimmune reactions. Possibly as a result of variations in sun exposure and vitamin D levels, MS is more prevalent in temperate countries than in tropical countries. Myelin breakdown, resulting in plaques or lesions in the central nervous system, is the hallmark of multiple sclerosis. MS causes damage to nerve fibers in addition to demyelination, which can result in neurodegeneration. The blood–brain barrier does not function properly, permitting immune cells to infiltrate the central nervous system and induce inflammation. Intervals of neurological worsening followed by intervals of remission are characteristic of relapsing-remitting multiple sclerosis (RRMS) (Cree 2007; Siffrin et al. 2010). RRMS is the first presentation of secondary progressive MS (SPMS), which subsequently progresses to constant development of disability.

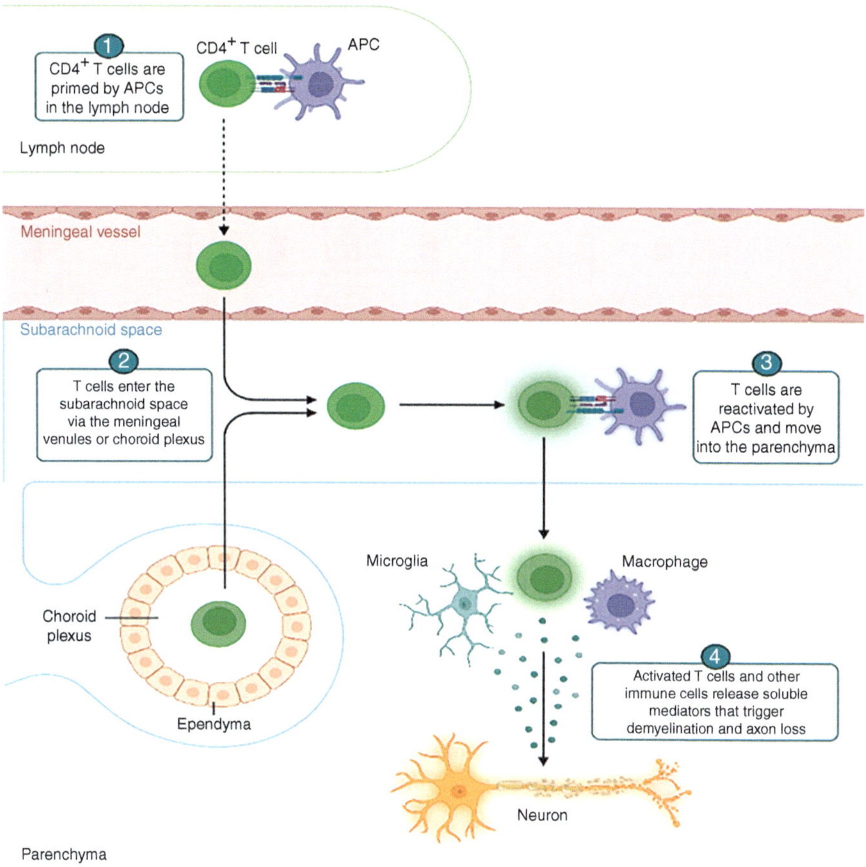

Fig. 1.2 Activation of central nervous system-reactive T cells in the pathogenesis of MS. (The images were created using Biorender.com)

The consistent progression of neurological impairment from the beginning is characteristic of primary progressive multiple sclerosis (PPMS).

1.3.3 Pathophysiology of T1D

Type 1 diabetes (T1D) is caused by complex interactions among environmental, immunological, and genetic factors (Ilonen et al. 2019; Zajec et al. 2022). Human leukocyte antigen (HLA) genes are among the genes that increase the risk of T1D. The risk is influenced by several genes, each of which has a minor impact. The hallmark of type 1 diabetes is an autoimmune reaction in which the immune system unintentionally targets pancreatic beta cells that produce insulin (James et al. 2023;

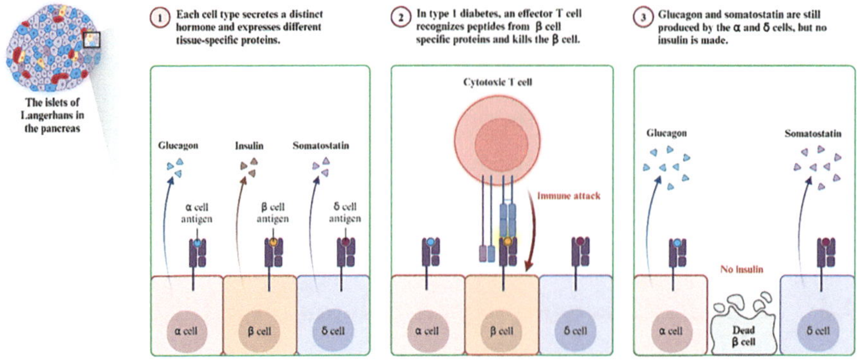

Fig. 1.3 Schematic representation of the immune response in T1D. (Image created with Biorender.com)

Zajec et al. 2022). T1D is characterized by the generation of autoantibodies or antibodies that target self-antigens. Dysregulation of T and B cells is a key factor in the pathophysiology of T1D (Fig. 1.3). In people who are genetically predisposed, environmental factors, such as infections, stress, and exposure to specific chemicals, can cause the development of T1D. The development of T1D may be influenced by environmental exposure from conception onward.

A trigger (such as infection) that activates the immune system frequently initiates the disease process. Chronic inflammation and tissue damage result from persistent immunological activation. The continuous reaction of the immune system gradually damages the beta cells that produce insulin. T1D causes a complete lack of insulin and raises blood sugar levels (hyperglycemia). Symptoms include exhaustion, unexplained weight loss, frequent urination (polyuria), and excessive thirst (polydipsia). Retinopathy, nephropathy, and neuropathy are some examples of microvascular complications. Heart disease and stroke are examples of macrovascular complications. Understanding the pathophysiology of autoimmune disease is essential to create tailored treatments and enhance patient outcomes.

1.4 Current Treatment Modalities

Autoimmune disease are complex and diverse; however, recent advancements have led to several promising treatment modalities (Table 1.1). Some of the current approaches are as follows:

Table 1.1 Current treatment strategies for autoimmune diseases

Autoimmune diseases	Categories of drugs	Drugs	Functions	References
Rheumatoid arthritis	Nonsteroidal antiinflammatory drugs (NSAIDs)	Ibuprofen and naproxen sodium	Ease pain and inflammation	Cox et al. (2025); Crofford (2013)
	Corticosteroids	Prednisone	Act quickly to control inflammation	Lim and Conn (2001)
	Disease-modifying antirheumatic drugs (DMARDs)	Methotrexate	Prevent joint and tissue damage	Lopez-Olivo et al. (2014)
	Biologics	Jakinibs or JAK inhibitors	Stop an overactive immune system, target specific steps in the inflammation process, and can slow, modify, or stop the disease	Rocha et al. (2021); Wang et al. (2022)
Multiple sclerosis	Corticosteroids	Oral prednisone and intravenous methylprednisolone	For the treatment of relapses	Alam et al. (1993); Horta-Hernández et al. (2019); Liu et al. (2017)
	Disease-modifying therapies (DMTs)	Interferon beta, glatiramer acetate, etc.	For relapsing-remitting MS, decrease inflammation and increase nerve growth	McGraw and Lublin (2013); Melendez-Torres et al. (2018)
	Bruton's tyrosine kinase (BTK) inhibitor	Evobrutinib, tolebrutinib, fenebrutinib (they are under clinical trial)	Altering the function of B cells, curtail disease progression by targeting immune cells	Krämer et al. (2023); Shulga et al. (2023)
Type 1 diabetes	Adjunctive therapies for T1D	Pramlintide (analogue of amylin), metformin	Marginal benefits at best in lowering A1c levels in T1D	Nally et al. (2019)
		Insulin	Improve glycemic control	Bolli (2006); Janež et al. (2020)

1.4.1 Immunosuppressive Drugs

Medications such as corticosteroids and methotrexate help reduce inflammation by suppressing the immune system. By weakening or inhibiting the immune system, immunosuppressive medications help prevent the body's tissues from being attacked (Meneghini et al. 2021; Stucker and Ackermann 2011; Wiseman 2016). To lower inflammation and inhibit immunological responses, medications, such as dexamethasone, methylprednisolone, and prednisone are frequently used. Multiple sclerosis, rheumatoid arthritis, and lupus can be treated effectively. However, they might not be appropriate for long-term use and can have serious adverse effects.

1.4.2 Biologic Therapies

To regulate autoimmune disease and reduce inflammation, biological therapies, also referred to as biological response modifiers, target specific immune system components (Johnston 2007; Sapkota et al. 2025). When traditional treatments, including disease-modifying antirheumatic drugs (DMARDs), are ineffective, these therapies are usually employed. These proteins are genetically modified to target specific components of the immune system. TNF inhibitors (adalimumab and infliximab) and interleukin inhibitors (anakinra and tocilizumab) are examples. Numerous autoimmune disease such as Crohn's disease, psoriasis, and rheumatoid arthritis are treated with biologics. Biologic therapies have drawbacks, including high costs, possible side effects, and the inconvenience of administration, even though they can have a major positive therapeutic impact and enhance patients' quality of life.

1.4.3 Targeted Synthetic DMARDs

A more recent class of drugs, called targeted synthetic disease-modifying antirheumatic drugs (DMARDs), targets specific immune response pathways (Frisell et al. 2023; Pang et al. 2022). Unlike traditional DMARDs, which have a broader impact on the immune system, targeted synthetic DMARDs offer a more precise approach for treating autoimmune disease. Small molecules, such as JAK inhibitors, selectively target specific pathways involved in the immune response. They offer a more refined approach than traditional immunosuppressants do. Compared with conventional DMARDs, targeted synthetic DMARDs have fewer adverse effects and a quicker onset of action.

1.4.4 Precision Cellular Therapies

New therapies are being developed to treat autoimmune disease, such as CAR T-cell therapy, which has shown promise for the treatment of some types of cancer (Li et al. 2024b; Schett et al. 2024). Chimeric autoantigen T-cell receptors (CATCR) have been developed to specifically eradicate self-reactive immune cells. By focusing on harmful immune cells, chimeric antigen receptor (CAR) T-cell therapy has been modified to treat autoimmune disease (Chung et al. 2024; Finkel and Radic 2023). Researchers are using gene editing to engineer T cells that can specifically eliminate self-reactive immune cells, offering a potential long-term solution without the need for continuous immunosuppression. Gene-edited T cells from healthy donors are used in allogeneic CAR T-cell therapy, which has several benefits over autologous CAR T-cell therapy, including instant availability and wider access. This method could reduce the cost and complexity of treatment, making it more accessible to patients

1.4.5 Lifestyle Modifications

Lifestyle changes can greatly enhance the quality of life of people with autoimmune disease and are essential for their management. Diet, exercise, and stress reduction may be very important to effectively treat autoimmune disease. Inflammation can be reduced by consuming meals rich in omega-3 fatty acids, antioxidants, and other antiinflammatory nutrients (Stromsnes et al. 2021; Wierenga and Pestka 2021). Fruits, vegetables, whole grains, lean meat, and healthy fats were included in this study. Understanding and avoiding these triggers are crucial because some meals may worsen the symptoms. Dairy products, processed meals, and gluten are the most common offenders. Maintaining adequate hydration can aid in symptom management and is crucial for general health. Walking, swimming, or yoga are examples of moderate exercise that can help lower inflammation, elevate mood, and improve general wellbeing. Increasing muscle strength helps ease pain associated with diseases such as rheumatoid arthritis and supports the joints. Deep breathing techniques, yoga, and mindfulness meditation are a few techniques that can help lower stress and enhance mental health. Consulting a mental health specialist can help them cope with chronic illness-related stress and anxiety. The performance of the immune system and general health depend on sufficient sleep. Improving sleep quality can be achieved by establishing a consistent sleep schedule and providing a comfortable sleeping space. Both symptoms and general health can be significantly improved by smoking cessation. Drinking too much alcohol may worsen symptoms and conflict with prescription drugs. These nonpharmacological methods can aid in relieving symptoms and enhancing general health.

1.4.6 Emerging Therapies

To provide more efficient and individualized treatments for autoimmune disease, ongoing research is investigating novel therapeutic targets and strategies, such as gene editing and personalized medicine. One innovative method that has a lot of potential for treating autoimmune disease is gene editing. With the help of this groundbreaking gene editing technology, scientists can precisely alter the genome and identify and fix genetic abnormalities that cause autoimmune disease. CRISPR/Cas9 can help modify immune responses and possibly create immune tolerance by modifying the genes involved in immune cell activation and control. By focusing on harmful immune cells, chimeric antigen receptor (CAR) T-cell therapy is being modified to treat autoimmune disease. To provide viable long-term treatment without the need for ongoing immunosuppression, researchers have employed gene editing to create T cells that can specifically destroy self-reactive immune cells. Precision medicine, which seeks to customize treatments for each patient according to their unique genetic and molecular profiles, includes gene editing as a fundamental element. This strategy may result in more individualized and successful autoimmune disease treatment. By lowering symptoms and stopping the disease course, these therapies seek to enhance the quality of life of those who suffer from autoimmune disease.

1.5 Nanotechnology in Autoimmune Disease Treatment

Utilizing the special qualities of materials and technology at the nanoscale, nanomedicine is transforming the treatment of autoimmune disease. Therapy for autoimmune disease is seeing promising new developments because of nanotechnology. The following are some significant developments and possible uses.

1.5.1 Targeted Drug Delivery

Targeted drug delivery is a cutting-edge technique that delivers medication directly to a particular location in the body, increasing the concentration of the drug while reducing its presence in nontarget locations. Drugs can be directly delivered by nanoparticles to specific immune cells that are implicated in autoimmune reactions. This strategy increases treatment effectiveness and reduces adverse effects. This strategy seeks to reduce the adverse effects and increase the effectiveness of the drug. Targeted drug delivery methods convey the medication to the intended site using carriers, such as polymer-drug conjugates, liposomes, or nanoparticles. When certain stimuli, such as pH shifts or enzymes in the target region, occur, these carriers can be used to release the medication. Notwithstanding its benefits, tailored drug delivery has drawbacks, including expensive development expenses, possible carrier toxicity, and difficulty in creating systems that can precisely target particular tissues.

1.5.2 Nanoparticle-Based Immunotherapy

An inventive method called nanoparticle-based immunotherapy boosts the body's immune response against disease, such as cancer, by using nanoparticles to deliver immunotherapeutic drugs directly to target cells. It is possible to design nanoparticles to deliver adjuvants, immune-modulating substances, or antigens to particular cells or organs. In response to particular stimuli, such as pH shifts or enzymes found in the tumor microenvironment, they can release their payload. Scientists are creating nanoparticles that can deactivate specific immune cells such as T and B cells, which attack the body's own tissues. Without complete inhibition of the immune system, this approach provides a more accurate means of managing autoimmune reactions. Notwithstanding its promise, nanoparticle-based immunotherapy has drawbacks such as the difficulty of creating nanoparticles that can precisely target particular tissues and release their payload. To prevent negative consequences, any harmful ingredients utilized to make nanoparticles must be biocompatible.

1.5.3 Inducing Immune Tolerance

By educating the immune system to ignore particular antigens, nanoparticles can be used to induce immunological tolerance, which is a possible treatment for autoimmune disease and allergies. Antigens and immunomodulatory substances can be delivered to antigen-presenting cells (APCs) via nanoparticles in a manner that encourages immunological tolerance as opposed to an immune response. Presenting the antigen without costimulatory cues that often set off an immune response is one way to do this. By exposing antigens in a way that trains the immune system to identify them as self-reducing autoimmune attacks, nanoparticles can be utilized to create immunological tolerance. This strategy is being investigated for several uses, including managing allergies, treating autoimmune conditions such as multiple sclerosis and rheumatoid arthritis, and preventing allograft rejection. Although promising, this strategy has several drawbacks, including the need to guarantee nanoparticle biocompatibility, achieve targeted distribution, and prevent unforeseen immunological reactions.

1.5.4 Enhanced Diagnostic Tools

Early detection and improved monitoring of autoimmune disease are made possible by the ability of nanotechnology to increase the sensitivity and accuracy of diagnostic instruments. Particularly for autoimmune disease, nanoparticles are transforming the diagnostic industry. High sensitivity and specificity are provided by nanoparticles, such as quantum dots (QDs) and gold nanoparticles (AuNPs), for the detection of biomarkers linked to autoimmune disease. Multiple detection molecules can be attached to them owing to their large surface area and compact size, which increases test accuracy. Autoantibodies, cytokines, and inflammatory indicators can be detected by nanoparticles in extremely low quantities, frequently prior to the onset of clinical symptoms. Better management of autoimmune disease and prompt intervention depend on early identification. The necessity for invasive procedures, such as biopsies, can be decreased by using nanoparticles in noninvasive diagnostic methods, such as imaging and biosensors. For instance, magnetic nanoparticles can be utilized as MRI contrast agents to visualize tissue damage and inflammation. Multiple biomarkers can be detected simultaneously in a single test thanks to nanoparticles. More precise diagnoses can result from a thorough examination of the illness condition, made possible by this multiplexing capacity. Point-of-care testing is made possible by the incorporation of nanoparticles into portable diagnostic instruments, which facilitates the diagnosis of autoimmune disease in settings with limited resources or in remote locations.

1.5.5 Gene Editing

By delivering gene-editing tools, such as CRISPR-Cas9, to particular cells, nanoparticles may be able to fix genetic abnormalities that lead to autoimmune disease. Within cells, CRISPR-Cas9 enables the precise editing of particular genes. Researchers hope to reduce or completely eradicate abnormal immune reactions by focusing on and altering the genes that cause autoimmune reactions. Numerous genes, particularly those involved in immune regulation and inflammatory responses, have been identified as potential targets for gene editing in autoimmune disease. For instance, research has been conducted on genes linked to multiple sclerosis, inflammatory bowel disease, rheumatoid arthritis, systemic lupus erythematosus, and type 1 diabetes. Even though gene editing is promising, there are obstacles to overcome, such as ensuring that the adjustments are safe and specific, preventing unintended consequences, and handling ethical issues regarding genetic alterations. Although these developments are still in the research and development stages, they have enormous potential to lead to more individualized and successful autoimmune disease treatments in the future.

1.6 Advancements in the Treatment of Autoimmune Diseases

Recently, several promising developments have been made in the management of autoimmune disease (Mallapaty 2024). Significant progress has been made in the treatment of autoimmune disease, raising the prospect of individualized and successful treatments. The following are some of the more encouraging developments.

1.6.1 Outsmarting the Immune System

In autoimmune disease, outwitting the immune system entails creative methods to alter or rewire the immune response as opposed to merely inhibiting it. Researchers are exploring ways to reprogram immune cells rather than suppress them. Recent advances in immunotherapy have aimed to reprogram immune cells to prevent them from attacking healthy tissues. This approach is being explored for type 1 diabetes, where researchers target the misguided immune response that destroys insulin-producing beta cells. A new subject called psychoneuroimmunology examines the connections between the immune system, neurological system, and psychological factors. Positive mental states such as mindfulness and relaxation may strengthen the immune system and alter the course of autoimmune disease. Adopting a healthy diet and lifestyle can help in managing autoimmune disease. Reducing stress, improving gut health, and maintaining a balanced diet can positively affect immune function and overall well-being. By preventing the immune system from targeting healthy cells, this strategy seeks to address the underlying cause of autoimmune disease, including type 1 diabetes.

1.6.2 Precision Medicine with AI

AI is used in medicine to enhance autoimmune disease-tailored treatments. AI-powered treatment solutions include improving medication development, finding biomarkers, improving diagnostics, and customizing treatment plans. AI can analyze vast amounts of patient data to improve medication development and customize treatments for specific individuals. More effective and individualized therapies are being developed using precision medicine techniques to customize treatments according to the unique characteristics of each patient.

1.6.3 Cellular Therapies

Developments in cellular therapy, including CAR T and TCR T-cell therapies, are being investigated for autoimmune disease. These treatments modify the immune response by focusing on specific chemicals found in autoreactive B and T cells. This cutting-edge treatment treats autoimmune disease using genetically modified donor immune cells (Mallapaty 2024). Patients in early trials reported better symptoms and remission, which is encouraging (Mallapaty 2024).

1.6.4 Partnerships for Precision Medicine

There have been exciting partnerships aimed at advancing precision medicine for autoimmune disease. One notable collaboration is between OMNY Health and Scientific Medicine. This partnership integrates transcriptomic data from Scipher's PrismRA Test into OMNY's extensive electronic health record (EHR) network, which includes data from over 80 million patients. This integration aims to provide deeper patient insights and drive the development of more targeted therapies.

1.6.5 Therapeutic Proteins

The therapeutic protein market is growing rapidly, with biologics, such as monoclonal antibodies and cytokines, offering precise and effective treatment options for autoimmune disease. There is optimism for better patient outcomes as a result of these developments, which are opening the door for more focused and efficient treatment of autoimmune disease. Millions of people suffering from autoimmune disease now have hope, thanks to these developments, which are opening the door to more individualized and efficient treatments.

1.7 Future Direction of Autoimmune Research

The future of autoimmune research is incredibly promising and is expected to bring significant advancements in the understanding, diagnosis, and treatment of these complex disease. The key areas of focus are as follows.

1.7.1 Precision Medicine

The goal of precision medicine for autoimmune disease is to customize the care for each patient according to their particular genetic, environmental, and lifestyle characteristics. Tailored Therapy Programs Clinicians can create individualized therapy programs that target particular pathways implicated in autoimmune disease by examining a patient's genetic composition. Identifying certain biomarkers aids in early autoimmune disease diagnosis and therapy response tracking. Highly specialized medicines are being designed using cutting-edge technologies, such as CRISPR-Cas9 gene editing and AI-driven methods. The effectiveness and safety of precision medicine techniques for autoimmune disease are currently being investigated in ongoing clinical trials. Precision medicine treatments are being developed more quickly owing to collaborations between research institutes and healthcare organizations. Researchers are attempting to develop individualized treatments based on a person's genetic and molecular profiles. This strategy seeks to improve treatment efficacy and minimize the adverse effects of customizing medicines for each patient's particular traits.

1.7.2 Genetics and Environment

Complex interactions between environmental and genetic factors influence the development of autoimmune disease (Ellis et al. 2014). The risk of autoimmune disease may be elevated by genetic predisposition (Ramos et al. 2015; Taneja 2020). For example, several autoimmune disease are closely linked to the human leukocyte antigen (HLA) locus. The majority of autoimmune disease are polygenic, which means that several genes, each with a minor impact, increase the risk. There is a hereditary component of many disease because they frequently run in families. In people who are genetically predisposed, environmental factors, including infections, smoking, exposure to specific chemicals, and even stress, can cause autoimmune reactions. Diet, exercise, and other lifestyle decisions can affect the development and course of autoimmune disease. Immune system regulation is greatly influenced by the gut microbiota, and dysbiosis or imbalances in this microbiota can lead to autoimmune disease. Personalized treatment plans and preventive measures can be developed with the aid of understanding how these factors interact. It is essential to understand how environmental influences and genetic predispositions interact with each other. Research is being conducted on the relationship between autoimmune disease and lifestyle choices, infections, and genetic factors.

1.7.3 Early Biomarkers

For autoimmune disease, early biomarkers are essential for prompt identification and treatment to avoid irreparable tissue damage and enhance patient outcomes. HLA-DR4 is a genetic marker linked to autoimmune disease that can aid in early identification (Bodis et al. 2018; Gough and Simmonds 2007). Before clinical symptoms appear, certain autoantibodies, such as anticyclic citrullinated peptide (anti-CCP) for rheumatoid arthritis and antinuclear antibodies (ANA) for lupus, can signal the existence of an autoimmune disease. Increased levels of cytokines, including tumor necrosis factor-alpha (TNF-α) and interleukin-6 (IL-6), can act as biomarkers for autoimmune and inflammatory processes (Aliyu et al. 2022; Kany et al. 2019; Tanaka et al. 2014). Some miRNAs, such as miR-155, can serve as early markers and are associated with autoimmune reactions (Cichalewska-Studzinska et al. 2024; O'Connell et al. 2010; Pashangzadeh et al. 2021). Immune dysregulation and early disease processes may be reflected by changes in certain metabolites in the blood or other bodily fluids. These biomarkers are being studied to assist customized therapy, improve early intervention, and increase diagnostic accuracy. However, issues such as validation gaps and disease heterogeneity need to be resolved. One of the main areas of research is identifying biomarkers that can forecast the development and course of the disease. Timely interventions and improved outcomes can result from early identification and management of autoimmune disease.

1.7.4 Targeted Therapies

Compared with conventional treatments, targeted medications for autoimmune disease provide a more sophisticated approach by precisely modifying specific immune system pathways or components. Proteins that are genetically modified from human genes are known as biologics. They target specific immune system components that cause inflammation, including B cells, interleukins (IL-6, IL-17, and IL-23), and tumor necrosis factor (TNF). Janus kinase (JAK) inhibitors are small-molecule inhibitors that target certain enzymes or signaling pathways involved in the immune response. Monoclonal antibodies, such as anti-CD20 for B-cell depletion therapy, target certain antigens on immune cells or inflammatory mediators. By genetically altering a patient's T cells, CAR T-cell therapy aims to target and eliminate particular cells implicated in autoimmune reactions. Costimulating chemical inhibitors regulate the immune response by targeting molecules necessary for T-cell activation and function. The goal of these medications is to maximize their effectiveness while reducing adverse effects, and research is still being conducted to identify novel targets and improve current medications.

1.7.5 Immunotherapy

Immunotherapy is one possible method for treating autoimmune disease (1). By modifying the immune system, it reduces its overactivity, which is the main cause of autoimmune disease. By retraining the immune system to stop attacking the body's own tissues or inhibiting its activity, immunotherapy can help control the immune system. Checkpoint inhibitors, cytokine therapy, and monoclonal antibodies are several forms of immunotherapy. It is being investigated for the treatment of inflammatory bowel disease, multiple sclerosis, lupus, and rheumatoid arthritis. The immune response must be balanced to prevent overstretching, which can result in infections or other problems. The safety and effectiveness of immunotherapy in patients with autoimmune disease have been studied in ongoing clinical trials. By modifying the immune response, immunotherapy innovations, such as checkpoint inhibitors and CAR T-cell therapy, are being modified to treat autoimmune disease.

1.7.6 Gene Editing

Gene editing, specifically CRISPR-Cas9 technology, offers a novel treatment option for autoimmune disease (Lee et al. 2022). Certain genes within the genome can be precisely edited using CRISPR-Cas9 technology. It has shown promise in altering gene expression and correcting genetic mutations. Targeting particular genes implicated in autoimmune reactions by gene editing may reduce or even eradicate unwarranted immune responses. Although promising, research on gene editing for autoimmune disease remains in its infancy. Addressing possible off-target consequences and guaranteeing treatment safety and effectiveness are challenging. The goal of ongoing research is to improve gene editing methods and fully investigate their potential in the treatment of autoimmune disease. Potential long-term remedies are being investigated to fix the genetic abnormalities that cause autoimmune disease using methods such as CRISPR-Cas9.

1.7.7 Microbiome Research

Research on the microbiome has revealed intriguing new information regarding autoimmune disease. The population of bacteria that reside in the intestine, known as the gut microbiome, is essential for preserving immunological homeostasis and balance (Joos et al. 2024; Ma et al. 2024). Autoimmune disease, such as rheumatoid arthritis, systemic lupus erythematosus, and type 1 diabetes, have been associated with dysbiosis or imbalances in the gut flora. The development and progression of autoimmune disease can be influenced by intricate relationships between the host immune system and gut bacteria (Jin et al. 2025; Song et al. 2024). As possible treatments for autoimmune disease, researchers are investigating microbiome-based therapeutics, including pharmacomicrobiomics and fecal microbiota transplantation (FMT). However, it is still difficult to use these discoveries in clinical settings, and further studies are required to identify reliable microbial targets and create efficient therapies. Novel therapeutic approaches and customized medicine may be made possible by understanding the involvement of microbiota in autoimmune disease. Millions of people suffering from autoimmune disease now have hope, thanks to these developments, which are opening the door to more individualized and efficient treatments.

1.8 Conclusion

Millions of people worldwide suffer from autoimmune disease, which are a serious global health concern that places a heavy strain on both individuals and the healthcare system. These conditions cause a variety of symptoms and consequences when the body's tissues are mistakenly attacked by the immune system. Rheumatoid arthritis, multiple sclerosis, and type 1 diabetes are examples of autoimmune disease. Over 80 different forms of autoimmune disease have been identified, and they are thought to affect 5–10% of the world's population. These disease are more common in women, and certain autoimmune disease are more prevalent in specific age groups and geographical regions. Autoimmune disease are chronic and often lead to long-term health issues, including pain, disability, and decreased quality of life. Individuals with autoimmune disease may experience additional health problems such as cardiovascular disease, infections, and mental health issues. Managing autoimmune disease involves significant healthcare expenses, including medications, doctor visits, and hospitalization. Chronic disease can lead to reduced work productivity, early retirement, and increased absenteeism, affecting both individuals and economies. Typical symptoms include fatigue, joint pain, skin rashes, and fever. Owing to their varied appearances, many disease can be difficult to diagnose. Hormonal effects, environmental factors, and genetics all play a role, although their precise causes are unknown. Autoimmune disease affect people's everyday lives and productivity and cause substantial morbidity, mortality, and financial expenses. Most autoimmune disease have no known cure, although therapies aim to control symptoms and lower inflammation by using drugs and altering lifestyle choices. More research, better diagnostics, efficient therapies, and higher public awareness are necessary to combat autoimmune disease. The quality of life of individuals affected by these difficult diseases can be enhanced by encouraging early diagnosis and prompt care.

Conflicts of Interest The authors declare that they have no competing interests.

References

Addissouky TA, Ali MMA, El Tantawy I, Sayed E, Wang Y (2024) Type 1 diabetes mellitus: retrospect and prospect. Bull Natl Res Cente 48:42

Alam SM, Kyriakides T, Lawden M, Newman PK (1993) Methylprednisolone in multiple sclerosis: a comparison of oral with intravenous therapy at equivalent high dose. J Neurol Neurosurg Psychiatry 56:1219–1220

Ali A, Rahul, Jori C, Kumar J, Kumar A, Kanika, Ansari MM, Ahmad A, Ali N, Yadav P, Parvez S, Navik U, Son YO, Khan R (2024) Sinapic acid-pullulan based inflammation responsive nanomicelles for the local treatment of experimental inflammatory arthritis. Int J Biol Macromol 278:134903

Aliyu M, Zohora FT, Anka AU, Ali K, Maleknia S, Saffarioun M, Azizi G (2022) Interleukin-6 cytokine: an overview of the immune regulation, immune dysregulation, and therapeutic approach. Int Immunopharmacol 111:109130

Barkhane Z, Elmadi J, Satish Kumar L, Pugalenthi LS, Ahmad M, Reddy S (2022) Multiple sclerosis and autoimmunity: a veiled relationship. Cureus 14:e24294

Bhamble S, Dighe S, Katari O, Yadav V, Ansari MM, Jain S (2024) Ionic liquid-based Ionogel: a novel strategy to improve dermal delivery and in vivo efficacy of etodolac in rheumatoid arthritis management. J Mol Liq 412:125841

Bodis G, Toth V, Schwarting A (2018) Role of human leukocyte antigens (HLA) in autoimmune diseases. Rheumatol Ther 5:5–20

Bolli GB (2006) Insulin treatment in type 1 diabetes. Endocr Pract 12(Suppl 1):105–109

Chung JB, Brudno JN, Borie D, Kochenderfer JN (2024) Chimeric antigen receptor T cell therapy for autoimmune disease. Nat Rev Immunol 24:830–845

Cichalewska-Studzinska M, Szymanski J, Stec-Martyna E, Perdas E, Studzinska M, Jerczynska H, Kulczycka-Wojdala D, Stawski R, Mycko MP (2024) The role of miR-155 in modulating gene expression in CD4+ T cells: insights into alternative immune pathways in autoimmune encephalomyelitis. Int J Mol Sci 25:11355

Colafrancesco S, Agmon-Levin N, Perricone C, Shoenfeld Y (2013) Unraveling the soul of autoimmune diseases: pathogenesis, diagnosis and treatment adding dowels to the puzzle. Immunol Res 56:200–205

Cox N, Mallen CD, Scott IC (2025) Pharmacological pain management in patients with rheumatoid arthritis: a narrative literature review. BMC Med 23:54

Cree B (2007) 20 – Demyelinating and inflammatory disorders of the central nervous system. In: Marshall RS, Mayer SA (eds) On call neurology, 3rd edn. W.B. Saunders, pp 266–293

Crofford LJ (2013) Use of NSAIDs in treating patients with arthritis. Arthritis Res Ther 15(Suppl 3):S2

Eid SA, Rumora AE, Beirowski B, Bennett DL, Hur J, Savelieff MG, Feldman EL (2023) New perspectives in diabetic neuropathy. Neuron 111:2623–2641

Elkon K, Casali P (2008) Nature and functions of autoantibodies. Nat Clin Pract Rheumatol 4:491–498

Ellis JA, Kemp AS, Ponsonby AL (2014) Gene-environment interaction in autoimmune disease. Expert Rev Mol Med 16:e4

Finckh A, Gilbert B, Hodkinson B, Bae SC, Thomas R, Deane KD, Alpizar-Rodriguez D, Lauper K (2022) Global epidemiology of rheumatoid arthritis. Nat Rev Rheumatol 18:591–602

Finkel TH, Radic M (2023) Chimeric receptors broaden the therapeutic landscape for autoimmune disease. Nat Rev Rheumatol 19:327–328

Frisell T, Bower H, Morin M, Baecklund E, Di Giuseppe D, Delcoigne B, Feltelius N, Forsblad-d'Elia H, Lindqvist E, Lindström U, Askling J (2023) Safety of biological and targeted synthetic disease-modifying antirheumatic drugs for rheumatoid arthritis as used in clinical practice: results from the ARTIS programme. Ann Rheum Dis 82:601–610

Gough SC, Simmonds MJ (2007) The HLA region and autoimmune disease: associations and mechanisms of action. Curr Genomics 8:453–465

Grant WB, Bhattoa HP, Pludowski P (2024) Chapter 56 – Determinants of vitamin D levels from sun exposure: a global perspective. In: Hewison M, Bouillon R, Giovannucci E, Goltzman D, Meyer M, Welsh J (eds) Feldman and pike's vitamin D, 5th edn. Academic Press, pp 97–113

Gregory GA, Robinson TIG, Linklater SE, Wang F, Colagiuri S, de Beaufort C, Donaghue KC, Harding JL, Wander PL, Zhang X, Li X, Karuranga S, Chen H, Sun H, Xie Y, Oram R, Magliano DJ, Zhou Z, Jenkins AJ, Ma RCW, Magliano DJ, Maniam J, Orchard TJ, Rai P, Ogle GD (2022) Global incidence, prevalence, and mortality of type 1 diabetes in 2021 with projection to 2040: a modelling study. Lancet Diabetes Endocrinol 10:741–760

Guan S-Y, Zheng J-X, Feng X-Y, Zhang S-X, Xu S-Z, Wang P, Pan H-F (2024) Global burden due to modifiable risk factors for autoimmune diseases, 1990–2021: temporal trends and socio-demographic inequalities. Autoimmun Rev 23:103674

Guo Q, Wang Y, Xu D, Nossent J, Pavlos NJ, Xu J (2018) Rheumatoid arthritis: pathological mechanisms and modern pharmacologic therapies. Bone Res 6:15

Hartley M, Hoare S, Lithander FE, Neale RE, Hart PH, Gorman S, Gies P, Sherriff J, Swaminathan A, Beilin LJ, Mori TA, King L, Black LJ, Marshall K, Xiang F, Wyatt C, King K, Slevin T, Pandeya N, Lucas RM (2015) Comparing the effects of sun exposure and vitamin D supplementation on vitamin D insufficiency, and immune and cardio-metabolic function: the Sun Exposure and Vitamin D Supplementation (SEDS) study. BMC Public Health 15:115

Horta-Hernández AM, Esaclera-Izquierdo B, Yusta-Izquierdo A, Martín-Alcalde E, Blanco-Crespo M, Álvarez-Nonay A, Torralba M (2019) High-dose oral methylprednisolone for the treatment of multiple sclerosis relapses: cost-minimisation analysis and patient's satisfaction. Eur J Hosp Pharm 26:280–284

Ilonen J, Lempainen J, Veijola R (2019) The heterogeneous pathogenesis of type 1 diabetes mellitus. Nat Rev Endocrinol 15:635–650

James EA, Joglekar AV, Linnemann AK, Russ HA, Kent SC (2023) The beta cell-immune cell interface in type 1 diabetes (T1D). Mol Metabol 78:101809

Janež A, Guja C, Mitrakou A, Lalic N, Tankova T, Czupryniak L, Tabák AG, Prazny M, Martinka E, Smircic-Duvnjak L (2020) Insulin therapy in adults with type 1 diabetes mellitus: a narrative review. Diabetes Ther 11:387–409

Jin J, Cai X, Rao P, Xu J, Li J (2025) Microbiota and immune dynamics in rheumatoid arthritis: mechanisms and therapeutic potential. Best Pract Res Clin Rheumatol 102035

Johnston SL (2007) Biologic therapies: what and when? J Clin Pathol 60:8–17

Joos R, Boucher K, Lavelle A, Arumugam M, Blaser MJ, Claesson MJ, Clarke G, Cotter PD, De Sordi L, Dominguez-Bello MG, Dutilh BE, Ehrlich SD, Ghosh TS, Hill C, Junot C, Lahti L, Lawley TD, Licht TR, Maguin E, Makhalanyane TP, Marchesi JR, Matthijnssens J, Raes J, Ravel J, Salonen A, Scanlan PD, Shkoporov A, Stanton C, Thiele I, Tolstoy I, Walter J, Yang B, Yutin N, Zhernakova A, Zwart H, Asnicar F, Typas A, Betsou F, Blottière H, Bork P, Boutron I, Carraturo F, Claesson M, Cordaillat-Simmons M, Druart C, Fasano A, Godoy Y, Haller D, Hassani Z, Hazenbrink DHMJ, Israelsen M, Iyappan A, Jarde A, Kampshoff S, Krag A, Kriaa A, Lavelle A, Metwaly A, Morozova V, Pinto F, Pons N, Prost P-L, Ravaud P, Rhimi M, Rodriguez J, Sarati A, Schierwagen R, Segata N, Serra D, Trebicka J, Vecchi C, Veiga P, Zitvogel L, Derosa L, Doré J, Ross RP, Human Microbiome Action C (2024) Examining the healthy human microbiome concept. Nat Rev Microbiol

Jori C, Ansari MM, Ahmad A, Ali N, Raza SS, Khan R (2024) Biomaterial-based combinatorial approach of aescin-comprised zein-coated gelatin nanoparticles alleviates synovial inflammation in experimental inflammatory arthritis. Nanoscale 16:7965–7975

Julián MT, Pérez-Montes de Oca A, Julve J, Alonso N (2024) The double burden: type 1 diabetes and heart failure-a comprehensive review. Cardiovasc Diabetol 23:65

Kahanovitz L, Sluss PM, Russell SJ (2017) Type 1 diabetes – a clinical perspective. Point Care 16:37–40

Kany S, Vollrath JT, Relja B (2019) Cytokines in inflammatory disease. Int J Mol Sci 20

Kim JW, Suh CH (2020) Systemic manifestations and complications in patients with rheumatoid arthritis. J Clin Med 9

Kondo N, Kuroda T, Kobayashi D (2021) Cytokine networks in the pathogenesis of rheumatoid arthritis. Int J Mol Sci 22
Krämer J, Bar-Or A, Turner TJ, Wiendl H (2023) Bruton tyrosine kinase inhibitors for multiple sclerosis. Nat Rev Neurol 19:289–304
Kropp M, Golubnitschaja O, Mazurakova A, Koklesova L, Sargheini N, Vo TKS, de Clerck E, Polivka J Jr, Potuznik P, Polivka J, Stetkarova I, Kubatka P, Thumann G (2023) Diabetic retinopathy as the leading cause of blindness and early predictor of cascading complications-risks and mitigation. EPMA J 14:21–42
Kumar A, Ali A, Kanika, Vyawahare A, Ahmad A, Mishra RK, Ansari MM, Nadeem A, Siddiqui N, Raza SS, Khan R (2023) Highly biocompatible smart injectable hydrogel for the management of rheumatoid arthritis. ACS Biomater Sci Eng 9:5312–5321
Lee MH, Shin JI, Yang JW, Lee KH, Cha DH, Hong JB, Park Y, Choi E, Tizaoui K, Koyanagi A, Jacob L, Park S, Kim JH, Smith L (2022) Genome editing using CRISPR-Cas9 and autoimmune diseases: a comprehensive review. Int J Mol Sci 23
Li D-P, Han Y-X, He Y-S, Wen Y, Liu Y-C, Fu Z-Y, Pan H-F, Cao F (2023) A global assessment of incidence trends of autoimmune diseases from 1990 to 2019 and predicted changes to 2040. Autoimmun Rev 22:103407
Li SJ, Wu YL, Chen JH, Shen SY, Duan J, Xu HE (2024a) Autoimmune diseases: targets, biology, and drug discovery. Acta Pharmacol Sin 45:674–685
Li YR, Lyu Z, Chen Y, Fang Y, Yang L (2024b) Frontiers in CAR-T cell therapy for autoimmune diseases. Trends Pharmacol Sci 45:839–857
Lim SS, Conn DL (2001) The use of low-dose prednisone in the management of rheumatoid arthritis. Bull Rheum Dis 50:1–4
Liu S, Liu X, Chen S, Xiao Y, Zhuang W (2017) Oral versus intravenous methylprednisolone for the treatment of multiple sclerosis relapses: a meta-analysis of randomized controlled trials. PLoS One 12:e0188644
Lopez-Olivo MA, Siddhanamatha HR, Shea B, Tugwell P, Wells GA, Suarez-Almazor ME (2014) Methotrexate for treating rheumatoid arthritis. Cochrane Database Syst Rev 2014:CD000957
Luo H (2024) Global burden and cross-country inequalities in six major immune-mediated inflammatory diseases from 1990 to 2021: a systemic analysis of the Global Burden of Disease Study 2021. Autoimmun Rev 23:103639
Ma H, Murphy C, Loscher CE, O'Kennedy R (2022) Autoantibodies – enemies, and/or potential allies? Front Immunol 13:953726
Ma Z, Zuo T, Frey N, Rangrez AY (2024) A systematic framework for understanding the microbiome in human health and disease: from basic principles to clinical translation. Signal Transduct Target Ther 9:237
Mallapaty S (2024) World-first therapy using donor cells sends autoimmune diseases into remission. Nature 634:519–520
McGraw CA, Lublin FD (2013) Interferon beta and glatiramer acetate therapy. Neurotherapeutics 10:2–18
McInnes IB, Buckley CD, Isaacs JD (2016) Cytokines in rheumatoid arthritis – shaping the immunological landscape. Nat Rev Rheumatol 12:63–68
Medhasi S, Chantratita N (2022) Human leukocyte antigen (HLA) system: genetics and association with bacterial and viral infections. J Immunol Res 2022:9710376
Melendez-Torres GJ, Armoiry X, Court R, Patterson J, Kan A, Auguste P, Madan J, Counsell C, Ciccarelli O, Clarke A (2018) Comparative effectiveness of beta-interferons and glatiramer acetate for relapsing-remitting multiple sclerosis: systematic review and network meta-analysis of trials including recommended dosages. BMC Neurol 18:162
Meneghini M, Bestard O, Grinyo JM (2021) Immunosuppressive drugs modes of action. Best Pract Res Clin Gastroenterol 54–55:101757
Miljković D, Spasojević I (2013) Multiple sclerosis: molecular mechanisms and therapeutic opportunities. Antioxid Redox Signal 19:2286–2334

Misra DP (2024) Clinical manifestations of rheumatoid arthritis, including comorbidities, complications, and long-term follow-up. Best Pract Res Clin Rheumatol 102020

Nally LM, Sherr JL, Van Name MA, Patel AD, Tamborlane WV (2019) Pharmacologic treatment options for type 1 diabetes: what's new? Expert Rev Clin Pharmacol 12:471–479

O'Connell RM, Kahn D, Gibson WS, Round JL, Scholz RL, Chaudhuri AA, Kahn ME, Rao DS, Baltimore D (2010) MicroRNA-155 promotes autoimmune inflammation by enhancing inflammatory T cell development. Immunity 33:607–619

Pang M, Sun Z, Zhang H (2022) Biologic DMARDs and targeted synthetic DMARDs and the risk of all-cause mortality in rheumatoid arthritis: a systematic review and meta-analysis. Medicine (Baltimore) 101:e29838

Pashangzadeh S, Motallebnezhad M, Vafashoar F, Khalvandi A, Mojtabavi N (2021) Implications the role of miR-155 in the pathogenesis of autoimmune diseases. Front Immunol 12:669382

Pisetsky DS (2023) Pathogenesis of autoimmune disease. Nat Rev Nephrol 19:509–524

Popoviciu MS, Kaka N, Sethi Y, Patel N, Chopra H, Cavalu S (2023) Type 1 diabetes mellitus and autoimmune diseases: a critical review of the association and the application of personalized medicine. J Pers Med 13

Qian Z, Li Y, Guan Z, Guo P, Zheng K, Du Y, Yin S, Chen B, Wang H, Jiang J, Qiu K, Zhang M (2023) Global, regional, and national burden of multiple sclerosis from 1990 to 2019: findings of global burden of disease study 2019. Front Public Health 11:1073278

Ramos PS, Shedlock AM, Langefeld CD (2015) Genetics of autoimmune diseases: insights from population genetics. J Hum Genet 60:657–664

Rocha CM, Alves AM, Bettanin BF, Majolo F, Gehringer M, Laufer S, Goettert MI (2021) Current jakinibs for the treatment of rheumatoid arthritis: a systematic review. Inflammopharmacology 29:595–615

Sapkota B, Makandar SN, Acharya S (2025) Biologic response modifiers (BRMs). StatPearls Publishing, StatPearls. Copyright © 2025, StatPearls Publishing LLC., Treasure Island (FL) ineligible companies. Disclosure: Shah Makandar declares no relevant financial relationships with ineligible companies. Disclosure: Saurav Acharya declares no relevant financial relationships with ineligible companies

Schett G, Müller F, Taubmann J, Mackensen A, Wang W, Furie RA, Gold R, Haghikia A, Merkel PA, Caricchio R, D'Agostino MA, Locatelli F, June CH, Mougiakakos D (2024) Advancements and challenges in CAR T cell therapy in autoimmune diseases. Nat Rev Rheumatol 20:531–544

Shah A, Panchal V, Patel K, Alimohamed Z, Kaka N, Sethi Y, Patel N (2023) Pathogenesis and management of multiple sclerosis revisited. Dis Mon 69:101497

Shulga O, Chabanova A, Kotsiuba O (2023) Bruton's tyrosine kinase inhibitors in the treatment of multiple sclerosis. Postep Psychiatr Neurol 32:23–30

Siffrin V, Vogt J, Radbruch H, Nitsch R, Zipp F (2010) Multiple sclerosis—candidate mechanisms underlying CNS atrophy. Trends Neurosci 33:202–210

Song Y, Li J, Wu Y (2024) Evolving understanding of autoimmune mechanisms and new therapeutic strategies of autoimmune disorders. Signal Transduct Target Ther 9:263

Sriwastava S, Khan E, Peterson S, Srivastava S, Lisak RP (2024) Chapter 1 – Introduction to multiple sclerosis: subtypes, pathogenesis, and diagnostic criteria. In: Sriwastava SK, Triantafylou-Bernitsas E (eds) Clinical aspects of multiple sclerosis essentials and current updates. Academic Press, pp 1–33

Stromsnes K, Correas AG, Lehmann J, Gambini J, Olaso-Gonzalez G (2021) Anti-inflammatory properties of diet: role in healthy aging. Biomedicine 9

Stucker F, Ackermann D (2011) Immunosuppressive drugs – how they work, their side effects and interactions. Ther Umsch 68:679–686

Tanaka T, Narazaki M, Kishimoto T (2014) IL-6 in inflammation, immunity, and disease. Cold Spring Harb Perspect Biol 6:a016295

Taneja V (2020) Chapter 26 – Genetic predisposition to autoimmune diseases conferred by the major histocompatibility complex: utility of animal models. In: Rose NR, Mackay IR (eds) The autoimmune diseases, 6th edn. Academic Press, pp 467–489

Tobón GJ, Youinou P, Saraux A (2010) The environment, geo-epidemiology, and autoimmune disease: rheumatoid arthritis. Autoimmun Rev 9:A288–A292

van Drongelen V, Holoshitz J (2017) Human leukocyte antigen-disease associations in rheumatoid arthritis. Rheum Dis Clin N Am 43:363–376

Vyawahare A, Ansari MM, Kumar A, Ahmad A, Mishra RK, Jori C, Nadeem A, Siddiqui N, Raza SS, Khan R (2023) Enzyme targeted delivery of sivelestat loaded nanomicelle inhibits arthritic severity in experimental arthritis. Life Sci 334:122206

Wang F, Tang X, Zhu M, Mao H, Wan H, Luo F (2022) Efficacy and safety of JAK inhibitors for rheumatoid arthritis: a meta-analysis. J Clin Med 11

Wang P, Zheng SG (2013) Regulatory T cells and B cells: implication on autoimmune diseases. Int J Clin Exp Pathol 6:2668–2674

Wang S, Yang N, Zhang H (2024) Metabolic dysregulation of lymphocytes in autoimmune diseases. Trends Endocrinol Metab 35:624–637

Wierenga KA, Pestka JJ (2021) Omega-3 fatty acids and inflammation – you are what you eat! Front Young Minds 9

Wiseman AC (2016) Immunosuppressive Medications. Clin J Am Soc Nephrol 11:332–343

World Health Organization (2023). Rheumatoid arthritis. https://www.who.int/news-room/factsheets/detail/rheumatoid-arthritis

Wu B, Zhou Y, Ni Q (2025) Global, regional and national burdens of chronic kidney disease due to T1DM and T2DM among adolescents and young adults aged 10-35 years from 1990-2021: a trend analysis based on the global burden of disease study 2021. Diabetes Res Clin Pract 220:111985

Wu D, Luo Y, Li T, Zhao X, Lv T, Fang G, Ou P, Li H, Luo X, Huang A, Pang Y (2022) Systemic complications of rheumatoid arthritis: focus on pathogenesis and treatment. Front Immunol 13:1051082

Wu GF, Alvarez E (2011) The immunopathophysiology of multiple sclerosis. Neurol Clin 29:257–278

Xiao F, Rui K, Shi X, Wu H, Cai X, Lui KO, Lu Q, Ballestar E, Tian J, Zou H, Lu L (2022) Epigenetic regulation of B cells and its role in autoimmune pathogenesis. Cell Mol Immunol 19:1215–1234

Yasmeen F, Pirzada RH, Ahmad B, Choi B, Choi S (2024) Understanding autoimmunity: mechanisms, predisposing factors, and cytokine therapies. Int J Mol Sci 25:7666

Zajec A, Trebušak Podkrajšek K, Tesovnik T, Šket R, Čugalj Kern B, Jenko Bizjan B, Šmigoc Schweiger D, Battelino T, Kovač J (2022) Pathogenesis of type 1 diabetes: established facts and new insights. Genes (Basel) 13

Zamanpoor M (2019) The genetic pathogenesis, diagnosis and therapeutic insight of rheumatoid arthritis. Clin Genet 95:547–557

Chapter 2
Genetic Predisposition and Epigenetic Modifications in Autoimmune Pathogenesis: Rheumatoid Arthritis, Type 1 Diabetes, and Multiple Sclerosis

Devendra Kumar Pandey and Shahnawaz

2.1 Introduction

Interactions between environmental and genetic variables lead to autoimmune disorders. The complex regulatory processes that contribute to the onset and evolution of such illnesses have been clarified by recent developments in genetic predisposition and epigenetic studies. The ability to tolerate oneself is essential for avoiding autoimmune diseases. The causes of this disintegration are not entirely known. The risk of autoimmune illnesses can be increased by a number of risk factors. Studies indicate that identical twins are more likely to have the same autoimmune conditions compared with fraternal twins or other family members, suggesting a genetic influence. However, since even identical twins do not always develop the same diseases, environmental exposures must also contribute. Unfortunately, we still have limited knowledge about these environmental influences.

Current findings demonstrate that environmental factors induce epigenetic changes, particularly in DNA methylation patterns, histone modifications, and noncoding RNAs that significantly impact the interaction between the environment and the host in autoimmune conditions. When epigenetic balance is disrupted, it may result in abnormal gene expression in particular cells and thus cause loss of immunological tolerance. These modified cells may contribute to autoimmunity in individuals with genetic risk, highlighting the significance of both genetic and epigenetic mechanisms in understanding certain autoimmune diseases.

D. K. Pandey (✉)
School of BioSciences and Biotechnology, Lovely Professional University, Phagwara, Punjab, India

Shahnawaz
Department of Biotechnology, Kashi Institute of Technology, AKTU, Varanasi, India

Y.-O. Son, M. Ghosh (eds.), *Cutting-Edge Strategies in Drug Delivery and Immunotherapy for Autoimmune Disorders*,
https://doi.org/10.1007/978-981-95-0520-3_2

This article briefly reviews the genetic contributions to rheumatoid arthritis (RA), multiple sclerosis (MS), and type 1 diabetes (T1DM), focusing on how epigenetic mechanisms may facilitate autoimmunity in predisposed individuals. In addition, it highlights the need for continued exploration in these vital areas.

2.2 Genetic and Epigenetic Mechanisms

2.2.1 Genetic Mechanism

Genetic polymorphisms are inherited changes in DNA. They can cause variations in traits and may increase the risk of certain disorders by affecting how genes work. Finding these genetic changes in rheumatic and autoimmune disorders helps us to explore how these diseases develop. Recent advances in science, such as gene testing and large-scale studies, have helped identify genes linked to immune problems in autoimmune disorders. Studies have found genes related to conditions like RA, MS, and T1DM. Researchers have also used animal models to study how some of these genes function in autoimmune disorders. However, studies of twins show that genetics alone do not cause autoimmune disorders. Other factors, such as environmental influences, are also important.

2.2.2 Epigenetic Mechanism

Epigenetics refers to heritable modifications in gene expression that occur without modification to the DNA sequence itself. These modifications are critical for regulating gene activity in eukaryotic organisms, as they modify chromatin structure and, accordingly, influence gene expression. Understanding these mechanisms can provide valuable insights from an evolutionary perspective. Prokaryotic organisms, such as bacteria, possess a few genes, smaller quantities of DNA, and lack a nucleus. Their regulation of gene transcription mainly occurs through transcription factors that bind to specific regions within the DNA. In contrast, eukaryotic organisms encompass multiple cell types, possess a greater number of genes, and contain significantly larger amounts of DNA, all organized within nuclei as chromatin. The roles of chromatin include packaging DNA within the nucleus and regulating gene expression. The nucleosome, made of DNA wrapped around histone proteins, is the fundamental unit of chromatin. These nucleosomes assemble into higher-order structures that enable the packaging of approximately 2–3 meters of human DNA into each cell nucleus. In its native form, chromatin is compact and inaccessible to transcription factors and transcription initiation machinery. Histone modifications, known as the "histone code," signal the remodeling of chromatin from a compact, inactive state to a more open, transcriptionally active form. For instance, the acetylation of lysine residues in histone tails promotes a transcriptionally active chromatin configuration, while deacetylation results in transcriptional silencing (Berger 2007; Dashwood and Ho 2007; Li 2002). The equilibrium between histone acetyltransferases and histone deacetylases governs the level of histone acetylation, rendering the process responsive to environmental factors that influence these enzymes and ultimately affect gene expression (Zhang and Reinberg 2001). A list of exogenous agents reported to alter DNA methylation with their detailed features is shown in Table 2.1.

Table 2.1 Genes associated with RA

Genetic factor	Specific variation	Cell type/ context	Effect/role in RA	Reference
HLA-DRB1	*HLA-DRB1*04:01, *HLA-DRB1*01:01	Antigen-presenting cells	Encodes the shared epitope associated with increased RA susceptibility; presents arthritogenic peptides to T cells.	
PTPN22	rs2476601 (R620W)	T cells, B cells	Alters T cell receptor signaling, leading to increased autoimmunity risk.	
STAT4	rs7574865	T cells, monocytes	Enhances Th1 and Th17 differentiation, promoting inflammatory responses.	
TRAF1-C5	rs10818488	Synovial fibroblasts, immune cells	Involved in TNF signaling; polymorphisms increase inflammation and joint damage.	
CTLA4	rs231775 (+49 A/G)	T cells	Reduces inhibitory signaling, increased T cell activation, and autoimmunity.	
PADI4	rs2240340	Neutrophils, monocytes	Involved in citrullination of proteins; polymorphisms increase production of anticitrullinated protein antibodies (ACPAs).	
TNFAIP3 (A20)	rs2230926	Various immune cells	Regulates NF-κB signaling; variants lead to prolonged inflammatory responses.	Plenge et al. (2007)
IL2RA	rs2104286	Regulatory T cells	Affects IL-2 signaling, influencing Treg function and contributing to autoimmunity.	
IL6R	rs2228145 (Asp358Ala)	Hepatocytes, immune cells	Modulates IL-6 signaling; variants can lead to increased inflammation.	
CCL21	rs2812378	Dendritic cells, T cells	Involved in lymphocyte trafficking; polymorphisms may enhance immune cell migration to joints.	
CD40	rs4810485	B cells, antigen-presenting cells	Influences B cell activation and autoantibody production.	

(continued)

Table 2.1 (continued)

Genetic factor	Specific variation	Cell type/ context	Effect/role in RA	Reference
TRAF6	rs540386	Various immune cells	Mediates signaling for TNF receptor and IL-1 receptor families; variants may enhance inflammatory responses.	
FCGR2A	rs1801274 (H131R)	Macrophages, neutrophils	Alters IgG binding affinity, affecting immune complex clearance and inflammation.	
IRF5	rs2004640	Dendritic cells, monocytes	Enhances type I interferon production, contributing to autoimmunity.	
PRDM1	rs548234	B cells, T cells	Regulates plasma cell differentiation; variants may affect antibody production.	
KIF5A	rs1678542	Neurons, synovial fibroblasts	Involved in intracellular transport; polymorphisms may influence synovial inflammation.	
AFF3	rs11676922	B cells, T cells	Associated with lymphocyte activation; variants may contribute to autoimmunity.	
SPRED2	rs934734	Synovial fibroblasts	Regulates ERK signaling pathway; polymorphisms may enhance fibroblast proliferation.	
CD28	rs1980422	T cells	Costimulatory molecule for T cell activation; variants may influence susceptibility to RA.	
TAGAP	rs212389	T cells	Involved in T cell activation; polymorphisms may affect immune responses.	Plenge et al. (2007)
IL21	rs6822844	T follicular helper cells	Influences B cell help and antibody production; variants may contribute to autoimmunity.	
CCR6	rs3093024	Th17 cells	Affects chemokine receptor expression, influencing Th17 cell migration to inflamed joints.	
ANKRD55	rs6859219	Various immune cells	Function unclear	

2.3 The Genetic and Epigenetic Landscape of RA

2.3.1 Genetic Mechanism of RA

RA affects nearly 3% of people worldwide. It is marked by persistent inflammation and damage to the synovial joints, which causes gradual joint degradation. From the studies on families and twins, a concordance rate of 12–30% is observed in monozygotic (identical) twins, emphasizing the strong genetic influence on susceptibility to RA. It is also more common among first-degree relatives of affected individuals. Interestingly, the prevalence of RA is lower in rural Africans in contrast to those who migrate to urban areas, highlighting the significance of environmental factors in the disease's development. The human leukocyte antigen (HLA) genes located at 6p21 show the strongest linkage to RA. Variants of the HLA-DRB1 allele are well documented, clearly indicating a predisposition to RA. However, the familial risk attributed to HLA genes is approximately 30%, implying that non-HLA genes might contribute to RA. The PTPN22 1858 T variant has been linked to RA, T1DM, and SLE. Despite extensive investigation, studies have produced inconsistent results regarding the association of the CTLA +49/G SNP with RA. Additionally, a single nucleotide polymorphism (SNP) haplotype marked by rs7574865 of the STAT4 gene has also been associated with susceptibility to both RA and SLE. Table 2.1 lists the genes associated with RA.

2.3.2 Epigenetic Mechanism of RA

The initial event is a T cell response to an environmental factor, though the exact identity of this trigger remains unclear. Evidence suggests that the abnormal gene expression in rheumatoid arthritis synovial fibroblasts (RASF) can occur independently of particular gene mutation, indicating that epigenetic mechanisms may significantly occur in the pathology of RA. Activated RASF exhibits DNA demethylation and an elevated level of normally methylated repetitive DNA elements, such as retrotransposon LINE-1. Additionally, there is hypermethylation of regulatory and death-associated proteins, alongside changes induced by histone deacetylases that affect the posttranslational activation of transcriptional activators and nuclear proteins. Within the framework of RA, T cell DNA also undergoes demethylation, which may contribute to the generation of autoreactive T and/or B cell clones, akin to the mechanisms observed in lupus. Furthermore, demethylation of the CD21 promoter has been observed in both peripheral blood mononuclear cells (PBMC) and synovial fluid cells of patients with RA. Support for the interaction of environmental and genetic factors in the pathogenesis of RA is provided by studies investigating the effects of cigarette smoking in conjunction with susceptibility genes present in RA patients. Notably, the occurrence of PTPN22 has been correlated with an increased vulnerability to RA. Table 2.2 lists the epigenetic mechanisms of RA. Figure 2.1 illustrates the key immune cells involved in RA.

Table 2.2 Epigenetic mechanisms of RA

Epigenetic mechanism	Target/change	Cell type/context	Effect/role in RA	Reference
DNA methylation	Hypermethylation of SFRP1 promoter	Fibroblast-like synoviocytes	Downregulation of SFRP1 leads to activation of Wnt signaling, promoting proliferation and invasion of FLS	Hu et al. (2020)
DNA methylation	Hypermethylation of Sirtuin 6 (SIRT6) promoter	Synovial tissue	Downregulation of SIRT6 contributes to increased inflammation and joint degradation	Zhang et al. (2019)
Histone modification	Increased KMT6 expression	Synovial fibroblasts	Upregulation leads to H3K27 methylation and generation of autoantibodies against H3K27me3 histones	Gupta et al. (2015)
Histone modification	Altered HDAC activity	Synovial tissue and fibroblasts	Increased HDAC expression but decreased activity affects gene expression related to inflammation	Gupta et al. (2015)
miRNAs	Upregulation of miR-146a	Peripheral blood mononuclear cells (PBMCs), macrophages, T and B cells	Modulates inflammatory responses by targeting key signaling pathways	Gupta et al. (2015)
miRNAs	UP miR-34a	Synovial fibroblasts	Promotes inflammation and joint destruction by targeting Axl	Hou et al. (2019)
miRNAs	DR miR-10a	Fibroblast-like synoviocytes	Elevate expression of proinflammatory cytokines and MMPs, enhancing FLS proliferation and invasion	Moran-Moguel et al. (2018)
miRNAs	DR miR-27a	Synovial tissue and fibroblasts	Enhances migration and invasion of FLS by increasing MMP expression	Moran-Moguel et al. (2018)
lncRNAs	UP of HOTTIP	Fibroblast-like synoviocytes	Recruits Dnmt3b to SFRP1 promoter, leading to hypermethylation and activation of Wnt signaling, promoting FLS proliferation and invasion	Hu et al. 2020

(continued)

Table 2.2 (continued)

Epigenetic mechanism	Target/change	Cell type/context	Effect/role in RA	Reference
lncRNAs	UP ZFAS1	Synovial tissue and fibroblasts	Enhances migration and invasion of FLS by sponging miR-27a	Ye et al. (2018)
lncRNAs	UP PVT1	Synovial tissue and fibroblasts	Induces methylation and downregulation of SIRT6, contributing to increased inflammation	Zhang et al. (2019)

Note: *UR* Upregulation; *DR* downregulation

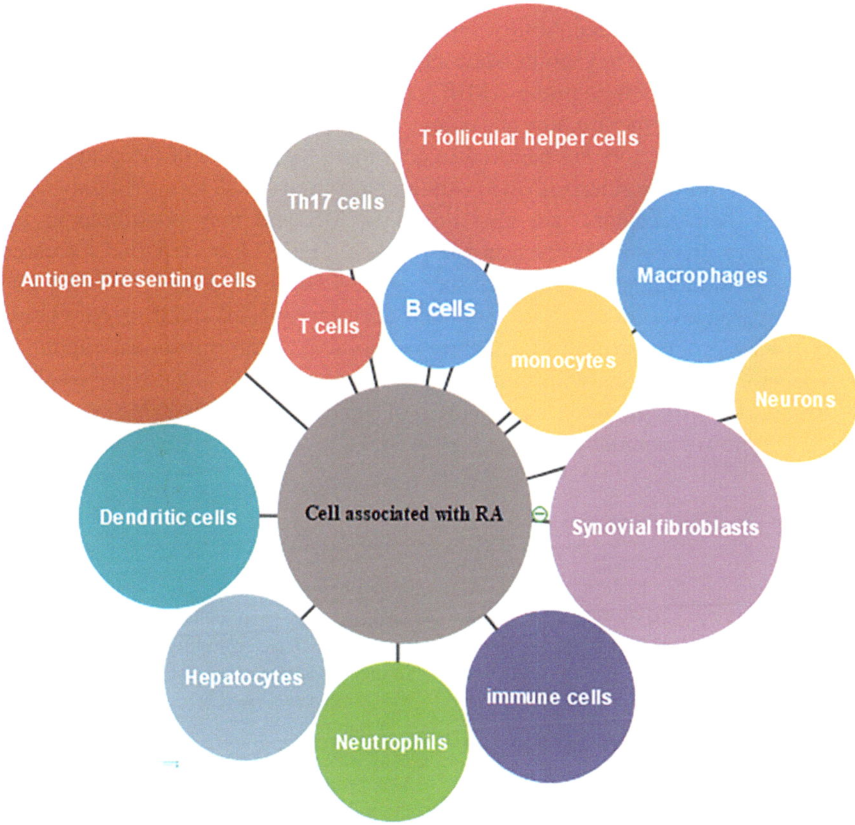

Fig. 2.1 Cells associated with RA

2.4 Genetic and Epigenetic Mechanisms of Multiple Sclerosis

2.4.1 Genetic Mechanism of MS

MS is a chronic inflammatory, neurodegenerative autoimmune disorder that is believed to be influenced by both genetic as well as environmental factors. Studies indicate a concordance rate of 30% for MS in monozygotic (identical) twins, compared with only 5% in dizygotic (fraternal) twins, emphasizing the strong genetic influence on susceptibility to the disease. The HLA gene cluster on chromosome 6p21.3 has been linked to multiple sclerosis (MS) through both candidate gene association studies and whole-genome linkage analyses. Certain HLA alleles—such as HLA-DR2, HLA-DRB11501, DQA10102, and DQB10602—show strong associations with MS susceptibility, with HLA-DRB11501 being the most consistently associated allele in genome-wide studies. T cells detect antigens presented by major histocompatibility complex (MHC) molecules, and changes in HLA-DRB1 residues can modify the structure and charge of the HLA molecule's peptide-binding site. HLA-related factors are believed to contribute to 20–60% of the genetic susceptibility to multiple sclerosis (MS). Additionally, non-MHC loci, including cytokines and their receptors, are also linked to MS and may contribute to the inflammatory processes observed in MS plaques. Studies have reported increased levels of proinflammatory cytokines such as IL-1, IL-4, IL-6, and TNF-α, along with decreased levels of the antiinflammatory cytokines IL-10 and IL-4, correlating with disease progression. Environmental factors (e.g., viral infections, vitamin D deficiency) also interact with genetic risk factors, contributing to the complex pathogenesis of MS. Table 2.3 contains a list of Genes associated with MS. Figure 2.2 illustrates the key immune cells involved in MS.

Table 2.3 Genes associated with MS

Genetic factor	Specific variation	Cell type/ context	Effect/role in MS	Reference
HLA-DRB1	*HLA-DRB1*15:01	Antigen-presenting cells	Strongly associated with increased MS susceptibility; influences antigen presentation to $CD4^{+}$ T cells, leading to autoimmunity.	Sawcer et al. (2011)
HLA-A	*HLA-A*02:01	Various immune cells	May confer protection against MS; possibly through modulation of CD8+ T cell responses.	
IL2RA (interleukin-2 receptor alpha chain)	rs2104286	Regulatory T cells	Polymorphisms affect IL-2 signaling, influencing regulatory T cell function and contributing to autoimmunity.	

(continued)

Table 2.3 (continued)

Genetic factor	Specific variation	Cell type/ context	Effect/role in MS	Reference
IL7R (interleukin-7 Receptor alpha chain)	rs6897932	T cells	Variations alter splicing and expression, impacting T cell development and homeostasis, thereby contributing to MS pathogenesis.	
TYK2 (tyrosine kinase 2)	rs34536443	Immune cells	Influences cytokine signaling pathways, affecting immune responses and contributing to MS susceptibility.	Ban et al. (2009)
CLEC16A (C-Type lectin domain family 16 member A)	rs12708716	Antigen-presenting cells	Involved in autophagy and antigen presentation; variants may affect immune tolerance mechanisms, increasing MS risk.	Dendrou et al. (2015)
CD40	rs6074022	B cells	Variants may influence B cell activation and antibody production, contributing to MS pathogenesis.	
CD58	rs2300747	Immune cells	Associated with altered immune cell adhesion and signaling; may influence T cell activation and MS susceptibility.	
EVI5 (ecotropic viral integration site5)	rs11154801	T cells	Variants may affect cell cycle regulation in T cells, influencing MS risk.	
TNFRSF1A (tumor necrosis factor receptor superfamily member 1A)	rs1800693	Immune cells	Polymorphisms may lead to altered TNF signaling, contributing to inflammation and MS susceptibility.	
IRF8 (interferon regulatory factor 8)	rs17445836	Dendritic cells	Variants may influence dendritic cell function and cytokine production, impacting MS risk.	
GPR65 (G protein-coupled receptor 65)	rs9523762	T cells	Associated with altered T cell responses and inflammation, contributing to MS pathogenesis.	
STAT3 (signal transducer and activator of transcription 3)	rs744166	T cells	Variants may affect Th17 cell differentiation and cytokine signaling, influencing MS susceptibility.	

(continued)

Table 2.3 (continued)

Genetic factor	Specific variation	Cell type/ context	Effect/role in MS	Reference
KIF21B (kinesin family member 21B)	rs12122721	Neurons	Polymorphisms may influence neuronal transport mechanisms, potentially impacting neurodegeneration in MS.	
ZFP36L1 (ZFP36 ring finger protein like 1)	rs7741164	Immune cells	Associated with regulation of mRNA stability for inflammatory cytokines, influencing MS risk.	
BACH2 (BTB domain and CNC homolog 2)	rs3757247	B and T cells	Variants may affect regulatory B and T cell function, contributing to immune dysregulation in MS.	
CXCR5 (C-X-C motif chemokine receptor 5)	rs630923	B and T cells	Involved in B and T cell migration to follicles; variants may influence MS susceptibility.	
IL12A (interleukin 12A)	rs4680534	Dendritic cells	Variants may affect IL-12 production, influencing Th	

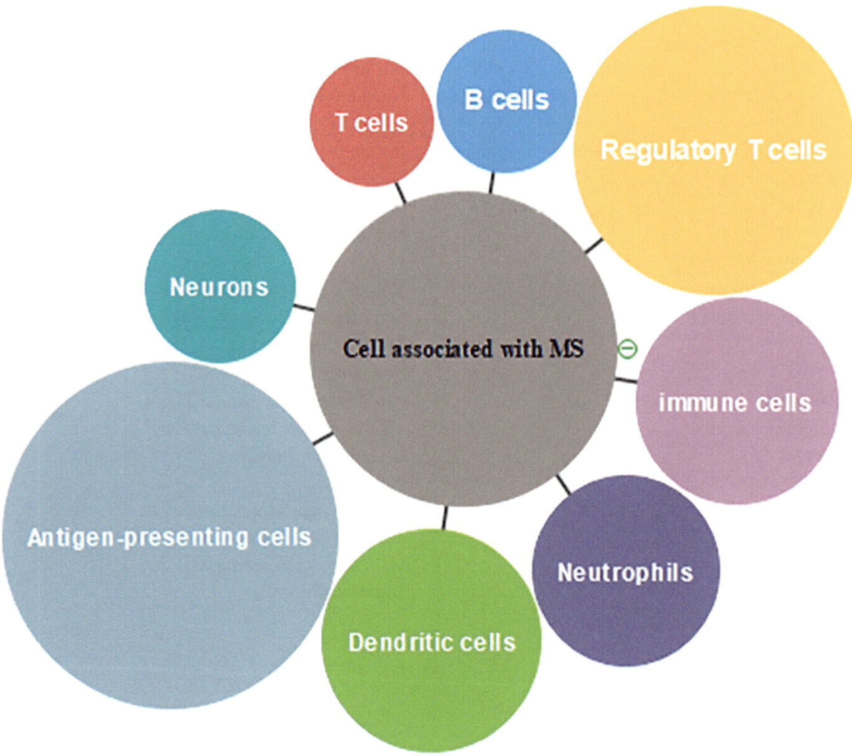

Fig. 2.2 Cell associated with MS

2.4.2 *Epigenetic Mechanism of MS*

A 70% discordance rate among monozygotic twins highlights the important role environmental factors play in disease development. Studies of individuals with comparable genetic backgrounds residing in different geographic regions have revealed notable variations in disease prevalence. For example, sunlight exposure seems to offer protective benefits against MS, possibly through the mediation of vitamin D3. Additionally, infection with the Epstein–Barr virus (EBV) has been associated with the inflammatory response seen in MS. Despite this, there is still a limited understanding of the epigenetics involved in MS. Citrullination of myelin basic protein (MBP) can lead to a loss of myelin stability in the brains of those with MS. This process is catalyzed by the enzyme peptide arginine deiminase 2 (PAD2). An increase in citrullinated MBP in the white matter of MS-affected brains may be attributed to hypomethylation of the PAD2 promoter, which leads to overexpression of PAD2. Table 2.4 contains a list of epigenetic mechanisms in MS.

Table 2.4 Epigenetic Mechanisms in MS

Epigenetic mechanism	Target/change	Cell type/context	Effect/role in MS	Reference
DNA methylation	Hypermethylation of HLA-DRB1 promoter	$CD4^+$ T cells	Alters HLA class II expression, influencing autoimmunity	Kumagai et al. (2012)
DNA methylation	Hypermethylation of SH2-containing protein tyrosine phosphatase-1 (SHP-1)	Peripheral blood mononuclear cells (PBMCs)	Reduces SHP-1 expression, leading to enhanced proinflammatory signaling	Kumagai et al. (2012)
DNA methylation	Hypermethylation of NDRG1 and BCL2L2	Normal-appearing white matter (NAWM) oligodendrocytes	Decreases expression of genes crucial for oligodendrocyte survival and function, impairing remyelination	Huynh et al. (2014)
Histone modification	Increased histone H3 acetylation	Frontal lobe white matter	Upregulates inhibitors of oligodendrocyte differentiation, hindering remyelination	Pedre et al. (2011)
Histone modification	Increased histone H3 citrullination	Normal-appearing white matter	Alters chromatin structure, potentially affecting gene expression related to myelination	Mastronardi et al. (2007)

(continued

Table 2.4 (continued)

Epigenetic mechanism	Target/change	Cell type/context	Effect/role in MS	Reference
miRNAs	Up miR-155, miR-34a, and miR-326	Active MS lesions	Downregulates CD47, enhancing macrophage-mediated myelin phagocytosis	Junker et al. (2009)
miRNAs	Up miR-17 and miR-20a	Whole blood	Modulates T cell activation genes, influencing immune responses	Cox et al. (2010)
miRNAs	Dr miR-338, miR-155, and miR-491	Cerebral white matter	Impairs neurosteroid synthesis, contributing to neurodegeneration	Noorbakhsh et al. (2011)
miRNAs	Dr miR-219 and miR-338-5p	Inactive MS lesions	Impedes oligodendrocyte precursor differentiation, affecting remyelination	Junker et al. (2009)

Note: *UR* Upregulation; *DR* downregulation

2.5 Genetics and Epigenetics of T1DM

2.5.1 Genetic Mechanism of T1DM

T1DM is influenced by both genetic as well as environmental triggers. Studies indicate a concordance rate of 30%–50% in monozygotic (identical) twins, compared with only 5% in dizygotic (fraternal) twins, emphasizing the strong genetic influence on susceptibility to the disease. In the Sardinian population, a noticeable prevalence of both T1DM and MS has been linked to common genetic factors. Whole-genome linkage scans have identified specific chromosome regions—6q26, 10q21.2, 20p12.3, and 22q11.22—that are associated with both T1DM and MS in the population. A comprehensive review has uncovered several additional genetic susceptibility loci for T1DM. The primary locus responsible for familial clustering of T1DM is found in the HLA region on chromosome 6p21, which accounts for approximately 50% of this clustering.

Notably, the HLA haplotypes DR4-DQ8 and DR3-DQ2 are present in 90% of T1DM patients, while the DR15-DR6 haplotype exhibits a protective association against the disease. Additionally, Table 2.5 outlines non-HLA loci connected to

Table 2.5 Genes associated with type 1 diabetes

Genetic factor	Specific variation	Cell type/ context	Effect/role in T1D	Reference
HLA class II genes	*HLA-DRB1*, *HLA-DQA1*, *HLA-DQB1* alleles	Antigen-presenting cells	Certain alleles (e.g., *HLA-DRB1* 03:01 and 04:01) increase susceptibility by presenting β-cell antigens to autoreactive T cells, leading to autoimmune responses.	
INS (insulin gene)	Variable number tandem repeats (VNTRs) in the promoter region	Pancreatic β-cells	Class I VNTRs are associated with reduced thymic expression of insulin, leading to inadequate immune tolerance and increased T1D risk.	Portuesi et al. (2013)
PTPN22 (protein tyrosine phosphatase nonreceptor type 22)	SNP rs2476601 (Arg620Trp)	T cells and B cells	The variant enhances negative regulation of T cell receptor signaling, impairing immune self-tolerance and increasing autoimmunity risk.	Portuesi et al. (2013)
IL2RA (interleukin-2 receptor alpha chain)	Polymorphisms affecting expression levels	Regulatory T cells	Altered IL-2 signaling affects regulatory T cell function, compromising immune regulation and promoting autoimmunity.	Portuesi et al. (2013)
CTLA4 (cytotoxic T-lymphocyte–associated protein 4)	Variants in the 3′ untranslated region	T cells	Modulates inhibitory signaling in T cells; certain variants reduce inhibitory function, leading to increased T cell activation and autoimmunity.	Portuesi et al. (2013)
IFIH1 (interferon induced with helicase C domain 1)	SNPs such as rs1990760	Innate immune cells	Variants can lead to an exaggerated antiviral response, potentially triggering autoimmune destruction of β-cells.	Nejentsev et al. (2009)

(continued)

Table 2.5 (continued)

Genetic factor	Specific variation	Cell type/ context	Effect/role in T1D	Reference
CLEC16A (C-type lectin domain family 16 member A)	SNPs like rs12708716	Antigen-presenting cells	Involved in autophagy and antigen presentation; variants may affect immune tolerance mechanisms.	Grant et al. (2009)
TYK2 (tyrosine kinase 2)	SNPs such as rs2304256	Immune cells	Affects cytokine signaling pathways; certain variants are associated with altered immune responses and T1D risk.	Diogo et al. (2015)
IL10 (interleukin 10)	Promoter polymorphisms	Regulatory T cells	Variants can lead to reduced IL-10 production, impairing antiinflammatory responses and promoting autoimmunity.	Eskdale et al. (1997)
STAT4 (signal transducer and activator of transcription 4)	SNPs like rs7574865	T cells	Influences Th1 cell differentiation; variants may enhance proinflammatory responses, contributing to autoimmunity.	Qu et al. (2009)
ERBB3 (Erb-B2 receptor Tyrosine kinase 3)	SNPs such as rs2292239	Pancreatic β-cells	Involved in β-cell development and survival; certain variants may impair β-cell function and increase T1D susceptibility.	Concannon et al. (2009)
SH2B3 (SH2B adaptor protein 3)	SNPs like rs3184504	Hematopoietic cells	Affects lymphocyte development and cytokine signaling; variants are linked to multiple autoimmune diseases, including T1D.	
GATA3 (GATA binding protein 3)	SNPs such as rs3824662	T-helper cells	Influences Th2 cell differentiation; variants may disrupt immune balance, favoring autoimmunity.	Hermann-Kleiter et al. (2012)
BACH2 (BTB domain and CNC homolog 2)	SNPs like rs3757247	B cells and T cells	Regulates immune tolerance; variants can impair regulatory T cell function, increasing autoimmunity risk.	Flanagan et al. (2014)

(continued)

Table 2.5 (continued)

Genetic factor	Specific variation	Cell type/ context	Effect/role in T1D	Reference
TNFAIP3 (TNF alpha induced protein 3)	SNPs such as rs2327832	Immune cells	Involved in NF-κB signaling and inflammation; certain variants are associated with heightened inflammatory responses.	Plenge et al. (2007)
PRKCQ (protein kinase C theta)	SNPs like rs4750316	T cells	Plays a role in T cell activation; variants may enhance T cell responses, contributing to autoimmunity.	Hirschhorn et al. (2003)
IL21 (interleukin 21)	SNPs such as rs6822844	T follicular helper cells	Influences B cell differentiation and antibody production; variants can affect immune regulation and T1D risk.	Smyth et al. (2023)

T1DM. The insulin gene (INS) locus at 11p15 is the second most significant susceptibility locus for T1DM. A region upstream of the insulin gene, containing a variable number of tandem repeats (VNTR), has been associated with T1DM. The size of the VNTR varies and correlates with the thymic expression of insulin mRNA. The short form of the insulin VNTR is linked to an increased risk of T1DM, while the long form is associated with a reduced risk of developing the disease. Fig. 2.3 illustrates the key immune cells involved in type 1 diabetes.

2.5.2 *Epigenetic Mechanism of T1DM*

Nongenetic factors may play a role in enhancing the autoimmune process that leads to the destruction of islet beta cells in T1DM. Proposed environmental factors include nutrition and viruses. Among these, the strongest evidence for a viral cause comes from congenital rubella syndrome. However, there is currently no direct

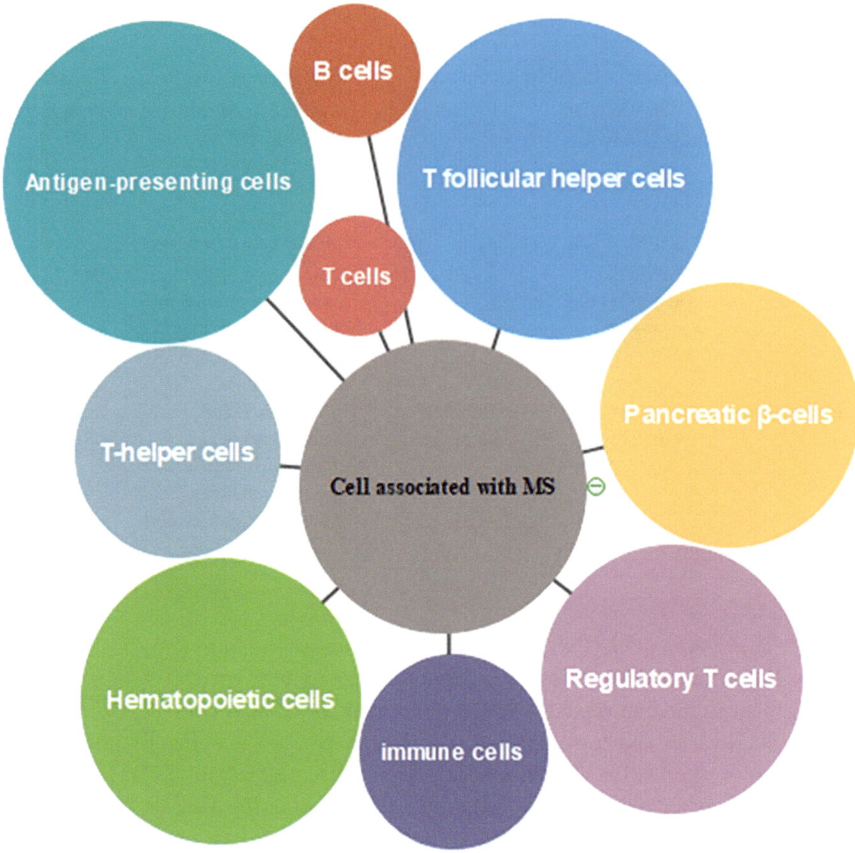

Fig. 2.3 Cells associated with type 1 diabetes

evidence linking epigenetic changes to T1DM. Nutrition is important since it provides methyl donors, such as methionine and choline, as well as essential cofactors like folic acid, vitamin B12, and pyridoxal phosphate, which are crucial for DNA and histone methylation. Therefore, it is possible that epigenetic mechanisms may also contribute to the development of T1DM. Table 2.6 summarizes key epigenetic mechanisms implicated in the pathogenesis of T1DM.

Table 2.6 Epigenetic mechanisms of T1DM

Epigenetic mechanism	Target/change	Cell type/ context	Effect/role in T1D	Reference
DNA methylation	Differential methylation at HLA-DQB1 and HLA-DRB1 genes	Monocytes	Altered expression of HLA class II molecules, influencing autoimmunity	Miao et al. (2012)
DNA methylation	Global DNA hypomethylation	CD4⁺ T cells	Enhanced gene expression leading to increased immune activation	Liu et al. (2015)
Histone modification	Increased H3K9Ac at HLA-DQB1	Monocytes	Upregulated HLA-DQB1 expression, contributing to autoimmune response	Miao et al. (2012)
Histone modification	Decreased H3 acetylation	CD4⁺ T cells	Reduced gene expression, potentially affecting immune regulation	Liu et al. (2015)
miRNAs	Upregulation of miR-326	Lymphocytes	Promotes Th17 differentiation, enhancing autoimmunity	Sebastiani et al. (2011)
miRNAs	Upregulation of miR-98, miR-23b, and miR-590-5p	CD8⁺ T cells	Suppression of proapoptotic pathways, leading to expansion of autoreactive T cells	de Jong et al. (2016)
miRNAs	Downregulation of miR-146	Mononuclear cells	Dysregulation of inflammatory responses	Yang et al. (2015)
miRNAs	Altered expression of miR-21, miR-34a, miR-146a, and miR-29	Pancreatic β-cells	Increased β-cell apoptosis through modulation of proinflammatory cytokines	Ruan et al. (2011)
miRNAs	Dysregulation of miR-375	Pancreatic islets	Impaired insulin secretion and β-cell function	
lncRNAs	Altered expression of various lncRNAs	Immune cells and β-cells	Modulation of gene expression affecting immune responses and β-cell viability	Hezova et al. (2010)

2.6 Summary

Although significant progress is being made in understanding the genetics of autoimmune disorders, the influence of environmental factors on disease development remains largely unknown. There is compelling evidence that the environment can modify the immune system through epigenetic mechanisms. This raises the possibility that epigenetics could also play a role in other autoimmune disorders. Further research into epigenetic abnormalities in diseases such as RA, MS, and T1DM could help clarify how environmental factors influence the immune system in these conditions and suggest potential ways to improve treatment.

1. Genetic and Environmental Susceptibility
2. Immune Activation (Preclinical Phase)
3. CNS Inflammation and Demyelination
4. Clinically Isolated Syndrome
5. Relapsing-Remitting MS
6. Secondary Progressive MS
7. Primary Progressive MS *(in ~10–15% of patients from onset)*
8. Advanced MS

References

Ban M, Booth DC, Heard X-Z et al (2009) Common variants in TYK2 and RELA are associated with multiple sclerosis. Nat Genet 41(12):1284–1289. https://doi.org/10.1038/ng.471

Berger SL (2007) The complex language of chromatin regulation during transcription. Nature 447:407–412

Concannon P, Rich SS, Todd JA (2009) Genetics of type 1 diabetes. N Engl J Med 360(16):1646–1654. https://www.nejm.org/doi/full/10.1056/NEJMra0808284

Cox MB, Cairns MJ, Gandhi KS et al (2010) MicroRNAs miR-17 and miR-20a inhibit T cell activation genes and are under-expressed in MS whole blood. PLoS One 5(8):e12132

Dashwood RH, Ho E (2007) Dietary histone deacetylase inhibitors: from cells to mice to man. Semin Cancer Biol 17:363–369

de Jong VM, Klinkenberg S, de Groot M et al (2016) miRNAs regulate human plasmacytoid dendritic cell function. J Immunol 196(12):5170–5180

Dendrou CA, Fugger L, Hauser SJ (2015) Neuroinflammation and multiple sclerosis: the role of genetic variation. Trends Immunol 36(3):144–158. https://doi.org/10.1016/j.it.2014.12.002

Diogo D, Kurreeman KA, Raychaudhuri S (2015) TYK2 gene associations with autoimmune and inflammatory diseases. Nat Genet 47(8):875–880. https://www.nature.com/articles/ng.3247

Eskdale J, Kube C, Tesch T, Gallagher G (1997) Mapping of the human IL10 gene and further characterization of the 5′ flanking sequence. Immunogenetics 46(2):120–128. https://doi.org/10.1007/s002510050254

Flanagan SE, Day RD, Sanjeevi ILB et al (2014) BACH2 variant is associated with type 1 diabetes and other immune-mediated diseases. Nat Genet 46(5):485–487. https://www.nature.com/articles/ng.2958

Grant SFA, Qu FA, Bradfield J et al (2009) Follow-up analysis of genome-wide association data identifies novel loci for type 1 diabetes. Diabetes 58(1):290–295. https://doi.org/10.2337/db08-1036

Gupta M, Maurya RN, Kumar A et al (2015) Epigenomics of autoimmune diseases. Immunol Cell Biol 93(4):271–276

Hermann-Kleiter N, Gruber B, Labi A et al (2012) The oncogene c-Rel regates GATA3 and Th2 polarization. Genes Immun 13:587–595. https://www.nature.com/articles/gene201210

Hezova R, Slaby O, Faltejskova P et al (2010) microRNA-342, microRNA-191 and microRNA-510 are differentially expressed in T regulatory cells of type 1 diabetic patients. Cell Immunol 260(2):70–74

Hirschhorn JN, Lohmueller GS, Byrne E, Hirschhorn K (2003) A comprehensive review of genetic association studies. Genet Med 4(2):45–61. https://www.nature.com/articles/ng1160

Hou C, Gu L, Guo Y et al (2019) Inhibition of miR-34a ameliorates collagen-induced arthritis in mice. J Cell Biochem 120(4):5644–5651

Hu F, Xu L, Zhao Y et al (2020) lncRNA HOTTIP regulates proliferation, apoptosis, and invasion of fibroblast-like synoviocytes through the Wnt signaling pathway in rheumatoid arthritis. Arthritis Res Ther 22(1):1–12

Huynh JL, Garg P, Thin TH et al (2014) Epigenome-wide differences in pathology-free regions of multiple sclerosis-affected brains. Nat Neurosci 17(1):121–130

Junker A, Krumbholz M, Eisele S et al (2009) MicroRNA profiling of multiple sclerosis lesions identifies modulators of the regulatory protein CD47. Brain 132(12):3342–3352

Kumagai N, Minohara M, Matsushita T et al (2012) Epigenetic regulation of SHP-1 expression in multiple sclerosis. J Neuroimmunol 244(1–2):69–72

Li E (2002) Chromatin modification and epigenetic reprogramming in mammalian development. Nat Rev Genet 3:662–673

Liu Y, Gao X, Miao Y et al (2015) The DNA methylation status of CD4$^+$ T cells in type 1 diabetes patients. Diabetes Res Clin Pract 108(1):89–96

Mastronardi FG, Noor A, Wood DD et al (2007) Increased citrullination of histone H3 in multiple sclerosis brain and animal models of demyelination. Mol Cell Proteomics 6(3):551–561

Miao F, Smith DD, Zhang L et al (2012) Lymphocytes from patients with type 1 diabetes display a distinct epigenetic signature. J Clin Endocrinol Metab 97(6):E1118–E1127

Moran-Moguel MC, Petarra-del Rio S, Mayorquin-Galvan EE et al (2018) Rheumatoid arthritis and miRNAs: a critical review through a functional approach. Front Immunol 9:1337

Nejentsev S, Walker E, Riches S et al (2009) Rare variants of IFIH1, a gene implicated in antiviral responses, protect against type 1 diabetes. Nature 458:960–963. https://www.nature.com/articles/ng.354

Noorbakhsh F, Ellestad KK, Maingat F et al (2011) Impaired neurosteroid synthesis in multiple sclerosis. Brain 134(9):2703–2721

Pedre X, Mastronardi F, Bruck W et al (2011) Changed histone acetylation patterns in normal-appearing white matter and early multiple sclerosis lesions. J Neurosci 31(9):3435–3445

Plenge RM, Seldin EM, Stahl BM et al (2007) TRAF1–C5 as a risk locus for rheumatoid arthritis: a Genomewide study. Nat Genet 39(12):1477–1482. https://www.nature.com/articles/ng2109

Portuesi R, Symeonidou DE, Wicker EC et al (2013) The role of the PTPN22 gene in autoimmunity: evidence from functional studies and genotype–phenotype associations. PLoS One 8(11):e79506. https://journals.plos.org/plosone/article?id=10.1371/journal.pone.0079506

Qu HQ, Marchand MJ, Grabs YS et al (2009) STAT4 association with type 1 diabetes in multiple populations. Genes Immun 10:424–428. https://www.nature.com/articles/gene200913

Ruan Q, Wang T, Kameswaran V et al (2011) The microRNA-21-PDCD4 axis prevents type 1 diabetes by blocking pancreatic beta cell death. Proc Natl Acad Sci 108(29):12030–12035

Sawcer S, Ban G, Hauser S (2011) The genetic basis of multiple sclerosis. Lancet Neurol 10(9):754–763. https://doi.org/10.1016/S1474-4422(11)70167-9

Sebastiani G, Ventriglia G, Stabilini A et al (2011) Increased expression of microRNA miR-326 in type 1 diabetic patients with ongoing islet autoimmunity. Diabetes Metab Res Rev 27(8):862–866

Smyth DJ, Cooper LJ, Concannon P et al (2023) IL-21 locus association with type 1 diabetes. Nat Genet

Yang M, Ye L, Wang B et al (2015) Decreased miR-146 expression in peripheral blood mononuclear cells from patients with type 1 diabetes mellitus. Arch Med Res 46(7):554–560

Ye Y, Xu Y, Lai Y et al (2018) Long non-coding RNA ZFAS1 promotes fibroblast-like synoviocyte proliferation and inflammatory response in rheumatoid arthritis via sponging miR-27a. Int Immunopharmacol 65:539–547

Zhang Y, Reinberg D (2001) Transcription regulation by histone methylation: interplay between different covalent modifications of the core histone tails. Genes Dev 15:2343–2360

Zhang L, Yang N, Song Y et al (2019) Long noncoding RNA s PVT1 promotes rheumatoid arthritis via inducing methylation of the Sirt6 promoter. Mol Med Rep 19(6):4984–4992

Chapter 3
Familial Aggregation and Hereditary Factors in Autoimmunity

Miraclin Prasanna A. and Priyankar Sen

Abbreviations

ADs	Autoimmune Diseases
AS	Ankylosing Spondylitis
ATD	Autoimmune Thyroid Disease
CD	Celiac Disease
CTLA4	Cytotoxic T-Lymphocyte Antigen 4
FDRs	First-Degree Relatives
FHS	Family History Score
GWAS	Genome-Wide Association Studies
HLA	Human Leukocyte Antigen
LD	Linkage Disequilibrium
MHC	Major Histocompatibility Complex
MS	Multiple Sclerosis
PBC	Primary Biliary Cirrhosis
PTPN22	Protein Tyrosine Phosphatase Nonreceptor Type 22
RA	Rheumatoid Arthritis
RSS	Relative Risk for Siblings
SLE	Systemic Lupus Erythematosus
SNPs	Single Nucleotide Polymorphisms
SRR	Sibling Recurrence Risk
SS	Sjögren's syndrome
SSc	Systemic Sclerosis
T1D	Type 1 Diabetes
TNF	Tumor Necrosis Factor
TNFAIP3	Tumor Necrosis Factor Alpha-Induced Protein 3
XCI	X Chromosome Inactivation

M. P. A. · P. Sen (✉)
Centre for Bio Separation Technology (CBST), School of Biosciences and Technology, VIT, Vellore, Tamil Nadu, India

Y.-O. Son, M. Ghosh (eds.), *Cutting-Edge Strategies in Drug Delivery and Immunotherapy for Autoimmune Disorders*,
https://doi.org/10.1007/978-981-95-0520-3_3

3.1 Introduction

Autoimmune diseases (ADs) represent a heterogeneous group of disorders with chronic conditions that affect specific target organs or multiple organ systems. ADs are chronic conditions that place a significant burden on the patient's medical care, economic costs, and diminish quality of life. It also has a long prodromal period before the appearance of clinically apparent symptoms (Tobon et al. 2012). According to previous studies, numerous autoantibodies can be serologically assessed far in advance of the onset of clinical illness. Identification of these markers and family history of autoimmunity can evaluate ADs' predictive value that may help in designing personalized medicine (Cardenas-Roldan et al. 2013), leading researchers to widely explore the pathological mechanisms of AD. The two main factors that influence the pathology of AD include genetic and environmental factors (Anaya 2006). Genetic contribution has the highest rate of concordance (from 15% to 60%) and with high recurrent risk ratios (λ_R) (Wandstrat and Wakeland 2001).

The genetic origin of AD is predisposed by individual and population variation. Population variations occur through migration, mutation, genetic drift, and natural selection; these genetic variations influence phenotypic expression in specific populations. Therefore, diving into population genetics, focusing on genetic variation, can elucidate human genetic diversity and the genetic factors causing disease (Torkamani et al. 2012). Each population embraces a mutational pool; in the case of individual mutations, such as polymorphisms, they often become unrecognized due to their mild effects, whereas mutations in conjunction with additional alleles would either prevent or favor autoimmune disorders. Based on the interaction between genetic variations, the quantifiable risk of acquiring an autoimmune phenotype will be altered. Due to the unpredictability of genetic variants, ADs are not inherited in a standard Mendelian way but rather possess a complicated or unidentified route of inheritance (Anaya 2006). Despite their complexity, ADs share a common origin, as proved by clinical observations that patients may develop multiple ADs over time or that multiple autoimmune conditions can coexist within an individual (polyautoimmunity) or among family members (familial autoimmunity). Similarly, common pathophysiological mechanisms have been identified among different ADs, reinforcing their interconnection. Further epidemiological studies highlight associations between specific ADs, bridging epidemiological findings with pathophysiological insights to enhance understanding of their shared immunogenetic and etiological factors. Lastly, genetic research provides strong evidence of overlapping hereditary components among ADs, emphasizing the likelihood of multiple autoimmune conditions occurring simultaneously in a single patient beyond statistical correlations (Castiblanco et al. 2015).

3.2 Genetic Factors Triggering Heritable Autoimmunity

The family of ADs is remarkably complex; despite affecting multiple organ systems, most ADs are known to share common underlying mechanisms. Additionally, significant gender and ethnic disparities are observed in the prevalence and progression of these conditions. The leading cause of death from AD is comparatively higher in women than in men; the influence of the X chromosome has been proposed as a contributing factor (Ramos et al. 2015). Ethnic/regional disparities tend to have different effects on AD susceptibility. For example, rheumatoid arthritis (RA) tends to be relatively low to moderate in low- and lower-middle-income countries, but these conditions seem to increase among immigrant populations residing in high-income countries. Notably, the burden of ADs is particularly high in affluent nations such as the UK, the Republic of Ireland, and New Zealand (Subramanian et al. 2020).

African Americans have a higher susceptibility to systemic lupus erythematosus (SLE) and scleroderma (systemic sclerosis [SSc]) compared with European Americans, often developing these conditions earlier in life with greater severity. However, they exhibit a lower risk for type 1 diabetes (T1D), thyroiditis, and multiple sclerosis (MS). Findings from research on AD have highlighted that despite their complexity, ADs share a certain degree of common genetic origins (Rajalingam and Ganjiwale 2024). Genetic factors that determine AD susceptibility and progression have been evidenced by a higher concordance rate observed in monozygotic twins compared with dizygotic twins with only a few exceptions, providing further evidence of a genetic basis for ADs (Morahan et al. 2008). The identification of genes involved in ADs is analyzed by three methods, as shown in Table 3.1. The first approach, identifying the candidate gene through a case-control approach, plays a crucial role in human genetic research. Numerous studies have sought to identify

Table 3.1 Methods used for identification of genes involved in ADs

Methods	Description	Strengths	Limitations
Candidate gene approach	Analyses specific candidate genes through case-control studies, testing a plausible hypothesis using small sets of SNPs.	Useful in hypothesis-driven research with targeted approach.	Limited statistical power, small sample sizes, and insufficient case-control matching.
Linkage analysis	Examines specific candidate genes through case-control studies. Tests a plausible hypothesis using small sets of SNPs.	Effective in detecting strong genetic components in families.	Limited power in identifying shared loci, requires multiple family samples.
Genome-wide association studies (GWAS)	Uses high-throughput SNP analysis to test thousands of hypotheses simultaneously, providing large-scale insights into genetic risk factors.	Highly effective in identifying genetic risk factors, large sample sizes, unbiased approach.	Requires large datasets, potential for false positives, limited in rare variant detection.

genes associated with autoimmunity by analyzing small sets of single nucleotide polymorphisms (SNPs) within one or a few specific candidate genes, thereby testing a plausible hypothesis. However, this approach has provided limited insights into the genetic basis of ADs, primarily due to inadequate statistical power, insufficient case-control matching, and small sample sizes. The second approach, linkage analysis, examines the transmission of two or more linked genes on a chromosome to the same daughter cell, where genetic loci are inherited together with a specific phenotypic trait within families. However, linkage analysis has limited power in identifying shared chromosomal loci or regions within families. This method is only effective when the genetic component of the disease is relatively strong and when genomic samples from multiple family members, particularly parents, are available.

The third approach gained popularity in the last few years. Advancements in molecular technology have enabled high-throughput analysis of SNPs, leading to significant progress in human genetics through large-scale, high-density GWAS. Like candidate gene studies, GWAS relies on association analysis but differs by simultaneously testing thousands of hypotheses rather than focusing on a single plausible hypothesis. This approach has proven particularly effective in identifying genetic risk factors for ADs (Invernizzi and Gershwin 2009; Hu and Daly 2012).

3.2.1 Role of MHC Locus Polymorphism in Autoimmune Heritability

The MHC is an important antigen-presenting molecule that helps T cells to recognize peptide fragments of the original antigen (Waithman et al. 2014). MHC genes play the most important role in predisposing to AD in both humans and animals. Located on the short arm of chromosome 6 (6p21.3), the MHC region spans approximately 3.6 megabase pairs (Mb) and is divided into three subregions: the telomeric class I, class III, and the centromeric class II regions. While the exact function of many associated MHC alleles remains unclear, certain alleles are believed to contribute to autoimmune responses by influencing immunogenicity. Recent molecular advancements have expanded the known MHC region to approximately 7.6 Mb of the human genome. Additionally, loci outside the well-defined HLA class I, II, and III regions have been identified, with 421 genes in the extended MHC, over 20% of which are believed to have immune-regulatory functions and may be linked to ADs (Horton et al. 2004. Certain genes within the MHC locus contain major disease-causing variants, including classical class I genes such as HLA-A, HLA-B, and HLA-C, as well as classical class II genes like HLA-DRB1, HLA-DQA1, and HLA-DQB1. Among these, HLA-B is the most polymorphic gene identified in the human genome (Fernando et al. 2008). The MHC's genetic structure exhibits high levels of linkage disequilibrium (LD) compared to the rest of the genome, posing technical challenges in identifying its SNPs, alleles, and amino acids. However, the recent availability of dense genotyping platforms has enhanced fine-mapping of the MHC locus by pinpointing genetic risk factors for AD, improving our understanding of disease associations, and identifying functional variants. Identifying genetic variations in the MHC is crucial for two main reasons: first, it provides insight into the underlying causes of autoimmunity and second, it enhances our understanding of the complexity of the human immune system. Ultimately, this knowledge will help develop better strategies to achieve more balanced patient-immune responses during treatment (Matzaraki et al. 2017).

ADs characterized by the presence of autoantibodies, such as RA, T1D, SLE, celiac disease (CD), and systemic sclerosis (SSc), are primarily linked to HLA class II alleles. In contrast, HLA class I alleles are associated with ADs like Behcet's disease, ankylosing spondylitis (AS), and psoriasis, where autoantibodies are not a defining feature. However, in certain cases, like Crohn's disease, which is associated with SNPs in the MHC region, classical HLA alleles do not show a significant association with the condition (Orozco and Rueda 2019).

3.2.2 Role of Non-MHC Locus Polymorphism in Autoimmune Heritability

There is substantial evidence indicating that genetic loci outside the HLA region play immune-modulating roles. These include cytotoxic T-lymphocyte antigen 4 (CTLA4), a member of the immunoglobulin superfamily expressed on helper T cells, which transmits inhibitory signals to T cells has been associated with several autoimmune disorders, such as autoimmune thyroid disease (ATD), T1D, and CD. Protein tyrosine phosphatase nonreceptor type 22 (PTPN22), primarily found in lymphoid tissues and involved in regulating T-cell receptor signaling pathways, and tumor necrosis factor (TNF) alpha–induced protein 3 (TNFAIP3), which suppresses NF-kappa B activation and TNF-mediated apoptosis, have been implicated in autoimmune conditions such as thyroid disease, T1D, and CD. Additionally, genetic loci associated with pathways governing B-cell and T-cell activation and differentiation, innate immunity, and cytokine signaling regulation contribute to the development of AD (Richard-Micelic and Criswell 2012; Ye et al. 2018). Non-HLA genes such as *CTLA4*, *PTPN22*, and *TNF-α* have been linked to the development of several ADs, including RA, ATD, T1D, MS, and SLE. These genes play a crucial role in immune regulation and contribute to the pathogenesis of autoimmunity by influencing T-cell activation, immune signaling, and inflammatory responses (Serrano et al. 2006). A comparative analysis of the role of MHC and non-MHC locus polymorphism in autoimmune heritability is shown in Table 3.2.

Table 3.2 Role of MHC versus non-MHC locus polymorphisms in autoimmune heritability

Aspect	MHC locus polymorphism	Non-MHC locus polymorphism	References
Genomic location	Located on chromosome 6 (6p21.3), spanning ~3.6–7.6 Mb.	Scattered across the genome outside the HLA region.	Sinkus et al. (2013), Brickner (2006)
Key genes involved	HLA-A, HLA-B, HLA-C (class I); HLA-DRB1, HLA-DQA1, HLA-DQB1 (class II).	CTLA4, PTPN22, TNFAIP3, and other regulating immune pathways.	Tokic et al. (2020), Houcken et al. (2018), and Hossen et al. (2023)
Role in autoimmunity	Involved in antigen presentation and immune response regulation.	Modulate immune signaling, T-cell activation, and inflammation.	Janeway Jr et al. (2001), Liang et al. (2012)
Associated ADs	HLA class II: RA, T1D, SLE, CD, SSc (autoantibody-associated). HLA class I: AS, Behcet's disease, psoriasis.	RA, ATD, T1D, MS, SLE, and other autoimmune diseases.	Bodis et al. (2018), Tizaoui et al. (2022)
Polymorphism impact	Highly polymorphic, especially HLA-B, leading to diverse immune responses.	Variants influence immune tolerance and inflammatory pathways.	Olson et al. (2020), Borysewicz-Sanczyk et al. (2020)
Genetic complexity	High linkage disequilibrium (LD) complicates SNP	Diverse genetic variants contribute independently to AD susceptibility.	Trowsdale and Knight (2013), Shamilov and Aneskievich (2018)
Technological advances	Dense genotyping platforms improve fine mapping of risk alleles.	Genome-wide association studies (GWAS) identify multiple risk loci.	Matzaraki et al. (2017), Lessard et al. (2012).
Clinical significance	Essential for understanding AD mechanisms and immune response balance.	Crucial for identifying immune-modulating pathways and potential therapeutic targets.	Fernando et al. (2008), Serrano et al. (2006)

3.2.3 *Sex-Biased Heritability*

The development of ADs has been shown to affect females and males differently, with females being more susceptible to these conditions. In fact, over 85% of individuals diagnosed with ADs such as thyroiditis, scleroderma, lupus, and Sjögren's syndrome (SS) are female (Billi et al. 2019). Sex bias in autoimmunity was initially attributed primarily to sex hormones, with estrogens in females and androgens in males influencing immune functions either positively or negatively in terms of maintenance, development, and effect. However, beyond hormonal influence, genetic mechanisms related to sex chromosomes have recently been recognized as contributors to sex-based differences in immune responses across various age groups. For instance, individuals with Klinefelter's syndrome (47,XXY) exhibit a susceptibility to lupus similar to that of women (46,XX), whereas women with Turner syndrome (45,X0) appear to have a protective advantage. This suggests that the X chromosome plays a role in the development of lupus, scleroderma, and SS (Bianchi et al. 2012). The sex bias of autoimmunity in females is also contributed to by the presence of two X chromosomes. Although one X chromosome is typically inactivated during early development, 15% of genes can escape X chromosome inactivation (XCI) under certain conditions. This escape from XCI has been linked to an increased risk of autoimmunity (Carrel and Willard 2005). Several important genes for autoimmune regulation are located on the X chromosome. In AD conditions, these genes are often overexpressed or exhibit hypomethylation in females but not in males. Notably, X-linked genes such as TLR7 and TLR8 have been shown to contribute to the development of SLE (Souyris et al. 2018).

3.3 Risk of Familial Aggregation Persevering as Heritable Autoimmunity

A primary characteristic of complex diseases, like ADs, is their tendency to cluster within families; this phenomenon is known as familial aggregation (recurrence risk or lambda (λ)). Familial aggregation is observed when a disease occurs more frequently among relatives of an affected individual compared with the general population (when $\lambda > 1.0$) (Anaya et al. 2007).

ADs occurring individually are rare; comparatively they generally occur in aggregates (clustering of different ADs), and collectively they affect approximately 5–9% of the population. However, the underlying reasons for their aggregation remain poorly understood (Cooper et al. 2009). Familial aggregation of AD likely reflects on clustering of certain pathological factors of different diseases, behaviors, or disorders within a given family (Niewold et al. 2011). The shared clinical symptoms, subphenotypes, pathophysiological mechanisms, and genetic factors exhibited in the affected individuals of ADs are collectively referred to as "autoimmune tautology." In clinical practice, two key phenomena support this concept: polyautoimmunity and familial autoimmunity. Polyautoimmunity occurs when an individual has two or more ADs, whereas familial autoimmunity refers to the occurrence of different ADs among relatives within a nuclear family (Cardenas-Roldan et al. 2013).

3.3.1 Etiology of Familial Aggregation

The clustering of diseases within families might be attributed to shared etiological factors, including environmental exposures, common genetic traits, or interactions between genetic and environmental influences. However, the relative contributions of genetic and environmental factors to disease coaggregation remain largely unclear. There is evidence supporting the genetic study that AD causative genes have a pleiotropic effect; some studies have suggested that several risk alleles have been identified with opposing effects across different disorders, increasing the risk for one disease while reducing susceptibility to another (Skov et al. 2020).

Although ADs develop from multigenic genes, they seem to share a clear correlational link. The patterns of aggregation across different autoimmune phenotypes suggest a shared genetic basis (Eisenbarth and Gottlieb 2004). For example, T1D is an immune-mediated disease that develops in genetically susceptible individuals, often alongside other ADs such as ATDs (e.g., Graves' disease or Hashimoto's thyroiditis) and CD, which are among the most prevalent co-occurring conditions. First-degree relatives (FDRs) of individuals with T1D have a higher predisposition to developing the disease and exhibit a greater prevalence of autoantibodies compared with the general population (Pugliese 2004). A study involving 265 multiplex North American families identified a functional variant (R620W) of the PTPN22 gene, which increases susceptibility to four ADs: T1D, RA, SLE, and ATD. However, no association was observed with MS. This genetic variant is a significant risk factor for multiple autoimmune phenotypes. In addition to genetic predisposition, environmental factors such as cigarette smoke exposure, crystalline silica, and Epstein-Barr virus infection have been linked to an increased risk of several ADs, including ATD, RA, and SLE. Despite these findings, the extent and significance of familial clustering across a broader range of autoimmune disorders remain uncertain. This is due to the relative rarity of many ADs and their heterogeneous phenotypes, which make diagnosis and confirmation challenging (Criswell et al. 2005).

3.3.2 Importance of Epidemiological Study to Evaluate Heritability in Familial Aggregation

Studies involving diverse populations, utilizing the epidemiological history of these groups, have proven to be one of the most effective methods for exploring genetic associations. Studies leveraging epidemiological history are a powerful approach for uncovering genetic associations, improving disease prediction, and advancing personalized medicine across different ethnic and geographical groups (Anaya et al. 2013). Epidemiological findings also highlight the significant role of nongenetic factors in disease clustering, as a stronger association between siblings, who share a more similar environment, is observed compared with parent-offspring pairs (Eaton et al. 2007). Epidemiological research on various ADs, like RA, SLE, T1D, and primary biliary cirrhosis (PBC), shows strong heritability due to strong genetic association; in contrast, certain ADs such as RA and MS heritability is based on environmental factors. Earlier studies on ADs heritability regarding genetic association were confirmed as ADs inheritance in monozygotic twins was observed compared with dizygotic twins. Despite this, the overall concordance rate in monozygotic twins remains well below 50%, suggesting that environmental and other nongenetic factors also play a crucial role in disease development (Seldin 2015). Findings from the most extensive analysis of shared genetic loci associated with ADs suggest that various population-level genetic factors, such as natural selection, coevolution with pathogens, random mutations, isolation, migration, and interbreeding, have contributed to the development of the diverse genetic loci that predispose individuals to autoimmunity. These factors have shaped genetic susceptibility in both similar and distinct environmental contexts (Ramos et al. 2011).

3.3.3 Statistical Methods to Assess Familial Aggregation through Epidemiological Studies

Disease aggregation is triggered by genetic, cultural, or environmental factors. Genetic epidemiology plays a key role in assessing whether the disease clusters in families. The field of genetic epidemiology focuses on identifying associations between genetic variants and phenotypic traits, with an emphasis on detecting genetic factors that increase susceptibility to diseases and related traits. However, correlation alone is insufficient to establish the heritability of ADs. For instance, a genetic variant prevalent in an isolated ethnic group may show a high correlation with a disease but could simply serve as a marker of ethnicity rather than a direct cause. To confirm a causal relationship between genetic variants and disease traits, robust statistical analyses must be applied. These analyses should incorporate epidemiological approaches that examine families or unrelated individuals while considering detailed disease-related data (Naj et al. 2012).

To test genetic hypotheses in families (Seiller-moiseiwitsch 2014). Two commonly used measures to assess familial aggregation are sibling recurrence risk (SRR) and relative risk for siblings (RSS). These metrics help quantify the likelihood of disease occurrence among siblings of affected individuals compared with the general population. Whereas SRR refers to the probability that a sibling of an affected individual (D) will also develop the disease. RRS measures the risk of disease among siblings of affected individuals (D) compared with the risk among siblings of unaffected individuals ($\check{D}$).

Mathematically, let D represent a diseased individual and $\check{D}$ denote a disease-free individual. The RSS is calculated as:

$$\mathrm{RSS} = \frac{P\left(\dfrac{D}{\text{Sibling of } D}\right)}{P\left(\dfrac{D}{\text{Sibling of D}}\right)} - \tag{3.1}$$

In Eq. 3.1, $P(D/\text{Sibling of } D)$ is the probability that a sibling of an affected individual also has the disease, and $P(D/\text{Sibling of } \check{D})$ is the probability that a sibling of a disease-free individual also has the disease. A higher RSS value indicates stronger familial aggregation, suggesting a greater genetic or shared environmental contribution to disease susceptibility.

The influence of familial aggregation of environmental risk factors on disease clustering, in the absence of genetic susceptibility, was examined by analyzing a risk factor with relative risk (R) and population exposure frequency (f), which was shared among siblings. Researchers evaluated how variations in R, f, and sibling exposure correlation (c) impact RRS and SRR. Findings indicate that both RRS and SRR increase with higher R and greater sibling exposure correlation (c), but tend to decrease when the exposure frequency (f) is high. Moreover, environmental factors with $R < 10$ result in low RRS and SRR, even when $c = 1$ (complete exposure correlation among siblings). These results suggest that environmental factor clustering alone is insufficient to explain the observed high RRS and SRR in many diseases, reinforcing the idea that genetic factors likely play a significant role in familial disease aggregation.

Relative risk (R or λ) in relatives of familial aggregation in AD is assessed using a pedigree approach and calculated with the formula below (Ganapati et al. 2019)

$$\text{RR} = \frac{\text{prevalence of AD in relatives of patients with AD}}{\text{highest prevalence of AD in general population}} \tag{3.2}$$

To assess familial aggregation, it is essential to account for factors beyond just parent-offspring relationships, including age distribution, family size, and the type of relatives considered in the analysis. Assessing aggregation in relatives beyond parents is quantified mainly in two ways. First, the number of affected FDRs is calculated by determining how many FDRs (parents, siblings, or children) are affected by the disease. It also takes into account the age of onset, distinguishing between early-onset (<55 years) and late-onset (≥55 years) cases. Second, the family history score (FHS) is calculated; this measure compares the observed number of disease cases (O) in a family to the expected number (E) based on general population incidence rates. The expected number (E) is calculated by multiplying the age- and sex-specific person-years in the family by the age- and sex-specific incidence rates in the general population. By incorporating these measures, researchers can more accurately evaluate familial disease risk and distinguish between genetic and environmental influences on disease clustering. The FHS is calculated using the formula:

$$\text{FHS} = \frac{O - E}{\sqrt{E}} \tag{3.3}$$

Based on the FHS value, familial disease risk is categorized as follows: Negative family history (FHS < 0.5), mild positive family history (0.5 ≤ FHS < 1.0), definite positive family history (1.0 ≤ FHS < 2.0), very strong family history (FHS ≥ 2.0). These classifications help quantify the degree of familial disease risk, allowing for a more structured assessment of genetic and environmental contributions to disease susceptibility.

Another method for evaluating excess familial risk relies on individual-level population risk estimates rather than family-level risk. This approach uses a test statistic that assesses whether the number of affected individuals in a family significantly deviates from what is expected based on population incidence rates. The test statistic is given by:

$$\sum_{i=1}^{n} \frac{\left(Y_i - p_i\right)}{p_i\left(1 - p_i\right)} \tag{3.4}$$

where Y_i represents the affection status of individual i (1 if affected, 0 if unaffected), p_i is the expected probability of disease for individual i, derived from population incidence rates, and n is the total number of individuals in the family. The reference distribution for this test statistic is obtained by computing its value across all 2^n possible family configurations, ranging from no affected members to all affected members. This approach allows researchers to determine if disease clustering within families is significantly higher than expected based on population risk estimates, providing insight into the potential influence of genetic or environmental factors on familial disease aggregation (Seiller-moiseiwitsch 2014).

The fundamental epidemiologic principles outlined above can be applied to identify familial aggregation and assess whether clustering of diseases is driven by genetic or environmental factors. These approaches are essential not only for detecting familial aggregation but also for establishing a critical foundation in understanding the genetic contributions to disease occurrence within the broader population (Khoury et al. 1993). Thus, the field of genetic epidemiology plays a crucial role in determining heritability by integrating genetics, epidemiology, and biostatistics. It aims to develop and apply optimal study designs and statistical methods to identify genes that influence the risk of complex and heterogeneous diseases. By combining these disciplines, genetic epidemiology enhances our understanding of disease mechanisms and contributes to more effective risk assessment, prevention, and therapeutic strategies (Duggal et al. 2019).

3.4 Conclusion

Familial aggregation and hereditary factors play a critical role in the development of ADs, highlighting the etiological complexity of AD due to its interplay between genetic predisposition and environmental influences. While significant progress has been made in identifying susceptibility loci through GWAS and epidemiological studies, much of the genetic basis of ADs remains to be explored. The variability in heritability across different autoimmune conditions suggests the need for further research into rare genetic variants, epigenetic modifications, and gene–environment interactions. Advances in genetic screening and personalized medicine hold promise for early diagnosis, risk prediction, and targeted interventions to improve patient welfare. Further understanding of the genetic landscape of ADs will pave the way for more effective prevention and therapeutic strategies in the future.

References

Anaya JM (2006) Autoimmunity and autoimmune disease: a Colombian construction beyond deconstructions. Autoimmun Rev 5(3):165–166

Anaya JM, Corena R, Castiblanco J, Rojas-Villarraga A, Shoenfeld Y (2007) The kaleidoscope of autoimmunity: multiple autoimmune syndromes and familial autoimmunity. Expert Rev Clin Immunol 3:623–635

Anaya JM et al (2013) Autoimmunity: from bench to bedside. El Rosario University Press, Bogota

Bianchi I, Lleo A, Gershwin ME, Invernizzi P (2012) The X chromosome and immune associated genes. J Autoimmun 38:187–192

Billi AC, Kahlenberg JM, Gudjonsson JE (2019) Sex bias in autoimmunity. Curr Opin Rheumatol 31(1):53–61

Bodis G, Toth V, Schwarting A (2018) Role of human leukocyte antigens (HLA) in autoimmune diseases. Rheumatol Ther 5:5–20

Borysewicz-Sanczyk H, Sawicka B, Wawrusiewicz-Kurylonek N et al (2020) Genetic association study of IL2RA, IFIH1, and CTLA-4 polymorphisms with autoimmune thyroid diseases and type 1 diabetes. Front Pediatr 8(481):1–11

Brickner AG (2006) Mechanisms of minor histocompatibility antigen immunogenicity: the role of infinitesimal versus structurally profound polymorphisms. Immunol Res 36:33–41

Cardenas-Roldan J, Rojas Villarraga A, Anaya JM (2013) How do autoimmune diseases cluster in families? A systematic review and meta-analysis. BMC Med 13(11):73

Carrel L, Willard HF (2005) X-inactivation profile reveals extensive variability in X-linked gene expression in females. Nature 434(7031):400–404

Castiblanco J, Sarmiento-Monroy JC, Mantilla RD, Rojas-Villarraga A, Anaya JM (2015) Familial aggregation and segregation analysis in families presenting autoimmunity, polyautoimmunity, and multiple autoimmune syndrome. J Immunol Res 2015:572353

Cooper GS, Bynum ML, Somers EC (2009) Recent insights in the epidemiology of autoimmune diseases: improved prevalence estimates and understanding of clustering of diseases. J Autoimmun 33(3–4):197–207

Criswell LA, Pfeiffer KA, Lum RF, Gonzalez B, Novitzke J, Kern M et al (2005) Analysis of families in the multiple autoimmune disease genetics consortium (MADGC) collection: the PTPN22 allele associates with multiple autoimmune phenotypes. Am J Hum Genet 76(5):61–71

Duggal P, Ladd-Acosta C, Ray D, Beaty TH (2019) The evolving field of genetic epidemiology: from familial aggregation to genomic sequencing. Am J Epidemiol 188(12):2069–2077

Eaton WW, Rose NR, Kalaydjian A, Pedersen MG, Mortensen PB (2007) Epidemiology of autoimmune diseases in Denmark. J Autoimmun 29:1–9

Eisenbarth GS, Gottlieb PA (2004) Medical progress: autoimmune polyendocrine syndromes (review). N Engl J Med 350(20):2068–2079

Fernando MMA, Stevens CR, Walsh EC et al (2008) Defining the role of the MHC in autoimmunity: a review and pooled analysis. PLoS Genet 4(4):1–9

Ganapati A et al (2019) Study of familial aggregation of autoimmune rheumatic diseases in Asian Indian patients with systemic lupus erythematosus. Rheumatol Int 39:2053–2060

Horton R, Wilming L, Rand V et al (2004) Gene map of the extended human MHC. Nat Rev Genet 5:889–899

Hossen MM, Ma Y, Yin Z, Xia Y, Du J, Huang JY, Huang JJ, Zou L, Ye Z, Huang Z (2023) Current understanding of CTLA-4: from mechanism to autoimmune diseases. Front Immunol 14:1–16
Houcken J, Degenhart C, Bender K, Konig J, Frommer L, Kahaly GJ (2018) PTPN22 and CTLA-4 polymorphisms are associated with polyglandular autoimmunity. J Clin Endocrinol Metab 103(5):1977–1984
Hu X, Daly M (2012) What have we learned from six years of GWAS in autoimmune diseases, and what is next? Curr Opin Immunol 24:571–575
Invernizzi P, Gershwin ME (2009) The genetics of human autoimmune disease. J Autoimmun 33:290–299
Janeway CA Jr, Travers P, Walport M et al (2001) Immunobiology: the immune system in health and disease. In: The major histocompatibility complex and its functions, 5th edn. Garland Science, New York
Khoury MJ, Beaty TH, and Cohen BH (1993) Epidemiologic approaches to familial aggregation. In: Fundamentals of genetic epidemiology. Oxford Academic, New York. (Online edn, 31 Oct. 2023)
Lessard CJ, Ice JA, Adrianto I, Wiley GB et al (2012) The genomics of autoimmune disease in the era of genome-wide association studies and beyond. Autoimmun Rev 11(4):267–275
Liang YL, Wu H, Shen X, Li PQ, Yang XQ, Liang L, Tian WH, Zhang LF, Xie XD (2012) Association of STAT4 rs7574865 polymorphism with autoimmune diseases: a meta-analysis. Mol Biol Rep 39(9):8873–8882
Matzaraki V, Kumar V, Wijmenga C, Zhernakova A (2017) The MHC locus and genetic susceptibility to autoimmune and infectious diseases. Genome Biol 18(76):1–21
Morahan G, Peeva V, Mehta M, Williams R (2008) Systems genetics can provide new insights in to immune regulation and autoimmunity. J Autoimmun 31:233–236
Naj AC, Park YS, Beaty TH (2012) Detecting familial aggregation. In: Elston R, Satagopan J, Sun S (eds) Statistical human genetics. Methods in molecular biology, vol 850. Humana Press
Niewold TB, Wu SC, Smith M, Morgan GA, Pachman LM (2011) Familial aggregation of autoimmune disease in juvenile dermatomyositis. Pediatrics 127(5):e1239–e1246
Olson E, Geng J, Raghavan M (2020) Polymorphisms of HLA-B: influences on assembly and immunity. Curr Opin Immunol 64:137–145
Orozco G, Rueda B (2019) Common genetic component in autoimmunity. J. In: Carmona F (ed) Genetics of rare autoimmune diseases. Rare diseases of the immune system. Springer, Cham
Pugliese A (2004) Genetics of type 1 diabetes. Endocrinol Metabol Clin North Am 33:1–16
Rajalingam A, Ganjiwale A (2024) Identification of common genetic factors and immune-related pathways associating more than two autoimmune disorders: implications on risk, diagnosis, and treatment. Genomics Inform 22(10):1–17
Ramos PS, Criswell LA, Moser KL, Comeau ME, Williams AH, Pajewski NM et al (2011) A comprehensive analysis of shared loci between systemic lupus erythematosus (SLE) and sixteen autoimmune diseases reveals limited genetic overlap. PLoS Genet 7:e1002406
Ramos PS, Shedlock AM, Langefeld CD (2015) Genetics of autoimmune diseases: insights from population genetics. J Hum Genet 60:657–664
Richard-Micelic C, Criswell LA (2012) Emerging patterns of genetic overlap across autoimmune disorders. Genome Med 4(1):6
Seiller-moiseiwitsch F (2014) Familal aggregation. Wiley, Hoboken. StatsRef: Statistics Reference Online
Seldin MF (2015) The genetics of human autoimmune disease: a perspective on progress in the field and future directions. J Autoimmun 64:1–12
Serrano NC, Millan P, Paez MC (2006) Non-HLA associations with autoimmune diseases. Autoimmun Rev 5:209–214
Shamilov R, Aneskievich BJ (2018) TNIP1 in autoimmune diseases: regulation of Toll-like receptor signaling. J Immunol Res 2018:1–13
Sinkus ML, Adams CE, Logel J, Freedman R, Leonard S (2013) Expression of immune genes on chromosome 6p21.3-22.1 in schizophrenia. Brain Behav Immun 32:51–62

Skov J, Eriksson D, Kuja-Halkola R et al (2020) Co-aggregation and heritability of organ-specific autoimmunity: a population-based twin study. Clinical study. Eur J Endocrinol 182(5):473–480

Souyris M, Cenac C, Azar P, Daviaud D, Canivet A, Grunenwald S et al (2018) TRL 7 escapes X chromosome inactivation in immune cells. Sci Immunol 3(19):eaap8855

Subramanian A et al (2020) Ethnicity-based differences in the incident risk of allergic diseases and autoimmune disorders: a UK-based retrospective cohort study of 4.4 million participants. Clin Exp Allergy 51(1):144–147

Tizaoui K, Shin JI, Jeong GH, Yang JW, Park S, Kim JH et al (2022) Genetic polymorphism of PTPN22 in autoimmune diseases: a comprehensive review. Medicina 58:1034

Tobon GJ et al (2012) Are autoimmune diseases predictable? Autoimmun Rev 11:259–266

Tokic S, Zizkova V, Stefanic M et al (2020) HLA-A, -B, -C, -DRB1, -DQA1, and –DQB1 allele and haplotype frequencies defined by next generation sequencing in a population of East Croatia blood donors. Sci Rep 10(5513):1–13

Torkamani A, Pham P, Libiger O, Bansal V, Zhang G, Scott-Van Zeeland AA et al (2012) Clinical implications of human population differences in genome-wide rates of functional genotypes. Front Genet 3:211

Trowsdale J, Knight JC (2013) Major histocompatibility complex genomics and human disease. Annu Rev Genomics Hum Genet 14:301–323

Waithman J, Moffat JM, Patterson NL, Van Beek AE, Mintern JD (2014) Antigen presentation. In: Reference module in biomedical sciences. Elsevier

Wandstrat A, Wakeland E (2001) The genetics of complex autoimmune diseases: non-MHC susceptibility genes. Nat Immunol 2(9):802–809

Ye J, Gillespie KM, Rodriguez S (2018) Unravelling the roles of susceptibility loci for autoimmune diseases in the post-GWAS Era. Genes (Basel) 9(8):377

Chapter 4
Decoding Environmental Triggers: Advancing Therapeutic Strategies and Preventive Measures for Rheumatoid Arthritis, Type 1 Diabetes, and Multiple Sclerosis

Jyotsna Bandi, Dokula Varaha Surya Prakash, Anand Kumar Nelapati, John Babu Dulla, and Thirupati Chinna Venkateswarulu

4.1 Introduction

Autoimmune disorders are a group of conditions in which the immune system mistakenly attacks the body's own tissues, leading to chronic inflammation, tissue damage, and impaired organ function. These diseases arise from a complex interplay of genetic susceptibility, epigenetic changes, and environmental factors that disrupt immune tolerance. While genetic variations in immune-regulatory genes lay the groundwork for susceptibility, environmental triggers such as UV radiation, infections, obesity, and dietary patterns play a crucial role in initiating or exacerbating the disease. This multifactorial nature highlights the importance of understanding both genetic predispositions and environmental influences in the development and progression of autoimmune conditions. Autoimmune diseases are categorized into organ-specific (e.g., type 1 diabetes [T1D]) and systemic (e.g., rheumatoid arthritis [RA], multiple sclerosis [MS]) forms, with both types causing widespread health burdens. Affecting approximately 5–8% of the global population, autoimmune diseases disproportionately impact women, likely due to the influence of hormonal factors such as estrogen (Albarbar and Aga 2024; Lassmann and Bradl 2017). This chapter explores the impact of environmental factors on the development and progression of autoimmune diseases (ADs), focusing on T1D, RA, and MS. While genetic factors are important, environmental triggers such as infections, pollutants,

J. Bandi · A. K. Nelapati (✉) · J. B. Dulla · T. C. Venkateswarulu
Department of Biotechnology, Vignan's Foundation for Science Technology and Research (Deemed to be University), Vadlamudi, Guntur, Andhra Pradesh, India

D. V. S. Prakash
Department of Biotechnology, Meerut Institute of Engineering and Technology, Meerut, Uttar Pradesh, India

Y.-O. Son, M. Ghosh (eds.), *Cutting-Edge Strategies in Drug Delivery and Immunotherapy for Autoimmune Disorders*,
https://doi.org/10.1007/978-981-95-0520-3_4

diet, smoking, and stress play a crucial role in shaping immune responses and influencing disease onset and progression (Anaya et al. 2016). T1D arises from autoimmune destruction of pancreatic β-cells, with strong genetic predisposition tied to HLA class II molecules (DQ, DR) and other loci. Environmental triggers like infections, dietary factors, and psychosocial influences also contribute, as evidenced by migration and socioeconomic studies (Atkinson and Eisenbarth 2001). The global rise in T1D, especially in developed nations, underscores the role of rapid environmental changes. Preventive measures targeting modifiable factors, while challenging to implement, hold promise for reducing incidence (Knip et al. 2005; Butalia et al. 2016). RA is a chronic autoimmune disease characterized by persistent synovial inflammation, joint pain, and progressive damage leading to functional impairment (McInnes and Schett 2011). Symptoms extend beyond joints, including fatigue and systemic organ involvement, severely impacting quality of life (Kim et al. 2025). The disease arises from a complex interplay of genetic factors, such as HLA-DR4 and HLA-DR1 alleles, and environmental triggers like smoking and infections, which promote autoantibody production and immune dysregulation (Smolen et al. 2016; klareskog et al. 2006b). Proinflammatory cytokines, including TNF-α and IL-6, drive chronic inflammation and joint destruction, marked by synovial hyperplasia and bone erosion (Firestein and McInnes 2017; Nygaard and Firestein 2020). Despite advances in biologics, RA affects 1% of the population, emphasizing the need for early diagnosis and personalized therapies to prevent long-term complications (McInnes and Schett 2011; Smolen et al. 2016). MS is characterized by neuroinflammation and axonal degeneration, with global prevalence influenced by latitude. Goldberg first identified that vitamin D deficiency, linked to reduced UVB exposure, has been identified as a potential risk factor (Abu Hajleh et al. 2020). Epidemiological evidence highlights associations between low vitamin D levels, Epstein-Barr virus (EBV) infection, and MS, particularly following infectious mononucleosis (Akeel Naji 2021). Smoking trends and environmental factors also modulate disease risk. Among the multifactorial causes of autoimmune diseases, environmental triggers play a pivotal role. These triggers, in conjunction with genetic predispositions and epigenetic modifications, contribute significantly to the disruption of immune tolerance (Zhang and Lu 2018). Environmental factors such as infections, dietary components, toxins, and microbiome alterations can influence the immune system, either by directly activating autoreactive immune cells or by creating an inflammatory milieu that lowers the threshold for autoimmunity (Bach 2002). The interplay between these environmental stimuli and an individual's genetic and epigenetic landscape determines susceptibility to autoimmune diseases (Baranzini and Oksenberg 2017). A deeper understanding of these triggers and their underlying mechanisms is critical for developing targeted therapies and preventive strategies to mitigate the onset and progression of autoimmunity. Autoreactive T cells play a central role in the development of autoimmune diseases. These cells can be activated through mechanisms such as molecular mimicry, dual TCRs, and chimeric TCRs. Upon activation, they differentiate into CD4+ and CD8+ T cells, leading to direct cellular attacks or the

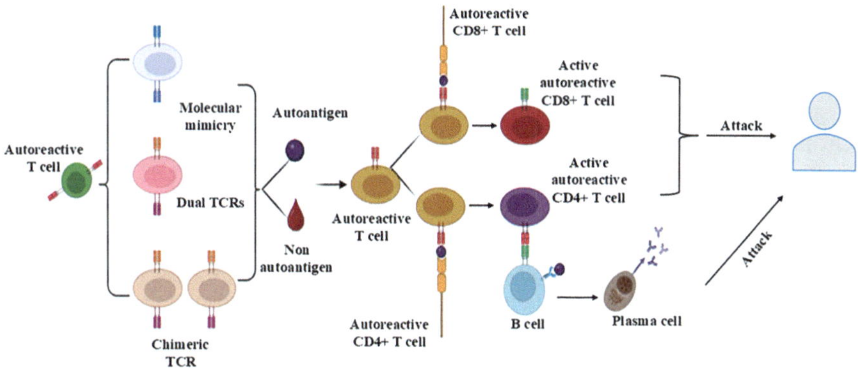

Fig. 4.1 Mechanisms of autoreactive T-cell activation and their role in autoimmune disease progression. (Source: Song et al. (2024))
The figure illustrates the mechanisms through which autoreactive T cells contribute to autoimmune diseases. Autoreactive T cells are activated via three pathways: molecular mimicry, where foreign antigens mimic autoantigens; dual TCRs, where one TCR recognizes nonautoantigens and the other targets autoantigens; and chimeric TCRs, which involve different combinations of Vα and Vβ regions enabling the recognition of both autoantigens and nonautoantigens. Upon encountering autoantigens, autoreactive T cells differentiate into CD4+ and CD8+ subsets. Activated CD8+ T cells directly attack target cells, while CD4+ T cells stimulate B cells to differentiate into plasma cells that produce autoantibodies, amplifying the immune response. This process ultimately leads to tissue damage and the development of autoimmune diseases

production of autoantibodies through B-cell stimulation, as illustrated in Fig. 4.1 (Song and Wu 2024).

4.2 Types of Environmental Triggers

4.2.1 Infections and Pathogens

Infections can play a significant role in triggering or exacerbating autoimmunity through various mechanisms, as evidenced in multiple autoimmune diseases. For T1D, infection-induced islet autoimmunity has been demonstrated in animal models. Acute viral infections, such as with EMC-D virus, or persistent low-grade infections, as seen with RIP-LCMV models, have been shown to trigger autoimmunity (Alves Abrantes et al. 2024). Molecular mimicry, where pathogen epitopes resemble islet antigens, also contributes; for instance, a shared sequence between GAD65 and Coxsackie B virus supports this hypothesis (Alves Abrantes et al. 2024). Human enteroviruses, particularly Coxsackie B viruses, have been implicated in T1D by inducing islet-specific autoantibodies, though results across studies remain inconsistent due to variations in research methodologies (Jun and Yoon 2003; Zorena et al. 2022). While no specific bacterial pathogen has been definitively associated with T1D, bacterial superantigens, such as those from *Mycobacterium tuberculosis* and *Streptococcus pyogenes*, may contribute to immune activation and potentially trigger autoimmunity (Peng and Hagopian 2007; Filippi and von Herrath 2008). In RA, infections and pathogens are well-recognized as potential environmental triggers in genetically predisposed individuals. For example, *Porphyromonas gingivalis*, a periodontal pathogen, facilitates autoantibody production against citrullinated peptides by promoting protein citrullination (Maresz et al. 2013). Epstein-Barr virus (EBV) has been associated with RA due to its ability to stimulate autoreactive B cells and induce rheumatoid factor production (Balandraud and Roudier 2018). Additionally, infections with *Escherichia coli* and *Proteus mirabilis* have been linked to RA via cross-reactivity between microbial and self-antigens (Ebringer and Rashid 2006). These findings underscore the importance of infections as triggers for immune dysregulation in RA. MS is another autoimmune disorder influenced by infections. Various bacterial and viral agents, including Herpesviridae, human endogenous retroviruses, and enterotoxins, have been implicated in MS onset or exacerbation (Al-Hussaniy 2023). Common infections in the respiratory, gastrointestinal, and urogenital tracts are thought to alter the neuroimmune system in genetically susceptible individuals, potentially leading to autoimmunity (Al-Hussaniy 2023). Viral infections, such as Theiler's murine encephalomyelitis in animal models, have been shown to induce demyelination, reflecting similar processes in MS (Al-hussaniy and AL-Biati 2022). Despite these findings, no specific pathogen has been conclusively identified as the cause of MS, possibly due to the transient nature of infections and the presence of autoreactive cells in both affected and unaffected individuals (Al-hussaniy and AL-Biati 2022). Furthermore, bacterial infections, such as *Helicobacter pylori*, have been linked to autoimmune diseases, including psoriasis and T1D, suggesting a broader role for microbial triggers in systemic autoimmune processes (Grasso and Chiarelli 2024).

4.2.2 Dietary Factors

Autoimmune diseases, such as T1D, RA, and MS, are chronic conditions where the body's immune system mistakenly attacks its own tissues. While genetics and environmental factors contribute significantly to the development of these diseases, emerging evidence suggests that dietary factors also play a crucial role in modulating immune responses and disease progression. Certain dietary patterns, nutrient deficiencies, or excesses can either promote or inhibit inflammation, which is a key feature in the pathogenesis of autoimmune disorders. Understanding how diet impacts immune function in these conditions is essential for developing dietary strategies that may complement conventional medical treatments and potentially reduce disease activity or even prevent the onset of autoimmune diseases in susceptible individuals. Below is an exploration of how specific dietary factors are linked to the development and progression of T1D, RA, and MS. T1D is a complex autoimmune condition where dietary factors, alongside genetic and environmental triggers, influence its onset and progression. Early-life exposure to cow's milk proteins has been associated with an increased risk of T1D in genetically predisposed individuals, likely due to molecular mimicry, where immune responses to milk proteins cross-react with pancreatic β-cells (Knip and Simell 2012). Similarly, high-gluten diets have been implicated in T1D development, as gluten may compromise gut integrity and promote immune dysregulation by increasing intestinal permeability (Lund-Blix et al. 2019). Vitamin D, essential for modulating immune function, plays a critical protective role; its deficiency during early life has been strongly linked to increased risk of T1D (Hyppönen et al. 2001). Omega-3 fatty acids, known for their antiinflammatory properties, may mitigate T1D development by reducing the production of inflammatory cytokines and improving immune tolerance (Poggioli et al. 2023). Furthermore, a diet high in processed foods and advanced glycation end products (AGEs) can exacerbate oxidative stress and systemic inflammation, potentially accelerating autoimmune processes (Uribarri et al. 2010). These findings suggest that dietary interventions focusing on reducing inflammatory foods and promoting antiinflammatory nutrients could play a complementary role in T1D prevention and management.

RA, another autoimmune disorder, is significantly influenced by dietary patterns that modulate inflammation. Diets high in red and processed meats are associated with increased levels of AGEs, which contribute to oxidative stress and immune dysregulation, exacerbating RA symptoms (Delrue et al. 2023). In contrast, the Mediterranean diet, rich in antiinflammatory components like fruits, vegetables, olive oil, and omega-3-rich fish, has demonstrated a protective effect, reducing RA incidence and improving disease outcomes (Sköldstam et al. 2003). High sugar consumption, which promotes obesity and increases inflammatory cytokines, is another dietary factor linked to worsened RA outcomes (Luo et al. 2023). Omega-3 fatty acids, particularly eicosapentaenoic acid (EPA) and docosahexaenoic acid (DHA), found in fatty fish, help suppress proinflammatory eicosanoids and cytokines, reducing disease activity (Innes and Calder 2018). For some individuals, gluten sensitivity may exacerbate RA symptoms, as gluten can trigger immune activation and inflammation in susceptible populations (Bruzzese et al. 2021). Vitamin D deficiency, which compromises immune regulation, has been implicated in both the increased risk and severity of RA, highlighting the importance of ensuring adequate levels for immune homeostasis (Song et al. 2012). MS is a chronic autoimmune condition where dietary influences impact immune regulation, neuroinflammation, and disease progression. High salt intake has been shown to promote the differentiation of proinflammatory Th17 cells, exacerbating the autoimmune response central to MS pathogenesis (Kleinewietfeld et al. 2013). Similarly, diets high in saturated fats from red and processed meats contribute to oxidative stress and inflammation, worsening disease activity (Esposito et al. 2018). Omega-3 fatty acids, particularly from fatty fish, offer significant antiinflammatory benefits by downregulating the production of proinflammatory cytokines, which may help slow disease progression (Calder 2020; Poggioli et al. 2023). Vitamin D deficiency, a well-established risk factor for MS, impairs immune tolerance and is linked to greater disease susceptibility and activity, emphasizing the importance of maintaining adequate levels (Ascherio et al. 2016). Gut microbiota composition, influenced by dietary fiber intake, also plays a key role in modulating immune responses in MS. The fermentation of dietary fiber into short-chain fatty acids (SCFAs) has protective effects, as SCFAs regulate inflammation and support gut barrier integrity (Jangi et al. 2016). Additionally, gluten sensitivity may exacerbate symptoms in some MS patients by increasing gut permeability and triggering systemic immune activation (Biesiekierski et al. 2013). These insights highlight the potential of targeted dietary strategies to reduce inflammation and support disease management.

4.2.3 Toxins and Chemicals

Exposure to toxins and chemicals is increasingly recognized as a significant factor in the development of autoimmune diseases, as these substances disrupt immune regulation and provoke inflammatory responses. Such environmental triggers often interact with genetic predispositions, facilitating the onset of diseases like T1D, RA, and MS. In the case of T1D, environmental toxins like nitrosamines found in processed foods and industrial emissions, and heavy metals such as arsenic and cadmium, contribute to β-cell damage through oxidative stress and inflammation. (Zorena et al. 2022). Additionally, prenatal and early-life exposure to pollutants like phthalates, bisphenol A (BPA), polychlorinated biphenyls (PCBs), and dioxins can impair immune development and increase susceptibility to autoimmune conditions (Bodin et al. 2015). Viral infections, particularly enteroviruses, may also act as molecular mimics, prompting autoimmune attacks on pancreatic β-cells (Akhbari et al. 2020). In RA, exposure to cigarette smoke promotes protein citrullination, triggering the formation of anticitrullinated protein antibodies (ACPAs) that lead to joint inflammation and damage (Chang et al. 2014). Silica dust, commonly encountered in industries like mining and construction, activates macrophages and releases inflammatory cytokines, aggravating immune dysregulation and increasing RA risk (Parks et al. 1999). Heavy metals like cadmium and mercury further contribute to chronic inflammation through their immunotoxic properties, enhancing the risk of autoimmune reactions (Klareskog et al. 2006a). In MS, organic solvents, pesticides, and persistent organic pollutants disrupt the blood-brain barrier, allowing immune cells to infiltrate the central nervous system and initiate demyelination, thereby increasing MS susceptibility (Ascherio and Munger 2016). Smoking exacerbates MS progression by promoting systemic inflammation and enhancing T-cell–mediated autoimmunity (Hedström et al. 2014), while air pollution, particularly particulate matter, worsens MS symptoms and accelerates disease progression through oxidative stress pathways (Kuntić et al. 2024). These findings underscore the need for further research to better understand how environmental factors contribute to autoimmune diseases and to develop more effective preventive strategies.

4.2.4 Medications and Vaccines

Medications and vaccines have been recognized as potential environmental triggers for autoimmune diseases, particularly in genetically predisposed individuals. While these interventions play crucial roles in treating diseases and preventing infections, they can sometimes disrupt immune regulation, leading to autoimmune responses such as T1D, RA, and MS. Certain medications and vaccines have been implicated in the development of autoimmune diseases, including T1D, RA, and MS. In T1D, immune checkpoint inhibitors (ICIs) used in cancer therapy, such as anti-PD-1 and anti-CTLA-4 antibodies, have been associated with the rapid onset of autoimmune diabetes in predisposed individuals. These medications may trigger autoimmunity by enhancing T-cell activity and reducing tolerance to self-antigens (Quandt et al. 2020). Similarly, glucocorticoids, while antiinflammatory, may contribute to immune dysregulation in rare cases, potentially exacerbating preexisting autoimmunity (Dendrou et al. 2015). For RA, biologic agents such as TNF inhibitors, used to treat autoimmune diseases, paradoxically induce autoimmunity in some cases. These medications can lead to the development of drug-induced lupus-like syndrome or the production of autoantibodies, potentially triggering new autoimmune conditions (Moulis et al. 2014). Additionally, statins, commonly prescribed for hyperlipidemia, have been reported to induce inflammatory myopathies, highlighting the potential for medications to modulate immune responses in susceptible individuals (Mammen 2016). In MS, disease-modifying therapies such as interferon-β and natalizumab can occasionally worsen autoimmune symptoms. While interferon-β is used to reduce MS relapse rates, it has been linked to the development of autoimmune thyroid disease in some patients (Antonelli et al. 2006). Natalizumab, an integrin inhibitor, can lead to immune reconstitution inflammatory syndrome (IRIS) upon discontinuation, exacerbating autoimmune activity (Yousry et al. 2006). Vaccines have also been scrutinized for their potential role in autoimmunity, although the evidence remains largely inconclusive. Some studies have suggested a possible link between vaccines, such as those for hepatitis B and HPV, and autoimmune diseases, including MS and RA, in genetically predisposed individuals. However, large-scale studies consistently demonstrate that the benefits of vaccination outweigh the minimal risk of autoimmune reactions (Langer-Gould et al. 2014). Overall, while some medications and vaccines may influence autoimmunity, their role is often context-dependent and requires further research for definitive conclusions.

4.2.5 Ultraviolet Radiation

Ultraviolet (UV) radiation plays a complex role in autoimmune diseases, acting as both a potential protective and triggering factor. While moderate UV exposure can regulate immune responses through vitamin D synthesis and immunomodulation, excessive or insufficient UV exposure may disrupt immune tolerance, influencing conditions like T1D, RA, and MS. Ultraviolet (UV) radiation has been implicated in the modulation of autoimmune diseases, including T1D, RA, and MS. Epidemiological studies have demonstrated a geographical correlation, with higher UV exposure linked to a lower prevalence of these conditions. For T1D, populations in areas with lower UV radiation are at increased risk, potentially due to UV-induced vitamin D production and its immunomodulatory effects, which include regulating cytokine expression and supporting regulatory T cells (Ponsonby et al. 2002). Similarly, in RA, UV radiation exposure has been associated with reduced disease severity, likely through its ability to suppress proinflammatory immune responses and reduce autoantibody production (Hart et al. 2011). MS, which shows a strong north-south gradient in prevalence, also appears inversely related to UV exposure, with vitamin D playing a key role in modulating immune pathways and protecting against neuroinflammation (Gombash et al. 2022). Despite the protective aspects of UV exposure, excessive exposure risks, such as skin damage and cancer, necessitate a balanced approach to utilizing UV's potential benefits.

The role of environmental triggers in the onset and progression of autoimmune diseases is multifaceted. Table 4.1 summarizes the major environmental triggers, their mechanisms of action, and the autoimmune diseases they influence, providing a consolidated overview of their impact.

4.3 Mechanisms of Environmental Triggers in Autoimmunity

4.3.1 Molecular Mimicry

Molecular mimicry occurs when foreign pathogens share structural similarities with self-antigens, leading to an immune response that mistakenly targets the body's own tissues. This mechanism is implicated in the development of several autoimmune diseases, including T1D, RA, and MS, where infections trigger immune reactions against self-proteins. Molecular mimicry is a key mechanism by which infections can trigger autoimmune diseases, including T1D, RA, and MS. In T1D, microbial

Table 4.1 Environmental triggers and associated autoimmune diseases

Trigger	Mechanism of action	Affected diseases	References
Infections (e.g., viruses, bacteria)	Molecular mimicry, bystander activation	T1D, RA, MS	Alves Abrantes et al. (2024), Coppieters et al. (2012)
Dietary factors (e.g., gluten, omega-3)	Inflammation modulation, gut microbiota changes	T1D, RA, MS	Poggioli et al. (2023), Bascuñán et al. (2025)
Toxins (e.g., silica dust, heavy metals)	Immune activation, oxidative stress	RA, MS, T1D	Parks et al. (1999), Klareskog et al. (2006a)
UV radiation	Immune regulation via vitamin D synthesis	T1D, RA, MS	Ponsonby et al. (2002), Hart et al. (2011)

antigens resembling pancreatic islet proteins can stimulate autoreactive T cells, as seen with Coxsackie B virus and glutamic acid decarboxylase (GAD65), which share structural similarities (Coppieters et al. 2012). For RA, *P. gingivalis*, a periodontal pathogen, can citrullinate host proteins, producing autoantigens that mimic self-proteins and drive autoantibody formation, contributing to joint inflammation (Muruganandam et al. 2024). Similarly, in MS, Epstein-Barr virus (EBV) epitopes can mimic myelin basic protein (MBP), triggering an immune response against myelin and leading to demyelination (Ramasamy et al. 2020). These examples underscore the pivotal role of molecular mimicry in disrupting immune tolerance and initiating autoimmune diseases.

4.3.2 Bystander Activation

Bystander activation refers to the phenomenon where immune cells, particularly T cells, are activated by inflammatory signals rather than specific recognition of their target antigen. This nonspecific activation can contribute to the development of autoimmune diseases, as seen in conditions like T1D, RA, and MS, where inflammation and cytokine release promote the activation of autoreactive T cells, triggering immune responses against self-tissues. Bystander activation refers to the phenomenon where immune cells, particularly T cells, are activated not by specific antigens but by inflammation or signals from nearby immune cells, potentially leading to autoimmune disease. In T1D, bystander activation is thought to contribute to β-cell destruction in the pancreas, particularly during viral infections or inflammation. Studies suggest that viral infections such as coxsackievirus can cause nonspecific activation of autoreactive T cells, which then attack insulin-producing cells (Shim et al. 2022). In RA, bystander activation may occur when inflammatory cytokines, such as TNF-α, induce the activation of autoreactive T cells, even in the absence of specific antigen recognition, driving joint inflammation and damage (Wehr et al. 2019). Finally, in MS, bystander activation plays a role in the pathogenesis of demyelination. In this case, inflammation in the central nervous system may lead to the activation of autoreactive T cells through local cytokine production, even in areas that are not directly infected (McCoy et al. 2006). These processes underscore the complex interplay between immune activation and autoimmunity in these diseases.

4.3.3 Epigenetic Modifications

Epigenetic modifications, including DNA methylation, histone modifications, and noncoding RNA regulation, are critical in influencing immune tolerance and the development of autoimmune diseases. In T1D, alterations in DNA methylation at key immunoregulatory genes such as *FOXP3* and *IL2RA* have been identified in individuals with T1D, suggesting epigenetic dysregulation in T regulatory cell function and autoimmunity (Zhang et al. 2021). Similarly, in RA, hypomethylation of proinflammatory genes like *TNF* and *IL6* has been observed in synovial fibroblasts, leading to persistent inflammation and joint damage (Yang et al. 2017). For MS, studies have shown differential DNA methylation patterns in immune-related pathways, including those regulating T-cell activation and differentiation, which may exacerbate neuroinflammation and autoimmunity (Chan 2020). Collectively, these findings underscore the role of epigenetic changes in disrupting immune homeostasis and driving the pathogenesis of autoimmune diseases.

4.3.4 Gut Microbiome Dysbiosis

The gut microbiome, a complex community of microorganisms residing in the gastrointestinal tract, is essential for maintaining immune homeostasis and tolerance. Dysbiosis, or an imbalance in the gut microbial composition, can disrupt this equilibrium, leading to heightened immune activation and the development of autoimmune diseases in genetically predisposed individuals. In T1D, a reduction in gut microbial diversity and a decrease in beneficial bacteria such as *Bifidobacterium* and *Lactobacillus* have been observed, contributing to intestinal barrier dysfunction and promoting islet autoimmunity (Kostic et al. 2015). Similarly, in RA, studies have linked the overrepresentation of *Prevotella copri* with disease onset, as it may trigger mucosal immune responses and systemic inflammation through molecular mimicry and dysregulated T-cell activation (Scher et al. 2013). In MS, gut microbiome dysbiosis characterized by increased levels of proinflammatory species like *Akkermansia muciniphila* and reduced *Prevotella* species has been associated with heightened neuroinflammation and altered peripheral immune regulation, exacerbating autoimmune processes (Cekanaviciute et al. 2017). These findings highlight the gut microbiome's significant influence on immune system regulation and its potential role as a therapeutic target for autoimmune diseases.

4.4 Case Studies and Emerging Evidence

4.4.1 Environmental Contributions to Type 1 Diabetes

Environmental factors play a pivotal role in triggering the onset of T1D in genetically predisposed individuals. For example, research has shown that early-life exposure to enteroviral infections, such as coxsackievirus B, can be associated with the development of β-cell autoimmunity, supporting the findings from the TEDDY study, which highlights enteroviral infections as significant triggers (Peng and Hagopian 2006). Additionally, cow's milk consumption in infancy has been implicated in T1D development. Studies suggest that early exposure to unhydrolyzed cow's milk formula may contribute to the risk of β-cell autoimmunity, as observed in longitudinal studies that align with findings from the TRIGR pilot study (Luopajärvi et al. 2008). Vitamin D deficiency during childhood has also been linked to the onset of T1D. Research has found significantly lower serum vitamin D levels in children who later developed T1D, supporting the hypothesis that adequate vitamin D intake may have a protective role (Savastio et al. 2016). Furthermore, gut microbiome dysbiosis, characterized by alterations in microbial diversity, has been associated with T1D development. Studies have shown that disruptions in gut microbiota, often triggered by factors such as repeated antibiotic use, may contribute to the onset of islet autoimmunity (Harbison et al. 2019). These examples collectively demonstrate the multifactorial nature of environmental triggers in the development of T1D.

4.4.2 Environmental Triggers in Rheumatoid Arthritis

Real-world case studies and research provide compelling evidence linking environmental triggers to RA. Smoking has long been identified as a major risk factor for the disease, with studies showing that prolonged smoking significantly increases the likelihood of developing RA. A Swedish cohort study revealed that individuals who smoked for over 20 years faced a doubled risk of RA (Källberg et al. 2011). Smoking-induced inflammation is known to drive the production of anticitrullinated protein antibodies (ACPAs), a hallmark of RA, which is supported by a case in which smoking was linked to the development of seropositive RA (Linn-Rasker et al. 2006). Occupational exposure to silica dust is another well-established environmental trigger for RA, with research showing that prolonged silica exposure can lead to progressive, seropositive RA in genetically predisposed individuals (Min et al. 2021). Chronic periodontal disease, particularly caused by the bacterium *P. gingivalis*, has been linked to RA, as bacterial-induced citrullination of proteins can initiate the autoimmune response. Studies suggest that individuals with poor oral health are at heightened risk for developing RA (Araújo et al. 2015). Furthermore, environmental pollution, especially air pollution from high-traffic areas, has been associated with an increased risk of RA. A case-control study in Vancouver found that individuals exposed to higher levels of nitrogen dioxide had elevated inflammatory cytokines and a higher likelihood of developing RA (Hart et al. 2013). Additionally, diet plays a significant role, with high sodium intake exacerbating inflammation through the promotion of Th17-mediated immune responses. Evidence suggests that individuals who switched to a low-sodium diet experienced significant improvement in symptoms, supporting the role of diet in RA management (Scrivo et al. 2019). These case studies underscore the complex interplay of environmental and genetic factors in the pathogenesis of RA.

4.4.3 Environmental Factors in Multiple Sclerosis

Environmental factors play a significant role in the onset and progression of MS, with case studies and emerging evidence shedding light on specific triggers. Smoking has been consistently identified as a risk factor for MS, with studies suggesting that smokers have a higher likelihood of developing the disease. A study found that smoking significantly increased the risk of MS in individuals with a genetic predisposition (Degelman and Herman 2017). Another environmental factor linked to MS is vitamin D deficiency, with low serum levels of vitamin D being associated with an increased risk of disease onset. Research has demonstrated that people living in regions with lower sunlight exposure have a higher prevalence of MS, and case studies have shown that vitamin D supplementation may help reduce relapse rates in patients (Ascherio et al. 2014). Viral infections, particularly Epstein-Barr virus (EBV), are also strongly associated with MS. A case study showed that EBV infection, along with a genetic predisposition, significantly increased the risk of MS development (Hedström et al. 2020). Furthermore, exposure to environmental toxins, such as pesticides, has been linked to MS risk. Studies have shown that individuals living in areas with higher pesticide exposure have a greater incidence of MS, supporting the role of environmental pollutants in the disease's onset (Parrón et al. 2011). These case studies and research findings highlight the complex interaction between environmental factors and genetic susceptibility in the development of MS.

4.5 Preventive Strategies and Risk Mitigation

4.5.1 Early Identification of Risk Factors

Preventive strategies and early identification of risk factors are crucial in mitigating the onset of autoimmune diseases such as T1D, RA, and MS. In T1D, early genetic screening and monitoring for environmental triggers, such as viral infections or cow's milk consumption, can help identify at-risk individuals. For example, studies suggest that genetic testing for HLA genotypes, combined with monitoring for enteroviral infections, may help in predicting T1D onset (Yahaya et al. 2023). In RA, early identification of environmental factors such as smoking and periodontal disease is essential for prevention. Smoking cessation programs and regular dental check-ups could be key preventive measures, as studies have shown that smoking increases the risk of RA, while periodontal disease contributes to disease progression through the induction of anticitrullinated protein antibodies (Roh 2019). In MS, early recognition of vitamin D deficiency and Epstein-Barr virus (EBV) exposure are important preventive strategies. Research has demonstrated that maintaining adequate vitamin D levels and avoiding EBV infection during childhood could reduce the risk of MS (Salzer et al. 2012). Additionally, lifestyle modifications such as smoking cessation and dietary interventions, including increased vitamin D intake, may play a role in reducing the risk of MS relapse (James et al. 2013). Early intervention and monitoring for these risk factors can significantly reduce the incidence and progression of these autoimmune diseases.

4.5.2 Lifestyle Modifications

Lifestyle modifications play a critical role in mitigating the risk of autoimmune diseases such as T1D, RA, and MS. In T1D, a balanced diet and exclusive breastfeeding during infancy, coupled with avoiding early introduction of cow's milk, have shown potential in reducing autoimmunity risk (Knip et al. 2011). For RA, smoking cessation is paramount, as smoking is strongly linked to the production of anticitrullinated protein antibodies (ACPAs) and disease development (Källberg et al. 2011). Similarly, adopting an antiinflammatory diet rich in omega-3 fatty acids and low in processed foods can reduce systemic inflammation and RA onset (Mazzucca et al. 2022). In MS, maintaining adequate levels of vitamin D through sunlight exposure or supplementation has been associated with reduced disease risk, given its immunomodulatory effects (Mokry et al. 2015). Physical activity, known to lower systemic inflammation, further contributes to reducing susceptibility across all three conditions (Warburton and Bredin 2017). These lifestyle interventions underline the importance of proactive health management in genetically predisposed individuals to prevent the onset of autoimmune diseases.

4.5.3 Public Health Interventions

Public health interventions aimed at early identification of risk factors for autoimmune diseases like T1D, RA, and MS play a vital role in mitigating their prevalence and impact. For T1D, newborn screening programs for genetic susceptibility, such as HLA typing, combined with regular monitoring of islet autoantibodies, enable early detection and intervention before clinical onset (Lee et al. 2014). In RA, community-based awareness campaigns emphasizing the risks associated with smoking and occupational hazards, such as silica exposure, can significantly reduce disease incidence, as highlighted in population studies (Källberg et al. 2011). MS prevention efforts have included widespread public health campaigns to promote adequate vitamin D intake through fortified foods, supplements, and safe sunlight exposure, given its protective role against MS onset (Mokry et al. 2015). Furthermore, global initiatives like vaccination programs are critical, as infections such as Epstein-Barr virus (EBV) have been linked to MS risk, with emerging evidence supporting the potential of EBV vaccines in prevention (Bjornevik et al. 2022). These public health measures underscore the importance of proactive surveillance, education, and preventive strategies in reducing autoimmune disease burden.

Table 4.2 Overview of environmental triggers, preventive strategies, and therapeutic implications for T1D, RA, and MS

Disease	Key environmental triggers	Preventive strategies	Therapeutic implications	References
T1D	Viral infections, cow's milk, vitamin D deficiency, gut microbiome dysbiosis	Genetic screening, monitoring autoantibodies, vitamin D supplementation, balanced microbiota through diet	Immunomodulators, antiviral therapies, nanoparticle-based delivery systems	TEDDY Study; Knip et al. (2011), Hyppönen et al. (2001), (Wang et al. 2024).
RA	Smoking, silica dust, periodontal infections, high sodium intake	Smoking cessation, workplace safety measures, oral health interventions, antiinflammatory diet	JAK inhibitors, TNF inhibitors, citrullinated protein antibody-targeted therapies	Källberg et al. (2011), Maresz et al. (2013), Scrivo et al. (2019)
MS	Vitamin D deficiency, Epstein-Barr virus (EBV), smoking, pesticides	Vitamin D supplementation, EBV vaccination, smoking cessation, monitoring toxin exposures	Monoclonal antibodies, vitamin D-based therapies, nanoparticle delivery systems	(Ascherio et al. 2016), Bjornevik et al. (2022), Ramasamy et al. (2020).

Effective preventive strategies are crucial to mitigating the impact of environmental triggers on autoimmune diseases. Table 4.2 outlines key preventive measures tailored to specific triggers and diseases, offering actionable insights for early intervention.

4.6 Therapeutic Implications

4.6.1 Targeting Environmental Triggers in Drug Development

Targeting environmental triggers in drug development has opened new therapeutic pathways for T1D, RA, and MS, focusing on mitigating disease progression by addressing modifiable risk factors. In T1D, drugs aimed at dampening immune responses triggered by viral infections, such as enteroviruses, are under investigation, with antiviral agents and vaccines showing promise in preclinical studies (Richardson and Morgan 2018). For RA, the role of smoking-induced inflammation has inspired therapies targeting citrullinated protein-specific pathways, including anticitrullinated protein antibody (ACPA)-neutralizing strategies, as well as interventions like JAK inhibitors to mitigate the downstream inflammatory cascade (klareskog et al. 2006b). In MS, therapies addressing Epstein-Barr virus (EBV) infection, a key trigger, are being explored, including anti-EBV vaccines and monoclonal antibodies like rituximab, which targets B cells implicated in EBV reactivation (Bjornevik et al. 2022). Additionally, vitamin D supplementation is being studied as both a preventive and adjunctive therapeutic strategy in MS, given its immunomodulatory effects on T-cell differentiation (Mokry et al. 2015). These advancements highlight the potential of integrating environmental risk factor targeting into precision medicine approaches for autoimmune diseases.

4.6.2 Personalized Medicine

Personalized medicine offers a transformative approach to the treatment of T1D, RA, and MS by tailoring therapies based on individual genetic, environmental, and immune profiles. In T1D, genetic screening for HLA risk alleles enables early identification of at-risk individuals, with immune-modulating therapies like teplizumab showing potential to delay disease onset in predisposed populations (Herold et al. 2019). For RA, advances in biomarker research, such as detecting anticitrullinated protein antibodies (ACPAs) and specific cytokine profiles, guide the use of targeted biologics like TNF inhibitors and JAK inhibitors, optimizing outcomes while minimizing adverse effects (Rein and Mueller 2017). MS treatment has similarly advanced through the integration of genomic and immune profiling, enabling stratification of patients for therapies like natalizumab or ocrelizumab, which target specific immune pathways implicated in disease progression (Hauser et al. 2017). These innovations underline the critical role of personalized medicine in improving efficacy and reducing the side effects of treatments for autoimmune diseases.

4.6.3 Nanoparticle-Based Drug Delivery Systems

Nanoparticle-based drug delivery systems (NDDS) are emerging as innovative therapeutic approaches for T1D, RA, and MS, offering targeted and efficient treatment with reduced systemic side effects. In T1D, nanoparticles are being utilized to deliver immunomodulatory agents such as rapamycin directly to pancreatic islets, effectively preventing β-cell destruction while minimizing systemic immune suppression (Nigam et al. 2022). For RA, liposomal formulations carrying methotrexate or TNF inhibitors enable precise delivery to inflamed joints, enhancing therapeutic efficacy and reducing gastrointestinal and hepatic toxicity (Logesh et al. 2023). Similarly, in MS, nanoparticles encapsulating agents like fingolimod or corticosteroids facilitate crossing the blood-brain barrier, targeting central nervous system lesions and reducing neuroinflammation with improved pharmacokinetics (Panghal and Flora 2024). These advancements demonstrate NDDS's potential to revolutionize autoimmune disease management by enhancing drug targeting and minimizing adverse effects.

Understanding and targeting environmental triggers have paved the way for novel therapeutic approaches. Table 4.3 highlights promising therapeutic

Table 4.3 Therapeutic implications targeting environmental triggers

Trigger	Therapeutic approach	Current status	References
Viral infections	Antiviral agents, vaccines	Preclinical trials	Richardson and Morgan (2018)
Smoking	Anticitrullinated protein therapies	Clinical trials	Klareskog et al. (2006a)
Vitamin D deficiency	Supplementation	Established adjunct therapy	Mokry et al. (2015)

interventions aimed at modulating the impact of these triggers, with an emphasis on current research progress and applications.

4.7 Future Directions and Challenges

Future research on environmental triggers and autoimmune responses in T1D, RA, and MS focuses on integrating advanced computational models, improving global monitoring, and translating discoveries into clinical applications. Computational models, such as machine learning algorithms, are increasingly used to predict environmental triggers and their complex interactions with genetic predisposition, enabling early risk stratification and personalized prevention strategies (Sadeghi et al. 2024). Enhanced global monitoring systems, such as integrating geospatial and longitudinal exposure data, aim to identify environmental hotspots linked to autoimmune diseases, providing actionable insights for public health interventions (Petit and Vuillerme 2024). Bridging basic research with clinical applications remains a priority, as advancements in understanding the mechanisms of autoimmunity are now guiding the development of targeted therapies, such as peptide-based immune modulators and tolerogenic vaccines (Yu et al. 2023). However, challenges remain in harmonizing multidisciplinary efforts and ensuring equitable access to emerging preventive and therapeutic approaches worldwide.

4.8 Conclusion

Environmental triggers are central to understanding the onset and progression of autoimmune diseases such as T1D, RA, and MS. While significant strides have been made in identifying these triggers—ranging from infections and pollutants to dietary and lifestyle factors—the integration of this knowledge into clinical and public health frameworks remains limited. Challenges persist in translating basic research findings into actionable therapies and prevention strategies due to the multifactorial nature of these diseases and interindividual variability. Key research gaps include the need for more longitudinal studies to establish causal relationships, refinement of computational models to predict environmental risk interactions, and global initiatives to harmonize data collection on environmental exposures. Addressing these gaps requires a multidisciplinary approach combining genomics, epidemiology, immunology, and public health efforts to develop precision therapies and targeted prevention strategies.

References

Abu Hajleh MN, Al-Samydai A, Al-Dujaili EAS (2020) Ginseng modulates steroid hormones action, production and metabolism; a narrative review on androgens, estrogens and corticosteroids. Indian Drugs 57:7–16

Akeel Naji H (2021) The psychosocial and economic impact of uveitis in Iraq. J Res Appl Basic Med Sci 7(4):207–215

Akhbari P, Richardson SJ, Morgan NG (2020) Type 1 diabetes: interferons and the aftermath of pancreatic beta-cell enteroviral infection. Microorganisms 8(9):1419

Albarbar B, Aga H (2024) A review on autoimmune diseases: recent advances and future perspectives. AlQalam J Med Appl Sci:718–729. https://doi.org/10.54361/ajmas.247337

Al-Hussaniy HA (2023) The development of molecular docking and molecular dynamics and their application in the field of chemistry and computer simulation. J Med Pharm Allied Sci 12:5552–5562. https://doi.org/10.55940/medphar20227

Al-hussaniy HA, AL-Biati HA (2022) The role of leptin hormone, neuropeptide Y, ghrelin and leptin/ghrelin ratio in Obesogenesis. Med Pharm J 1(2):52–63. https://doi.org/10.55940/medphar20227

Alves Abrantes J, Veríssimo De Azevedo J, Fernandes F, Almeida V, Custódio De Oliveira L, Ferreira De Oliveira M, Galvão De Araújo J, Lanza D, Bezerra F, Andrade V, De Medeiros A, Fernandes T, Fernandes J (2024) Viruses as a potential environmental trigger of type 1 diabetes mellitus (review). Biomed Rep 20:81. https://doi.org/10.3892/br.2024.1770

Anaya JM, Ramirez-Santana C, Alzate MA, Molano-Gonzalez N, Rojas-Villarraga A (2016) The autoimmune ecology. Front Immunol 7:139. https://doi.org/10.3389/fimmu.2016.00139

Antonelli A, Ferri C, Fallahi P, Ferrari SM, Ghinoi A, Rotondi M, Ferrannini E (2006) Thyroid disorders in chronic hepatitis C virus infection. Thyroid 16(6):563–572. https://doi.org/10.1089/thy.2006.16.563

Araújo VMA, Melo IM, Lima V (2015) Relationship between periodontitis and rheumatoid arthritis: review of the literature. Mediat Inflamm 2015(1):259074. https://doi.org/10.1155/2015/259074

Ascherio A, Munger KL (2016) Epidemiology of multiple sclerosis: from risk factors to prevention—an update. In: Seminars in neurology, vol 36, No. 02. Thieme Medical Publishers, pp 103–114. https://doi.org/10.1055/s-0036-1579693

Ascherio A, Munger KL, White R, Köchert K, Simon KC, Polman CH et al (2014) Vitamin D as an early predictor of multiple sclerosis activity and progression. JAMA Neurol 71(3):306–314. https://doi.org/10.1001/jamaneurol.2013.5993

Atkinson MA, Eisenbarth GS (2001) Type 1 diabetes: new perspectives on disease pathogenesis and treatment. Lancet 358:221–229. https://doi.org/10.1016/S0140-6736(01)05415-0

Bach JF (2002) The effect of infections on susceptibility to autoimmune and allergic diseases. N Engl J Med 347(12):911–920. https://doi.org/10.1056/NEJMra020100

Baranzini SE, Oksenberg JR (2017) The genetics of multiple sclerosis: from 0 to 200 in 50 years. Trends Genet 33(12):960–970

Bascuñán KA, Araya M, Rodríguez JM, Roncoroni L, Elli L, Alvarez JDPL, Valenzuela R (2025) Interplay of n-3 polyunsaturated fatty acids, intestinal inflammation, and gut microbiota in celiac disease pathogenesis. Nutrients 17(4):621

Balandraud N, Roudier J (2018) Epstein-Barr virus and rheumatoid arthritis. Joint Bone Spine 85(2):165–170

Biesiekierski JR, Muir JG, Gibson PR (2013) Is gluten a cause of gastrointestinal symptoms in people without celiac disease? Curr Allergy Asthma Rep 13:631–638. https://doi.org/10.1007/s11882-013-0386-4

Bjornevik K, Cortese M, Healy BC, Kuhle J, Mina MJ, Leng Y et al (2022) Longitudinal analysis reveals high prevalence of Epstein-Barr virus associated with multiple sclerosis. Science 375(6578):296–301. https://doi.org/10.1126/science.abj8222

Bodin J, Stene LC, Nygaard UC (2015) Can exposure to environmental chemicals increase the risk of diabetes type 1 development? Biomed Res Int 2015(1):208947

Bruzzese V, Scolieri P, Pepe J (2021) Efficacy of gluten-free diet in patients with rheumatoid arthritis, vol 72, p 213

Butalia S, Kaplan GG, Khokhar B, Rabi DM (2016) Environmental risk factors and type 1 diabetes: past, present, and future. Can J Diabetes 40:586–593. https://doi.org/10.1016/j.jcjd.2016.05.002

Calder PC (2020) n-3 PUFA and inflammation: from membrane to nucleus and from bench to bedside. Proc Nutr Soc 79(4):404–416

Cekanaviciute E, Yoo BB, Runia TF, Debelius JW, Singh S, Nelson CA et al (2017) Gut bacteria from multiple sclerosis patients modulate human T cells and exacerbate symptoms in mouse models. Proc Natl Acad Sci 114(40):10713–10718. https://doi.org/10.1073/pnas.1711235114

Chan VSF (2020) Epigenetics in multiple sclerosis. In: Epigenetics in allergy and autoimmunity, pp 309–374. https://doi.org/10.1007/978-981-15-3449-2_12

Chang K, Yang SM, Kim SH, Han KH, Park SJ, Shin JI (2014) Smoking and rheumatoid arthritis. Int J Mol Sci 15(12):22279–22295. https://doi.org/10.3390/ijms151222279

Coppieters KT, Wiberg A, von Herrath MG (2012) Viral infections and molecular mimicry in type 1 diabetes. APMIS 120(12):941–949. https://doi.org/10.1111/apm.12011

Degelman ML, Herman KM (2017) Smoking and multiple sclerosis: a systematic review and meta-analysis using the Bradford Hill criteria for causation. Mult Scler Relat Disord 17:207–216. https://doi.org/10.1016/j.msard.2017.07.020

Delrue C, Speeckaert R, Delanghe JR, Speeckaert MM (2023) The potential influence of advanced glycation end products and (s) RAGE in rheumatic diseases. Int J Mol Sci 24(3):2894. https://doi.org/10.3390/ijms24032894

Dendrou CA, Fugger L, Friese MA (2015) Immunopathology of multiple sclerosis. Nat Rev Immunol 15(9):545–558. https://doi.org/10.1038/nri3871

Ebringer A, Rashid T (2006) Rheumatoid arthritis is an autoimmune disease triggered by Proteus urinary tract infection. Journal of Immunology Research 13(1):41–48.

Esposito S, Bonavita S, Sparaco M, Gallo A, Tedeschi G (2018) The role of diet in multiple sclerosis: a review. Nutr Neurosci 21(6):377–390. https://doi.org/10.1080/1028415X.2017.1303016

Filippi CM, von Herrath MG (2008) Viral trigger for type 1 diabetes: pros and cons. Diabetes 57:2863–2871. https://doi.org/10.2337/db07-1023

Firestein GS, McInnes IB (2017) Immunopathogenesis of rheumatoid arthritis. Immunity 46(2):183–196

Gombash SE, Lee PW, Sawdai E, Lovett-Racke AE (2022) Vitamin D as a risk factor for multiple sclerosis: immunoregulatory or neuroprotective? Frontiers in neurology 13:796933

Grasso EA, Chiarelli F (2024) Type 1 diabetes and other autoimmune disorders in children. Pediatr Diabetes (1):5082064

Harbison JE, Roth-Schulze AJ, Giles LC, Tran CD, Ngui KM, Penno MA et al (2019) Gut microbiome dysbiosis and increased intestinal permeability in children with islet autoimmunity and type 1 diabetes: a prospective cohort study. Pediatr Diabetes 20(5):574–583. https://doi.org/10.1111/pedi.12865

Hart PH, Gorman S, Finlay-Jones JJ (2011) Modulation of the immune system by UV radiation: more than just the effects of vitamin D? Nat Rev Immunol 11(9):584–596. https://doi.org/10.1038/nri3045

Hart JE, Källberg H, Laden F, Bellander T, Costenbader KH, Holmqvist M et al (2013) Ambient air pollution exposures and risk of rheumatoid arthritis: results from the Swedish EIRA case–control study. Ann Rheum Dis 72(6):888–894

Hauser SL, Bar-Or A, Comi G, Giovannoni G, Hartung HP, Hemmer B et al (2017) Ocrelizumab versus interferon beta-1a in relapsing multiple sclerosis. N Engl J Med 376(3):221–234. https://doi.org/10.1056/NEJMoa1601277

Hedström AK, Hillert J, Olsson T, Alfredsson L (2014) Alcohol as a modifiable lifestyle factor affecting multiple sclerosis risk. JAMA neurology 71(3):300–305

Hedström AK, Huang J, Brenner N, Butt J, Hillert J, Waterboer T et al (2020) Smoking and Epstein–Barr virus infection in multiple sclerosis development. Sci Rep 10(1):10960. https://doi.org/10.1038/s41598-020-67883-w

Herold KC, Bundy BN, Long SA, Bluestone JA, DiMeglio LA, Dufort MJ et al (2019) An anti-CD3 antibody, teplizumab, in relatives at risk for type 1 diabetes. N Engl J Med 381(7):603–613. https://doi.org/10.1056/NEJMoa1902226

Hyppönen E, Läärä E, Reunanen A, Järvelin MR, Virtanen SM (2001) Intake of vitamin D and risk of type 1 diabetes: a birth-cohort study. Lancet 358(9292):1500–1503

McInnes IB, Schett G (2011) The Pathogenesis of Rheumatoid Arthritis New England Journal of Medicine 365(23):2205–19 https://doi.org/10.1056/NEJMra1004965

Innes JK, Calder PC (2018) Omega-6 fatty acids and inflammation. Prostaglandins Leukot Essent Fat Acids 132:41–48. https://doi.org/10.1016/j.plefa.2018.03.004

James E, Dobson R, Kuhle J, Baker D, Giovannoni G, Ramagopalan SV (2013) The effect of vitamin D-related interventions on multiple sclerosis relapses: a meta-analysis. Multiple Sclerosis Journal 19(12):1571–79

Jangi S, Gandhi R, Cox LM, Li N, Von Glehn F, Yan R et al (2016) Alterations of the human gut microbiome in multiple sclerosis. Nat Commun 7(1):12015. https://doi.org/10.1038/ncomms12015

Jun H, Yoon J (2003) A new look at viruses in type 1 diabetes. Diabetes Metab Res 19:8–31. https://doi.org/10.1002/dmrr.337

Källberg H, Ding B, Padyukov L, Bengtsson C, Rönnelid J, Klareskog L et al (2011) Smoking is a major preventable risk factor for rheumatoid arthritis: estimations of risks after various exposures to cigarette smoke. Ann Rheum Dis 70(3):508–511. https://doi.org/10.1136/ard.2009.120899

Kim D, Ansari Md M, Ghosh M, Heo Y, Choi Ki-C, Son Y (2025) Implications of obesity-mediated cellular dysfunction and adipocytokine signaling pathways in the pathogenesis of osteoarthritis. Mol Asp Med 103:101361. https://doi.org/10.1016/j.mam.2025.101361

Klareskog L, Padyukov L, Rönnelid J, Alfredsson L (2006a) Genes, environment and immunity in the development of rheumatoid arthritis. Curr Opin Immunol 18(6):650–655. https://doi.org/10.1016/j.coi.2006.06.004

Klareskog L, Stolt P, Lundberg K, Källberg H, Bengtsson C, Grunewald J et al (2006b) A new model for an etiology of rheumatoid arthritis: smoking may trigger HLA–DR (shared epitope)–restricted immune reactions to autoantigens modified by citrullination. Arthritis Rheumatism 54(1):38–46. https://doi.org/10.1002/art.21575

Kleinewietfeld M, Manzel A, Titze J, Kvakan H, Yosef N, Linker RA et al (2013) Sodium chloride drives autoimmune disease by the induction of pathogenic TH17 cells. Nature 496(7446):518–522. https://doi.org/10.1038/nature11868

Knip M, Simell O (2012) Environmental triggers of type 1 diabetes. Cold Spring Harb Perspect Med 2(7):a007690

Knip M, Veijola R, Virtanen SM, Hyöty H, Vaarala O, Åkerblom HK (2005) Environmental triggers and determinants of type 1 diabetes. Diabetes 54:S125–S136. https://doi.org/10.2337/diabetes.54.suppl_2.S125

Knip M, Virtanen SM, Becker D, Dupre J, Krischer JP, Åkerblom HK, TRIGR Study Group (2011) Early feeding and risk of type 1 diabetes: experiences from the Trial to Reduce Insulin-dependent diabetes mellitus in the Genetically at Risk (TRIGR). Am J Clin Nutr 94:S1814–S1820. https://doi.org/10.3945/ajcn.110.000711

Kostic AD, Gevers D, Siljander H, Vatanen T, Hyötyläinen T, Hämäläinen AM et al (2015) The dynamics of the human infant gut microbiome in development and in progression toward type 1 diabetes. Cell Host Microbe 17(2):260–273

Kuntić M, Hahad O, Münzel T, Daiber A (2024) Crosstalk between oxidative stress and inflammation caused by noise and air pollution—implications for neurodegenerative diseases. Antioxidants 13(3):266

Langer-Gould A, Qian L, Tartof SY, Brara SM, Jacobsen SJ, Beaber BE et al (2014) Vaccines and the risk of multiple sclerosis and other central nervous system demyelinating diseases. JAMA Neurol 71(12):1506–1513

Lassmann H, Bradl M (2017) Multiple sclerosis: experimental models and reality. Acta Neuropathol 133:223–244. https://doi.org/10.1007/s00401-016-1631-4

Lee HS, Burkhardt BR, McLeod W, Smith S, Eberhard C, Lynch K et al (2014) Biomarker discovery study design for type 1 diabetes in The Environmental Determinants of Diabetes in the Young (TEDDY) study. Diabetes Metab Res Rev 30(5):424–434. https://doi.org/10.1002/dmrr.2510

Linn-Rasker SP, Van Der Helm-Van Mil AHM, Van Gaalen FA, Kloppenburg M, De Vries RRP, le Cessie S et al (2006) Smoking is a risk factor for anti-CCP antibodies only in rheumatoid arthritis patients who carry HLA-DRB1 shared epitope alleles. Ann Rheum Dis 65(3):366–371. https://doi.org/10.1136/ard.2005.041079

Logesh K, Raj B, Bhaskaran M, Thirumaleshwar S, Gangadharappa HV, Osmani RA, Spandana KA (2023) Nanoparticulate drug delivery systems for the treatment of rheumatoid arthritis: a comprehensive review. J Drug Deliv Sci Technol 81:104241. https://doi.org/10.1016/j.jddst.2023.104241

Lund-Blix NA, Dong F, Mårild K, Seifert J, Barón AE, Waugh KC et al (2019) Gluten intake and risk of islet autoimmunity and progression to type 1 diabetes in children at increased risk of the disease: the Diabetes Autoimmunity Study in the Young (DAISY). Diabetes Care 42(5):789–796. https://doi.org/10.2337/dc18-2315

Luo TT, Wu YJ, Yin Q, Chen WG, Zuo J (2023) The involvement of glucose and lipid metabolism alteration in rheumatoid arthritis and its clinical implication. J Inflamm Res Volume 16:1837–1852

Luopajärvi K, Savilahti E, Virtanen SM, Ilonen J, Knip M, Åkerblom HK, Vaarala O (2008) Enhanced levels of cow's milk antibodies in infancy in children who develop type 1 diabetes later in childhood. Pediatr Diabetes 9(5):434–441. https://doi.org/10.1111/j.1399-5448.2008.00413.x

Mammen AL (2016) Statin-associated autoimmune myopathy. N Engl J Med 374(7):664–669. https://doi.org/10.1056/NEJMra1515161

Maresz KJ, Hellvard A, Sroka A, Adamowicz K, Bielecka E, Koziel J et al (2013) Porphyromonas gingivalis facilitates the development and progression of destructive arthritis through its unique bacterial peptidylarginine deiminase (PAD). PLoS Pathog 9(9):e1003627. https://doi.org/10.1371/journal.ppat.1003627

Mazzucca CB, Scotti L, Cappellano G, Barone-Adesi F, Chiocchetti A (2022) Nutrition and rheumatoid arthritis onset: a prospective analysis using the UK Biobank. Nutrients 14(8):1554. https://doi.org/10.3390/nu14081554

McCoy L, Tsunoda I, Fujinami RS (2006) Multiple sclerosis and virus induced immune responses: autoimmunity can be primed by molecular mimicry and augmented by bystander activation. Autoimmunity 39(1):9–19. https://doi.org/10.1080/08916930500484799

Min YS, Kim MG, Ahn YS (2021) Rheumatoid arthritis in silica-exposed workers. Int J Environ Res Public Health 18(23):12776. https://doi.org/10.3390/ijerph182312776

Mokry LE, Ross S, Ahmad OS, Forgetta V, Smith GD, Leong A et al (2015) Vitamin D and risk of multiple sclerosis: a Mendelian randomization study. PLoS Med 12(8):e1001866. https://doi.org/10.1371/journal.pmed.1001866

Moulis G, Sommet A, Lapeyre-Mestre M, Montastruc JL (2014) Is the risk of tumour necrosis factor inhibitor-induced lupus or lupus-like syndrome the same with monoclonal antibodies and soluble receptor? A case/non-case study in a nationwide pharmacovigilance database. Rheumatology 53(10):1864–1871. https://doi.org/10.1093/rheumatology/keu214

Muruganandam A, Migliorini F, Jeyaraman N, Vaishya R, Balaji S, Ramasubramanian S et al (2024) Molecular mimicry between gut microbiome and rheumatoid arthritis: current concepts. Med Sci 12(4):72. https://doi.org/10.3390/medsci12040072

Nigam S, Bishop JO, Hayat H, Quadri T, Hayat H, Wang P (2022) Nanotechnology in immunotherapy for type 1 diabetes: promising innovations and future advances. Pharmaceutics 14(3):644. https://doi.org/10.3390/pharmaceutics14030644

Nygaard G, Firestein GS (2020) Restoring synovial homeostasis in rheumatoid arthritis by targeting fibroblast-like synoviocytes. Nat Rev Rheumatol 16(6):316–333. https://doi.org/10.1038/s41584-020-0413-5

Panghal A, Flora SJS (2024) Nano-based approaches for the treatment of neuro-immunological disorders: a special emphasis on multiple sclerosis. Discov Nano 19(1):171. https://doi.org/10.1186/s11671-024-04135-0

Parks CG, Conrad K, Cooper GS (1999) Occupational exposure to crystalline silica and autoimmune disease. Environ Health Perspect 107(suppl 5):793–802. https://doi.org/10.1289/ehp.99107s5793

Parrón T, Requena M, Hernández AF, Alarcón R (2011) Association between environmental exposure to pesticides and neurodegenerative diseases. Toxicology and applied pharmacology 256(3):379–85

Peng H, Hagopian W (2006) Environmental factors in the development of Type 1 diabetes. Rev Endocr Metab Disord 7:149–162. https://doi.org/10.1007/s11154-006-9024-y

Peng H, Hagopian W (2007) Environmental factors in the development of Type 1 diabetes. Rev Endocr Metab Disord 7:149–162. https://doi.org/10.1007/s11154-006-9024-y

Petit P, Vuillerme N (2024) Leveraging administrative health databases to address health challenges in farming populations: a scoping review and bibliometric analysis (1975-2024). JMIR Public Health Surveill 11:e62939

Poggioli R, Hirani K, Jogani VG, Ricordi C (2023) Modulation of inflammation and immunity by omega-3 fatty acids: a possible role for prevention and to halt disease progression in autoimmune, viral, and age-related disorders. European Review for Medical & Pharmacological Sciences 27(15)

Ponsonby AL, McMichael A, Van Der Mei I (2002) Ultraviolet radiation and autoimmune disease: insights from epidemiological research. Toxicology 181:71–78. https://doi.org/10.1016/S0300-483X(02)00257-3

Quandt Z, Young A, Anderson M (2020) Immune checkpoint inhibitor diabetes mellitus: a novel form of autoimmune diabetes. Clin Exp Immunol 200(2):131–140. https://doi.org/10.1111/cei.13424

Ramasamy R, Mohammed F, Meier UC (2020) HLA DR2b-binding peptides from human endogenous retrovirus envelope, Epstein-Barr virus and brain proteins in the context of molecular mimicry in multiple sclerosis. Immunol Lett 217:15–24. https://doi.org/10.1016/j.imlet.2019.10.017

Rein P, Mueller RB (2017) Treatment with biologicals in rheumatoid arthritis: an overview. Rheumatol Ther 4(2):247–261. https://doi.org/10.1007/s40744-017-0073-3

Richardson SJ, Morgan NG (2018) Enteroviral infections in the pathogenesis of type 1 diabetes: new insights for therapeutic intervention. Curr Opin Pharmacol 43:11–19

Roh S (2019) Smoking as a preventable risk factor for rheumatoid arthritis: rationale for smoking cessation treatment in patients with rheumatoid arthritis. J Rheumatic Dis 26(1):12–19. https://doi.org/10.4078/jrd.2019.26.1.12

Sadeghi P, Karimi H, Lavafian A, Rashedi R, Samieefar N, Shafiekhani S, Rezaei N (2024) Machine learning and artificial intelligence within pediatric autoimmune diseases: applications, challenges, future perspective. Expert Rev Clin Immunol 20:just-accepted. https://doi.org/10.1080/1744666X.2024.2359019

Salzer J, Hallmans G, Nyström M, Stenlund H, Wadell G, Sundström P (2012) Vitamin D as a protective factor in multiple sclerosis. Neurology 79(21):2140–2145. https://doi.org/10.1212/WNL.0b013e3182752ea8

Savastio S, Cadario F, Genoni G, Bellomo G, Bagnati M, Secco G et al (2016) Vitamin D deficiency and glycemic status in children and adolescents with type 1 diabetes mellitus. PLoS One 11(9):e0162554. https://doi.org/10.1371/journal.pone.0162554

Scher JU, Sczesnak A, Longman RS, Segata N, Ubeda C, Bielski C et al (2013) Expansion of intestinal Prevotella copri correlates with enhanced susceptibility to arthritis. elife 2:e01202. https://doi.org/10.7554/eLife.01202

Scrivo R, Perricone C, Altobelli A, Castellani C, Tinti L, Conti F, Valesini G (2019) Dietary habits bursting into the complex pathogenesis of autoimmune diseases: the emerging role of salt from experimental and clinical studies. Nutrients 11(5):1013. https://doi.org/10.3390/nu11051013

Shim CH, Cho S, Shin YM, Choi JM (2022) Emerging role of bystander T cell activation in autoimmune diseases. BMB Rep 55(2):57. https://doi.org/10.5483/BMBRep.2022.55.2.183

Sköldstam L, Hagfors L, Johansson G (2003) An experimental study of a Mediterranean diet intervention for patients with rheumatoid arthritis. Ann Rheum Dis 62(3):208–214

Smolen JS, Breedveld FC, Burmester GR, Bykerk V, Dougados M, Emery P et al (2016) Treating rheumatoid arthritis to target: 2014 update of the recommendations of an international task force. Ann Rheum Dis 75(1):3–15

Song GG, Bae SC, Lee YH (2012) Association between vitamin D intake and the risk of rheumatoid arthritis: a meta-analysis. Clin Rheumatol 31:1733–1739. https://doi.org/10.1007/s10067-012-2080-7

Song Y, Li J, Wu Y (2024) Evolving understanding of autoimmune mechanisms and new therapeutic strategies of autoimmune disorders. Signal Transduct Target Ther 9(1):263. https://doi.org/10.1038/s41392-024-01952-8

Uribarri J, Woodruff S, Goodman S, Cai W, Chen XUE, Pyzik R et al (2010) Advanced glycation end products in foods and a practical guide to their reduction in the diet. J Am Diet Assoc 110(6):911–916. https://doi.org/10.1016/j.jada.2010.03.018

Wang J, Mei L, Hao Y, Xu Y, Yang Q, Dai Z, Ji Y (2024) Contemporary perspectives on the role of vitamin D in enhancing gut health and its implications for preventing and managing intestinal diseases. Nutrients 16(14):2352

Warburton DE, Bredin SS (2017) Health benefits of physical activity: a systematic review of current systematic reviews. Curr Opin Cardiol 32(5):541–556. https://doi.org/10.1097/HCO.0000000000000437

Wehr P, Purvis H, Law SC, Thomas R (2019) Dendritic cells, T cells and their interaction in rheumatoid arthritis. Clin Exp Immunol 196(1):12–27. https://doi.org/10.1111/cei.13256

Yahaya TO, Liman UU, Obadiah CD, Zakari ZI, Anyebe D, Clement BG, Muhammad BM (2023) Environmental triggers' involvement in the development of type 1 diabetes mellitus. Explor Res Hypothesis Med 8(2):150–156. https://doi.org/10.14218/ERHM.2022.00051

Yang F, Zhou S, Wang C, Huang Y, Li H, Wang Y et al (2017) Epigenetic modifications of interleukin-6 in synovial fibroblasts from osteoarthritis patients. Sci Rep 7(1):43592. https://doi.org/10.1038/srep43592

Yousry TA, Major EO, Ryschkewitsch C, Fahle G, Fischer S, Hou J et al (2006) Evaluation of patients treated with natalizumab for progressive multifocal leukoencephalopathy. N Engl J Med 354(9):924–933. https://doi.org/10.1056/NEJMoa054693

Yu X, Mai Y, Wei Y, Yu N, Gao T, Yang J (2023) Therapeutic potential of tolerance-based peptide vaccines in autoimmune diseases. Int Immunopharmacol 116:109740. https://doi.org/10.1016/j.intimp.2023.109740

Zhang P, Lu Q (2018) Genetic and epigenetic influences on the loss of tolerance in autoimmunity. Cellular & molecular immunology 15(6):575–85

Zhang J, Chen LM, Zou Y, Zhang S, Xiong F, Wang CY (2021) Implication of epigenetic factors in the pathogenesis of type 1 diabetes. Chin Med J 134(09):1031–1042

Zorena K, Michalska M, Kurpas M, Jaskulak M, Murawska A, Rostami S (2022) Environmental factors and the risk of developing type 1 diabetes—old disease and new data. Biology 11:608. https://doi.org/10.3390/biology11040608

Zorena K, Jaskulak M, Michalska M, Mrugacz M, Vandenbulcke F (2022) Air pollution, oxidative stress, and the risk of development of type 1 diabetes. Antioxidants 11(10):1908.

Chapter 5
Environmental Triggers and Autoimmune Responses

Subin Ok and Siyoung Yang

5.1 Introduction

5.1.1 Introduction to Autoimmune Diseases (ADs)

ADs are a diverse group of disorders in which the immune system mistakenly targets its own cells and tissues. This results in chronic inflammation, tissue damage, and a wide array of clinical manifestations, depending on the affected organs. Examples include systemic lupus erythematosus (SLE), which involves multiple organ systems, and type 1 diabetes (T1D), in which immune cells selectively attack the pancreatic beta cells responsible for insulin production (Ahmed and Lambert 2014).

The immune system typically distinguishes between self and nonself through intricate tolerance mechanisms. Central tolerance occurs during immune cell development in the thymus or bone marrow, ensuring the deletion of autoreactive cells. Peripheral tolerance further safeguards against autoimmune responses through regulatory T cells and mechanisms, such as anergy and apoptosis. If these processes fail, autoreactive lymphocytes proliferate and initiate autoimmunity.

S. Ok · S. Yang (✉)
Department of Biological Sciences, Sungkyunkwan University, Suwon, Republic of Korea
e-mail: dhrtnqls55@g.skku.edu; yangsy@skku.edu

Y.-O. Son, M. Ghosh (eds.), *Cutting-Edge Strategies in Drug Delivery and Immunotherapy for Autoimmune Disorders*,
https://doi.org/10.1007/978-981-95-0520-3_5

5.1.2 The Rising Prevalence of ADs

ADs have become increasingly prevalent worldwide. Conditions such as multiple sclerosis (MS), rheumatoid arthritis (RA), and inflammatory bowel disease (IBD) are now more frequently diagnosed, particularly in industrialized nations (Bach 2002). The incidence of T1D has been rising annually by approximately 3–5%, with a pronounced increase in younger populations (Patterson et al. 2009). This trend cannot be attributed solely to genetic factors, as evolutionary changes have occurred over millennia. Instead, the growing prevalence points to environmental and lifestyle factors as the key contributors.

The "hygiene hypothesis" provides a compelling explanation for this increase, suggesting that reduced microbial exposure during early childhood alters immune system development and predisposes individuals to ADs. Improved sanitation, decreased family size, and the widespread use of antibiotics may reduce immune challenges and skew the immune system toward a proautoimmune state (Bach 2018).

5.1.3 The Importance of Understanding Triggers

Identifying the triggers of ADs is crucial for their effective prevention, diagnosis, and management. Genetic predisposition plays a fundamental role, particularly in the human leukocyte antigen (HLA) system, which governs antigen presentation to immune cells (Trowsdale and Knight 2013). HLA-DR4 is strongly associated with RA, whereas HLA-DQ2 and HLA-DQ8 are critical markers for celiac disease (Sollid and Jabri 2013). However, genetics alone cannot fully explain disease onset. Environmental factors, such as infections, dietary habits, chemical exposure, and stress act as catalysts that interact with genetic susceptibility to disrupt immune tolerance (Zhernakova et al. n.d.).

Infections: Pathogens, such as the Epstein-Barr virus (EBV), can mimic host antigens and trigger cross-reactive immune responses, as seen in SLE and MS. Viral and bacterial infections also promote ADs via bystander activation and epitope spread (James and Robertson 2021).

Dietary Influences: Diet profoundly affects immune regulation, with certain foods acting as triggers in predisposed individuals (Lerner and Matthias n.d.). Gluten induces a pathological immune response in patients with celiac disease carrying the HLA-DQ2 or HLA-DQ8 alleles. Diets rich in processed foods can also disrupt the gut microbiota, leading to increased intestinal permeability and promotion of autoimmune conditions (Rook n.d.).

Environmental Toxins: Exposure to substances, such as silica, smoking, and industrial chemicals, is linked to an increased autoimmune risk. Silica dust has been associated with SLE, whereas smoking exacerbates RA by inducing posttranslational modifications, such as protein citrullination (Hutchinson et al. 2001).

Psychological and Physical Stress: Stress influences immune function by altering cytokine profiles and regulatory T-cell activity. Chronic stress is a recognized risk factor for autoimmune flares in conditions, such as RA and lupus (Quan and Banks n.d.).

5.1.4 Bridging Genetic and Environmental Insights

Understanding the interplay between genetic and environmental factors is the key to revealing the mechanisms underlying ADs. T1D is often triggered by viral infections in individuals carrying HLA-DQ8, in which molecular mimicry and epitope spreading lead to beta-cell destruction (TEDDY Study Group, 2008). Similarly, smoking-induced citrullination amplifies the autoimmune response in RA, particularly in patients with a genetic (Hutchinson et al. 2001).

Advances in molecular technologies, such as genomics and microbiomics, have revolutionized autoimmune research (Lessard et al. n.d.). Studies, such as the Environmental Determinants of Diabetes in the Young (TEDDY) project, have explored how infections, diet, and other environmental factors interact with genetic susceptibility to influence disease onset. These efforts have paved the way for precision medicine approaches that are tailored to individual risk profiles.

ADs are complex disorders driven by a convergence of genetic predispositions and environmental influences. Their increasing prevalence highlights the urgent need for a deeper understanding of their triggers and mechanisms of action. By identifying the modifiable environmental factors and integrating this knowledge into clinical practice, researchers and clinicians can develop targeted prevention and management strategies (Fig. 5.1). Addressing this intricate interplay is promising for reducing the disease burden and improving patient outcomes in the era of precision medicine (Topol n.d.).

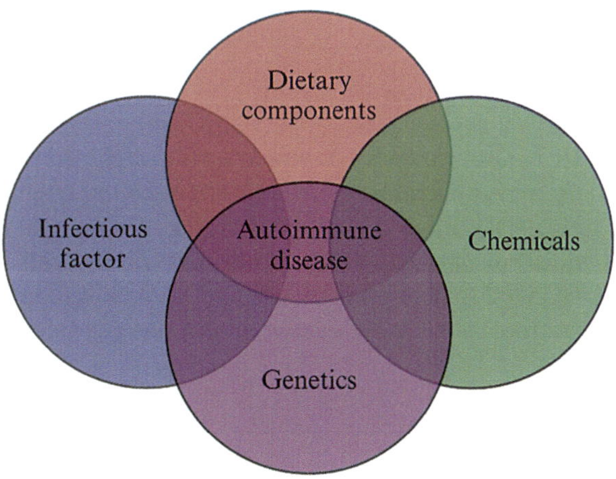

Fig. 5.1 Venn diagram illustrating the interplay of genetic, dietary, chemical, and infectious factors in the development of autoimmune diseases, highlighting the multifactorial nature of their pathogenesis

5.2 Part I: Understanding Autoimmune Responses

5.2.1 *Fundamentals of Autoimmune Pathogenesis*

ADs result from the immune system's failure to distinguish self from nonself, leading to the destruction of body tissues. This pathological immune response arises from the complex interplay among genetic predispositions, environmental triggers, and failures in immune regulatory mechanisms. Understanding the fundamental processes of autoimmunity is essential for the development of effective treatment and prevention strategies. This section discusses the mechanisms underlying autoimmune pathogenesis, focusing on genetic influences, environmental factors, and breakdown of self-tolerance (Marrack et al. n.d.).

5.2.2 *Mechanisms of Autoimmunity*

ADs develop when the mechanisms that maintain immune tolerance are compromised. The immune system, designed to protect the body from pathogens, mistakenly targets self-antigens and initiates chronic inflammation and tissue damage.

5.2.2.1 The Role of Genetics in Autoimmunity

Genetic predisposition is the backbone of susceptibility to ADs. Studies have consistently demonstrated that certain genetic loci, particularly within the HLA system, are critical for determining autoimmune risk.

HLA System and Autoimmune Risk

HLA molecules located on chromosome 6 are central to antigen presentation and immune recognition. HLA allele variations influence the susceptibility to specific ADs.

- HLA-DR4: Associated with RA, with allelic variants increasing the susceptibility to citrullinated protein autoantigens (Gregersen et al. n.d.).
- HLA-DQ2/DQ8: Essential for celiac disease pathogenesis, where gluten peptides bind strongly to these alleles and drive T-cell activation.
- HLA-B27: Strongly associated with ankylosing spondylitis and other spondyloarthropathies, likely because of its molecular mimicry mechanisms (Khan n.d.).

Non-HLA Genetic Contributions

Beyond the HLA system, numerous non-HLA genes influence autoimmune susceptibility.

PTPN22: Encodes a protein tyrosine phosphatase involved in T-cell receptor signaling. Polymorphisms in PTPN22 are associated with an increased risk of RA, T1D, and SLE (Bottini and Peterson n.d.).

CTLA4: Functions as an immune checkpoint, inhibiting T-cell activation. Variants in CTLA4 are linked to autoimmune thyroid diseases and T1D (Ueda et al. 2003).

- IL2RA (CD25): Regulates regulatory T-cell (Treg) development. Mutations impair Treg function and contribute to diseases, such as MS (Sharp et al. 2015).

Epigenetic Modifications

Epigenetic changes, such as DNA methylation and histone modifications, play a role in the modulation of gene expression. Environmental factors, including diet and stress, can alter epigenetic markers and influence the risk of AD without altering the DNA sequence itself (Feinberg 2018; Liu et al. 2013).

5.2.2.2 Environmental Factors as Catalysts

Although genetics lays the foundation for autoimmune susceptibility, environmental factors often act as catalysts that initiate disease onset or exacerbate the existing conditions.

Pathogens and Molecular Mimicry

Pathogen exposure can trigger autoimmunity through molecular mimicry, bystander activation, and epitope spreading.

- EBV: In SLE and MS, EBV-infected B cells express viral antigens that mimic self-antigens and elicit autoreactive immune response.
- Enteroviruses: Linked to T1D, these viruses are thought to initiate pancreatic beta-cell destruction through molecular mimicry and inflammation-induced epitope spreading (TEDDY Study Group, 2008).

Smoking and Environmental Pollutants

- Smoking: A well-established risk factor for RA. Smoking promotes protein citrullination, creating neoantigens that elicit immune responses in genetically susceptible individuals, particularly HLA-DR4 carriers (Hutchinson et al. 2001).
- Silica dust: Silica exposure induces chronic inflammation and macrophage activation, contributing to immune dysregulation (Parks et al. 2016).

Dietary Factors

- Gluten: In genetically predisposed individuals, gluten triggers celiac disease through immune-mediated intestinal damage (Sollid and Jabri 2013).
- Western Diets: High amounts of processed foods and low fiber, which promote gut dysbiosis, increasing the risk of IBD and other autoimmune conditions (Albenberg and Wu n.d.).

5.2.2.3 Breakdown of Self-Tolerance

Failure of self-tolerance is central to the pathogenesis of AD. Self-tolerance mechanisms are designed to prevent autoreactive immune cells from attacking host tissues (Klein et al. 2014).

Central Tolerance

Central tolerance occurs in the primary lymphoid organs (thymus for T cells and bone marrow for B cells), where autoreactive lymphocytes are eliminated during development.

Thymic selection: T cells undergo positive and negative selection in the thymus. The failure to eliminate autoreactive T cells contributes to autoimmunity.

Peripheral Tolerance

Peripheral tolerance mechanisms suppress autoreactive cells that escape central tolerance. These mechanisms include:

- Tregs: Tregs inhibit autoreactive T-cell activity and secrete antiinflammatory cytokines, such as IL-10 and TGF-β. Impaired Treg function has been implicated in diseases, such as MS and SLE (Sakaguchi et al. n.d.).

Immune Checkpoints: Molecules, such as CTLA-4 and PD-1, dampen T-cell activation, preventing excessive immune responses. Checkpoint dysfunction also contributes to uncontrolled inflammation (Fife and Bluestone n.d.).

Immune Privilege

Certain tissues, such as the brain, eyes, and testes, are considered immune privileged because of their physical barriers and local immunosuppressive environments. Breaches to these barriers, as observed in trauma or infections, can expose sequestered antigens and trigger autoimmunity. Understanding autoimmune responses requires a comprehensive examination of the genetic, environmental, and immunological factors that drive disease. Genetic predisposition creates a foundation for susceptibility, whereas environmental exposure acts as a catalyst, triggering or exacerbating disease onset. The failure of self-tolerance mechanisms further perpetuates chronic inflammation and tissue damage. By unraveling these mechanisms, researchers and clinicians can identify therapeutic targets and develop interventions that restore the immune balance, offering hope for improved outcomes in autoimmune disease management.

5.2.3 Common ADs

ADs are a diverse group of disorders characterized by immune dysregulation and self-reactivity. These diseases range from systemic conditions, that involve multiple organ systems, to organ-specific diseases that target specific tissues. The triggers, clinical presentations, and severities of these diseases vary widely, reflecting the complex interplay between genetic, environmental, and immunological factors. This section explores several common ADs, including their pathogenesis, clinical features, and triggers, and emphasizes the diversity of their manifestations.

5.2.4 RA

RA is a chronic systemic autoimmune disease primarily affecting synovial joints. It is characterized by persistent inflammation, synovial hyperplasia, and subsequent joint destruction.

5.2.4.1 Pathogenesis

RA develops because of a combination of genetic susceptibility and environmental triggers, leading to immune-mediated synovial inflammation.

- Genetic factor HLA-DR4, a specific allele within the HLA-DR family, is strongly associated with RA. This allele enhances the presentation of citrullinated peptides to T cells, perpetuating immune activation (Hutchinson et al. 2001).

Environmental Triggers: Smoking is a major risk factor that induces protein citrullination and creates neoantigens that elicit autoreactive immune responses. Additionally, exposure to silica and air pollutants has been implicated in increasing RA risk (Parks et al. 2016).

5.2.4.2 Clinical Features

RA commonly presents as symmetrical polyarthritis, primarily affecting the small joints of the hands and feet. The systemic manifestations include fatigue, weight loss, and subcutaneous rheumatoid nodules. Advanced disease may involve extra-articular organs, including the lungs (interstitial lung disease) and cardiovascular system.

5.2.4.3 Mechanisms of Damage

- Immune Cell Infiltration: Synovial inflammation is driven by $CD4^+$ T cells, B cells, and macrophages, which release proinflammatory cytokines, such as TNF-α and IL-6.

Pannus Formation: Hyperplastic synovium or pannus invades the cartilage and bone, leading to joint destruction (McInnes and Schett n.d.).

5.2.5 SLE

SLE is a prototypical systemic autoimmune disease characterized by the production of autoantibodies against nuclear components. It affects multiple organ systems and has varied clinical courses.

5.2.5.1 Pathogenesis

- Autoantibody Production: The hallmark of SLE is the presence of antinuclear antibodies (ANA) targeting DNA, histones, and ribonucleoproteins. These autoantibodies form immune complexes that are deposited in the tissues and trigger inflammation (Tsokos n.d.).
- Triggers: Environmental factors, such as UV radiation, can induce keratinocyte apoptosis and release nuclear antigens that fuel autoantibody production. EBV

infections and certain medications have also been implicated (James and Robertson 2021).

5.2.5.2 Clinical Features

SLE manifests with diverse symptoms, including:

- Cutaneous: Photosensitive rash, discoid lesions, and malar rash (butterfly rash).
- Renal: Lupus nephritis, which is characterized by proteinuria and hematuria, is the leading cause of morbidity.
- Neurological disorders: Seizures, psychosis, and cognitive dysfunction.
- Hematological examination revealed anemia, thrombocytopenia, and leukopenia (D'Cruz et al. n.d.).

5.2.5.3 Disease Mechanisms

- Immune complexes activate the complement system, leading to tissue damage and inflammation.
- TLR activation by nucleic acid-containing immune complexes perpetuates the inflammatory cycle.

5.2.6 MS

MS is a chronic demyelinating disease of the central nervous system (CNS). It is driven by autoreactive T cells that target myelin and the protective sheath surrounding the neurons.

5.2.6.1 Pathogenesis

- Genetic Susceptibility: Polymorphisms in the HLA-DRB1 allele are strongly associated with MS and influence T-cell responses to myelin antigens.
- Environmental factors: EBV infection is a key trigger, with nearly all patients with MS showing evidence of prior EBV exposure. Vitamin D deficiency and smoking were the other risk factors (Bjornevik et al. 2022).

5.2.6.2 Clinical Features

MS symptoms vary depending on the location of demyelination:

- Sensory Disturbances: Numbness, tingling, or pain.

- Motor Symptoms: Weakness, spasticity, and gait disturbances.
- Visual Impairment: Optic neuritis, presenting as vision loss or pain during eye movement.
- Cognitive and Psychological: Memory issues and depression (Filippi et al. n.d.).

5.2.6.3 Mechanisms of Damage

- T-Cell Activation: Autoreactive $CD4^+$ and $CD8^+$ T cells infiltrate the CNS and target oligodendrocytes.
- Microglial Activation: Resident CNS macrophages contribute to inflammation and neuronal damage (Hohlfeld and Wekerle n.d.).

5.2.7 T1D

T1D is an organ-specific autoimmune disease characterized by the destruction of insulin-producing beta cells in the pancreas, resulting in insulin deficiency and hyperglycemia.

5.2.7.1 Pathogenesis

- Genetic Factors: HLA-DQ8 and HLA-DQ2 alleles confer susceptibility to T1D (Noble and Erlich n.d.).
- Environmental Triggers: Viral infections, such as those caused by coxsackievirus, are believed to initiate autoimmunity through molecular mimicry, in which viral antigens resemble beta-cell proteins (TEDDY Study Group, 2008).

5.2.7.2 Clinical Features

T1D typically presents in childhood or adolescence with symptoms, such as polyuria, polydipsia, weight loss, and fatigue. Diabetic ketoacidosis is a life-threatening complication that often marks disease onset.

5.2.7.3 Mechanisms of Damage

- T-Cell–mediated Destruction: $CD8^+$ cytotoxic T cells attack beta cells (Bluestone et al. 2010).
- Autoantibodies antiinsulin and anti-GAD65 are disease biomarkers.

5.2.8 Diversity in Symptoms and Severity

ADs display significant variability in their clinical manifestations and severity:

- Systemic Versus Organ-specific: RA and SLE affect multiple systems, whereas T1D and MS target specific organs.
- Acute versus Chronic Presentation: T1D often presents acutely, whereas MS and RA have chronic, relapsing-remitting courses.
- Heterogeneity within Diseases: Even within a single disease, symptoms and progression rates vary widely among patients, necessitating personalized approaches for diagnosis and treatment.

Understanding the common ADs and their diverse manifestations highlights the complexity of immune dysregulation. Each disease represents a unique interplay among genetic predispositions, environmental triggers, and immune mechanisms. This diversity underscores the need for individualized approaches to manage and treat ADs, with a focus on early detection, targeted therapies, and patient-centered care.

5.3 Part II: Environmental Triggers

5.3.1 Environmental Factors Influencing Autoimmunity

ADs arise from an interplay between genetic predisposition and environmental exposure, which collectively disrupt immune tolerance and lead to self-reactivity. Environmental factors, ranging from chemical exposure and dietary habits to infection and stress, significantly influence the onset, progression, and severity of ADs. This section explores these triggers in detail, highlighting their mechanisms and impacts.

5.3.1.1 Chemical Exposures

Exposure to pollutants, pesticides, and industrial chemicals represents an important category of environmental triggers of ADs. These chemicals contribute to immune dysregulation via oxidative stress, chronic inflammation, and antigen modification.

Silica Dust and SLE

Silica dust, which is commonly encountered in mining and construction industries, is strongly associated with an increased risk of SLE. Epidemiological studies have demonstrated that miners and industrial workers exposed to silica are disproportionately affected by SLE compared with that in the general population. The proposed mechanism involves macrophage activation and chronic inflammation. Silica particles are phagocytosed by macrophages, leading to the release of proinflammatory cytokines, such as IL-1β and TNF-α. This process induces apoptosis and release of nuclear antigens that form immune complexes and drive systemic inflammation (Parks et al. 2016).

Pesticides and RA

Pesticides, particularly organophosphates, have been implicated in ADs such as RA. These chemicals disrupt immune tolerance by altering antigen presentation and inducing oxidative stress. Miller revealed that agricultural workers exposed to high levels of pesticides exhibited high rates of RA, highlighting the occupational risks associated with these exposures (Chittrakul et al. 2022).

Smoking and ADs

Smoking is a well-established risk factor of ADs, particularly RA. Smoking induces protein citrullination, a posttranslational modification in which arginine residues are converted into citrulline. These citrullinated proteins act as neoantigens and elicit an autoimmune response. Smokers carrying the HLA-DR4 allele are at a significantly higher risk of developing RA due to this mechanism. Additionally, smoking exacerbates disease severity by increasing systemic inflammation and impairing Treg function (Klareskog et al. n.d.).

5.3.1.2 Dietary Factors

Diet plays a fundamental role in modulating immune responses and maintaining gut health, with specific dietary components acting as triggers of ADs.

Gluten and Celiac Disease

Gluten is a primary dietary trigger for celiac disease, a condition characterized by the immune-mediated destruction of the small intestinal mucosa. Genetically predisposed individuals carrying the HLA-DQ2 or HLA-DQ8 alleles are highly susceptible to gluten-induced inflammation. Upon ingestion, gluten peptides are deamidated by tissue transglutaminases, generating antigens that activate $CD4^{+}$ T cells. This immune cascade results in villous atrophy, malabsorption, and systemic complications (Sollid and Jabri 2013).

Western-Style Diets and IBD

The transition to a Western-style diet rich in processed foods, sugar, and saturated fats has been associated with an increasing incidence of IBD. These diets disrupt gut microbiota diversity and increase intestinal permeability, allowing microbial products to enter circulation and trigger immune responses. Vaarala (2008) demonstrated that a high-fat diet impairs gut barrier function, promotes systemic inflammation, and exacerbates Crohn's disease and ulcerative colitis.

Probiotic and Prebiotic Interventions

Restoring gut health through probiotic- and prebiotic-rich diets has shown promise in mitigating autoimmune conditions. Foods, such as yogurt, kefir, and fermented vegetables, support the growth of beneficial gut microbes, enhance Treg function, and reduce inflammation.

5.3.1.3 Infections

Viruses and bacteria are potent environmental triggers of ADs. These infectious agents disrupt the immune regulation through molecular mimicry, epitope spreading, and bystander activation.

EBV

EBV is strongly associated with systemic ADs, such as SLE and MS. EBV-infected B cells express autoantigens that mimic host proteins, triggering autoreactive T- and B-cell responses. James et al. (2021) demonstrated that patients with high titers of anti-EBV antibodies have a significantly increased risk of developing SLE, suggesting a causal relationship mediated by molecular mimicry.

Enteroviruses and T1D

Enteroviruses have also been implicated in the pathogenesis of T1D. The TEDDY Study Group (2008) reported that children exposed to enteroviruses early in life exhibit increased autoantibody production against pancreatic beta cells. This process involves epitope spreading and bystander activation, leading to the autoimmune destruction of insulin-producing cells.

Helicobacter pylori and Autoimmune Gastritis

Chronic *H. pylori* infections have been linked to autoimmune gastritis, a condition characterized by immune-mediated damage to the stomach lining. *H. pylori* infection triggers cross-reactive immune responses that target gastric antigens, contributing to disease progression (Kusters et al. n.d.).

5.3.1.4 Stress

Stress is increasingly recognized as a significant environmental factor influencing the development and exacerbation of AD (Fig. 5.2). Chronic psychological and physiological stress disrupts immune regulation, creating a proinflammatory environment conducive to autoimmunity (Whitacre et al. 1994).

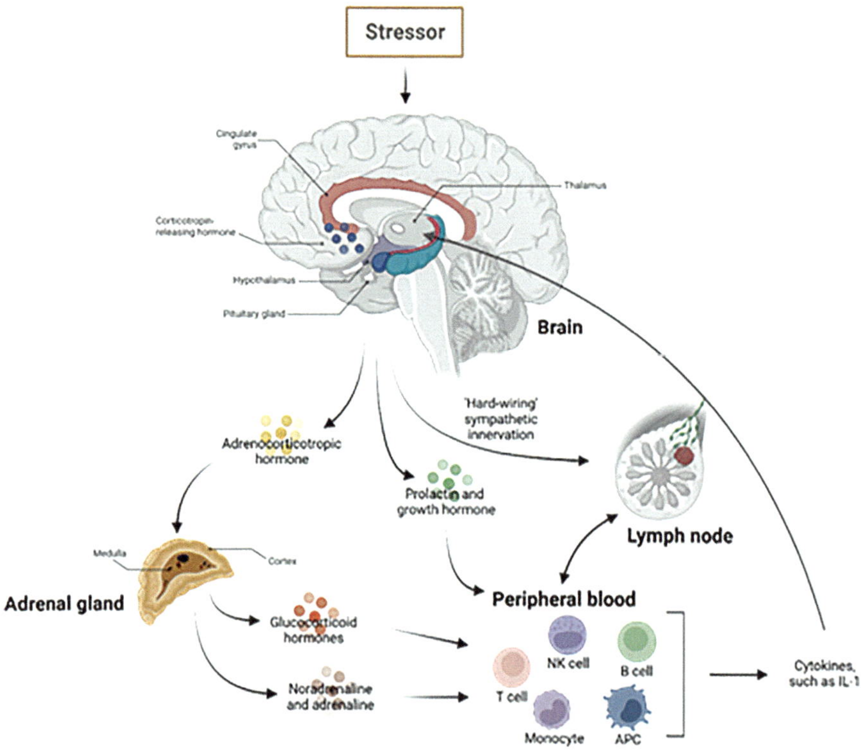

Fig. 5.2 Chronic stress acts as a significant environmental trigger, disrupting the hypothalamic-pituitary-adrenal (HPA) axis and immune homeostasis, leading to increased proinflammatory cytokine production and immune dysregulation, which collectively contribute to the onset and exacerbation of autoimmune diseases

Impact on Cytokine Profiles

Chronic stress alters cytokine production, increasing levels of proinflammatory cytokines, such as IL-6 and TNF-α, while reducing antiinflammatory cytokines, such as IL-10. These imbalances impair immune tolerance and exacerbate ADs, such as RA and SLE (Marshall Jr. n.d.).

Reduction in Tregs

Stress decreases the activity of Tregs, which are critical for maintaining immune homeostasis. This reduction promotes the activation of autoreactive T and B cells, thereby contributing to disease flares.

Environmental triggers play a central role in the development and progression of ADs. By understanding the specific mechanisms by which chemical exposure, dietary factors, infections, and stress influence immune function, researchers and clinicians can develop targeted interventions to mitigate their effects. Addressing these environmental factors through lifestyle modifications, public health policies, and clinical strategies is essential for reducing the global burden of ADs (Table 5.1).

Table 5.1 Summary of key environmental triggers, their mechanisms, and associated autoimmune diseases, highlighting the interplay between environmental factors and immune dysregulation

Environmental trigger	Examples	Mechanism of action	Associated autoimmune diseases
Chemical exposures	Pollutants, pesticides (e.g., organophosphates), industrial chemicals (e.g., silica dust)	Induction of oxidative stress, immune dysregulation, and neoantigen formation	Rheumatoid arthritis (RA), systemic lupus erythematosus (SLE)
Dietary factors	Gluten, dairy, processed foods	Disruption of gut microbiota, increased intestinal permeability ("leaky gut"), activation of immune responses	Celiac disease, inflammatory bowel disease (IBD)
Infections	Epstein-Barr virus (EBV), coxsackievirus, *Helicobacter pylori*	Molecular mimicry, epitope spreading, bystander activation	SLE, type 1 diabetes (T1D), RA
Stress	Chronic psychological stress	Dysregulation of cytokine profiles, reduced Treg activity, and heightened proinflammatory states	RA, SLE, Multiple sclerosis (MS)

5.3.2 *The Microbiome and Autoimmunity*

The gut microbiome, an intricate ecosystem comprising bacteria, viruses, fungi, and other microorganisms, plays a pivotal role in immune regulation and maintenance of systemic homeostasis. The relationship between gut health and autoimmunity has gained increasing attention, as dysbiosis or microbial imbalance has been implicated as a major contributor to the development and exacerbation of ADs. This chapter explores the complex interactions between the gut microbiota and immune system, delving into the mechanisms through which dysbiosis influences autoimmune conditions, and highlights the therapeutic implications and future research directions.

5.3.3 *Gut Health and Immune Regulation*

The gut microbiota profoundly influences the immune system through direct interactions with immune cells, metabolite production, and maintenance of the intestinal barrier. These interactions are central to maintaining the balance between proinflammatory and antiinflammatory immune responses.

5.3.3.1 Role of Microbial Diversity in Immune Tolerance

Microbial diversity ensures the robustness of gut-immune system interactions. Beneficial microbes, such as *Faecalibacterium prausnitzii* and *Akkermansia muciniphila,* play critical roles in modulating immune function by producing metabolites that promote tolerance. In contrast, the loss of diversity due to diet, infection, or antibiotic use creates a proinflammatory milieu that triggers autoimmune responses (Belkaid and Hand n.d.).

5.3.3.2 Short-Chain Fatty Acids (SCFAs) and Tregs

SCFAs, including acetate, propionate, and butyrate, are fermentation products of dietary fiber produced by commensal bacteria. SCFAs enhance Treg differentiation through mechanisms involving histone deacetylase inhibition and G protein-coupled receptor signaling. Tregs suppress autoreactive immune responses and maintain immune tolerance (Arpaia et al. 2013).

5.3.3.3 Gut Barrier Integrity

The gut epithelium, reinforced by tight junctions, serves as a physical barrier to microbial translocation. Dysbiosis weakens this barrier, leading to increased intestinal permeability, commonly referred to as a "leaky gut." This condition allows microbial antigens, such as lipopolysaccharides (LPS) to enter the systemic circulation, triggering inflammation and exacerbating ADs, such as SLE (Zhang et al. 2020).

5.3.4 Dysbiosis and ADs

Dysbiosis has been implicated in numerous ADs, with specific microbial alterations linked to distinct conditions.

In RA, dysbiosis is characterized by the over-representation of *Prevotella copri*, which has been associated with disease onset. *P. copri* enhances Th17 cell activity, which is a key driver of inflammation in RA. Reduced levels of SCFA-producing bacteria exacerbate inflammation by impairing Treg function (Scher et al. 2013).

Longitudinal studies on T1D have demonstrated that dysbiosis precedes the development of islet autoantibodies in genetically predisposed children. A reduction in *Bacteroides* species and SCFA-producing microbes has been linked to weakened gut barrier function, facilitating beta-cell destruction.

Patients with MS exhibit a distinct microbial signature, with reduced levels of *Butyricimonas* and *Bacteroides fragilis*. These alterations shift the balance between antiinflammatory Tregs and proinflammatory Th17 cells, thereby contributing to neuroinflammation. Experimental models have shown that reintroducing beneficial microbes can restore the immune balance and ameliorate symptoms (Berer et al. n.d.).

In Crohn's disease and ulcerative colitis, dysbiosis is characterized by decreased levels of *F. prausnitzii* and increased levels of *Escherichia coli*. These shifts disrupt the gut homeostasis, perpetuate inflammation, and cause tissue damage.

5.3.5 Mechanisms Linking Dysbiosis to Autoimmunity

5.3.5.1 Molecular Mimicry

Pathogenic microbes produce antigens that mimic host proteins, leading to cross-reactive immune responses. *Klebsiella pneumoniae* antigens resemble HLA-B27, contributing to ankylosing spondylitis development (Rojas et al. n.d.).

5.3.5.2 Epitope Spreading

Dysbiosis-induced inflammation exposes host antigens, broadening the scope of autoimmune responses. In T1D, microbial dysbiosis facilitates exposure to pancreatic beta-cell antigens, promoting autoantibody production (Simmons and Michels n.d.).

5.3.5.3 Bystander Activation

Microbial antigens can nonspecifically activate autoreactive T and B cells. LPS from Gram-negative bacteria activates toll-like receptor 4 (TLR4), amplifying inflammatory cascades in diseases, such as RA and MS.

5.3.6 Therapeutic Implications

5.3.6.1 Probiotics and Prebiotics

Probiotics, such as *Lactobacillus rhamnosus* and *Bifidobacterium longum*, can reduce inflammation and restore microbial balance. Prebiotics, including inulins and fructooligosaccharides, enhance the growth of beneficial bacteria and improve SCFA production (Saxelin et al. n.d.).

5.3.6.2 Fecal Microbiota Transplantation (FMT)

FMT has been successfully used for IBD treatment by restoring microbial diversity and symptom alleviation. Ongoing trials are exploring its efficacy in RA, MS, and T1D treatment (Moayyedi et al. n.d.).

5.3.6.3 Dietary Interventions

High-fiber diets promote SCFA production and support gut barrier integrity and Treg function. Gluten-free diets are essential to prevent immune activation and intestinal damage in patients with celiac disease (Brown et al. n.d.).

5.3.6.4 Precision Microbiome-Based Therapies

Advances in metagenomics and metabolomics have enabled the development of personalized microbiome-based interventions. Tailoring therapies for individual microbial signatures has significant potential for improving the outcomes of Ads (Cho and Blaser 2012).

Research on the role of the microbiome in autoimmunity continues to expand, with several emerging areas of focus.

Identification of Microbial Biomarkers

Disease-specific microbial signatures have been investigated as diagnostic tools and predictors of therapeutic responses.

Engineered Microbes

Synthetic biological approaches have been used to develop engineered probiotics that deliver therapeutic molecules directly to the gut.

Microbiome-Host Interaction Studies

Advanced technologies, such as spatial transcriptomics and single-cell RNA sequencing, have been employed to map the interactions between gut microbes and host immune cells.

Longitudinal Cohort Studies

Large-scale studies such as the TEDDY study are crucial for understanding the temporal relationship between dysbiosis and autoimmune disease development.

5.4 Part III: Case Studies and Research Findings

5.4.1 *Case Studies and Epidemiological Evidence*

5.4.1.1 The Present Case Studies Illustrate Specific Instances in Which Environmental Triggers Have Been Linked to the Onset or Exacerbation of AD

Environmental factors are pivotal in triggering and exacerbating ADs, particularly in genetically predisposed individuals. The interaction between genetic susceptibility and environmental exposure is complex, and case studies provide critical insights into the specific mechanisms and pathways involved. This section highlights several key examples in which environmental triggers, including chemical exposure, dietary factors, infections, and lifestyle behaviors, have been directly linked to the onset or progression of ADs.

5.4.2 *Silica Exposure and SLE*

Silica exposure is a well-documented environmental trigger of SLE. Workers in the mining, construction, and manufacturing industries are particularly vulnerable to prolonged exposure to silica dust. Epidemiological studies have consistently demonstrated a strong association between silica exposure and the risk of SLE.

A meta-analysis by Parks et al. (2016) reviewed multiple occupational studies and found a significantly elevated risk of SLE among silica-exposed populations, with an odds ratio of 1.78. The pathophysiological mechanism involves silica-induced inflammation through macrophage activation and the production of proinflammatory cytokines, such as IL-1β and TNF-α. These inflammatory signals promote apoptosis and the release of nuclear antigens, which serve as autoantigenic triggers.

Further investigations have revealed that silica exposure compromises the function of Tregs and enhances the activation of autoreactive T and B cells. Subsequent immune dysregulation contributes to the formation of immune complexes that are deposited in tissues, causing systemic inflammation characteristic of SLE. These findings underscore the critical importance of occupational safety measures, including stringent exposure limits and protective equipment, to mitigate autoimmune risks in industrial workers.

5.4.3 Smoking as a Risk Factor for RA

Smoking is another extensively studied environmental factor in the etiology of AD, particularly RA. Case-control studies have consistently shown that smoking increases the risk of RA, particularly in individuals carrying the HLA-DR4 allele.

The mechanism underlying this association involves smoking-induced protein citrullination, a posttranslational modification in which arginine residues are converted into citrulline. This modification generates neoantigens that are recognized as foreign by the immune system, triggering an autoimmune response. Hutchinson et al. (2001) demonstrated that smokers with the HLA-DR4 genotype exhibited a 20-fold increased risk of developing RA compared with that of nonsmokers without this allele.

In addition to promoting citrullination, smoking alters immune cell behavior by increasing proinflammatory cytokine production and impairing Treg function. These immune alterations exacerbate synovial inflammation, leading to progressive joint destruction. Therefore, smoking cessation programs are vital, particularly for individuals with a genetic predisposition for RA.

5.4.4 EBV and Systemic ADs

EBV, a ubiquitous human herpesvirus, has been implicated in several systemic ADs, including SLE and MS. Case studies have highlighted that EBV is a potent environmental trigger that interacts with genetic susceptibility to drive autoimmune responses.

James et al. (2021) reported that individuals with a history of EBV infection exhibit a significantly higher risk of developing MS, particularly when infection occurs during adolescence or early adulthood. EBV infects B cells, leading to the expression of viral proteins that mimic host antigens. This molecular mimicry triggers autoreactive T- and B-cell responses, resulting in chronic inflammation and tissue damage.

In SLE, EBV-infected B cells overexpress Sm and Ro antigens, which are targeted by lupus-specific autoantibodies. The persistent presence of these autoantibodies exacerbates the disease severity and contributes to systemic manifestations. These findings suggest that EBV vaccination may serve as a preventive measure against certain ADs.

5.4.5 Gluten as a Trigger for Celiac Disease

Celiac disease provides a clear example of how dietary environmental factors interact with a genetic predisposition to trigger autoimmunity. Gluten, a protein found in wheat, barley, and rye, serves as the primary environmental trigger in individuals carrying HLA-DQ2 or HLA-DQ8 alleles.

Sollid and Jabri (2013) described how gluten peptides resist enzymatic digestion in the gastrointestinal tract, leading to deamidation by tissue transglutaminase. These modified peptides are then presented by HLA-DQ2 or HLA-DQ8 molecules to $CD4^{+}$ T cells, initiating an immune cascade. The resulting inflammation damages the intestinal epithelium and causes villous atrophy and malabsorption.

Case studies have demonstrated that strict adherence to a gluten-free diet effectively reverses symptoms and prevents complications in most patients with CD. These findings highlight the importance of an early diagnosis and dietary management in genetically predisposed populations.

5.4.6 Chemical Exposures and Autoimmune Skin Disorders

Occupational exposure to chemicals, such as phenolic compounds and pesticides, has been associated with autoimmune skin disorders, including vitiligo and psoriasis. Phenolic compounds, which are commonly used in the production of plastics, dyes, and adhesives, are structurally similar to melanin precursors. This similarity is hypothesized to induce cross-reactive immune responses that target melanocytes, thereby leading to their destruction.

A case control study by Schallreuter et al. (2008) found that individuals with prolonged occupational exposure to phenolic compounds had a significantly increased risk of developing vitiligo (Njoo and Westerhof n.d.). These findings emphasize the need for regulatory measures to limit chemical exposure in industrial settings and prevent autoimmune complications.

5.4.7 Occupational Exposure to Pesticides and Systemic ADs

Pesticides, particularly organophosphates, have been implicated in the development of systemic ADs, such as SLE and RA. Selmi et al. (2012) investigated the prevalence of ADs among agricultural workers exposed to pesticides and found a significantly higher incidence than that in nonexposed populations.

The proposed mechanism involves pesticide-induced oxidative stress and epigenetic changes that alter immune gene expression and disrupt homeostasis. These findings underscore the importance of adopting safer agricultural practices and promoting the use of nontoxic alternatives to mitigate the autoimmune risks associated with pesticide exposure (Table 5.2).

5.4.7.1 Review of Epidemiological Studies Highlighting Trends in Autoimmune Disease Incidence Related to Environmental Changes

ADs are a growing public health challenge with increasing global incidence rates. While genetic predisposition plays a central role, the rapid rise in ADs cannot be explained solely by genetic factors, as they have evolved over millennia. Environmental changes have emerged as critical drivers of autoimmune disease pathogenesis. Epidemiological studies have shed light on these trends and identified key environmental triggers, including urbanization, pollution, dietary shifts, and microbial exposure. This review synthesizes evidence from epidemiological studies to illustrate how environmental factors contribute to the increasing prevalence of autoimmune conditions.

Table 5.2 Overview of environmental triggers linked to autoimmune diseases, highlighting their associated conditions, key studies, and mechanisms of action

Environmental trigger	Associated disease(s)	Case study/cited evidence	Mechanism of action
Silica exposure	Systemic lupus erythematosus (SLE)	Parks et al. (2016): Meta-analysis of silica-exposed populations	Macrophage activation, cytokine production (IL-1β, TNF-α), immune complex deposition
Smoking	Rheumatoid arthritis (RA)	Hutchinson et al. (2001): 20-fold increased RA risk in smokers	Protein citrullination, neoantigen formation, Treg dysfunction
Epstein-Barr virus (EBV)	SLE, multiple sclerosis (MS)	James et al. (2021): Increased MS risk after adolescent infection	Molecular mimicry, autoantibody production, chronic inflammation
Gluten	Celiac disease	Sollid and Jabri (2013): HLA-DQ2/DQ8 linked to gluten sensitivity	Presentation of deamidated gluten peptides by HLA molecules, intestinal damage
Phenolic compounds	SLE, RA	Schallreuter et al. (2008): Increased vitiligo risk in exposed workers	Cross-reactive immune responses targeting melanocytes
Pesticides	Vitiligo, psoriasis	Selmi et al. (2012): Higher autoimmune incidence in farmers	Oxidative stress, epigenetic changes, immune gene dysregulation

Urbanization and Autoimmune Disease Trends

The global shift toward urbanization has been accompanied by an increase in ADs, particularly in developed and rapidly developing countries. Urbanization is associated with environmental changes, including reduced exposure to the natural environment, increased pollution, and lifestyle shifts, all of which affect immune function.

A comprehensive review by Bach (2002) proposed the "hygiene hypothesis," which suggests that reduced microbial exposure during early childhood in urban settings skews immune responses toward autoimmunity. This hypothesis is supported by epidemiological data showing higher rates of ADs, such as T1D and MS, in urban populations than that in rural populations. A large cohort study in Europe reported a 30% higher incidence of T1D in children living in urban areas than that in their rural counterparts.

Similarly, MS prevalence correlated with urbanization and latitude. This trend is partly attributable to lower sunlight exposure and subsequent vitamin D deficiency, which are known risk factors for MS.

Pollution and Its Role in AD Incidence

Air pollutants, industrial emissions, and particulate matter are increasingly recognized as contributors to ADs. Fine particulate matter (PM2.5) and nitrogen dioxide (NO_2), commonly found in urban and industrial areas, are associated with systemic inflammation and immune dysregulation. A Swedish population-based study by Hart et al. (2013) examined the association between air pollution and RA and found that individuals living in areas with high PM2.5 levels had a significantly increased risk of RA. The study proposed that air pollutants induce oxidative stress, alter antigen presentation, and exacerbate systemic inflammation, thereby creating a proautoimmune environment. Another example is SLE, in which exposure to ozone and diesel exhaust particles is linked to disease exacerbation. A US-based cohort study found that women exposed to high levels of traffic-related air pollution had a 20% higher risk of developing SLE than those exposed to low levels of pollution. These findings emphasize the urgent need for environmental regulations to reduce pollution and mitigate the risk of *AD*.

Dietary Shifts and AD Patterns

The global transition toward Western-style diets, characterized by high fat, sugar, and processed food consumption, has been implicated in the rising incidence of ADs, such as IBD and T1D. Epidemiological studies have highlighted the influence of dietary changes on the gut health and systemic immune responses. Ananthakrishnan (2015) reviewed the impact of diet on IBD prevalence and noted that populations adopting Western diets experienced higher rates of Crohn's disease and ulcerative colitis. These diets reduce gut microbiota diversity and impair intestinal barrier function, allowing microbial products to enter the circulation and trigger inflammation. Similarly, a prospective study in Finland found that children consuming a high-fat diet had an elevated risk of developing T1D, emphasizing the role of diet in early immune programming.

Infections and Molecular Mimicry in Autoimmunity

Infections have long been implicated as environmental triggers of ADs and epidemiological studies have provided robust evidence. Pathogens, such as viruses and bacteria, can initiate autoimmune responses through molecular mimicry, epitope spreading, and bystander activation. Similar trends have been observed in rheumatic fever, where *Streptococcus pyogenes* infections lead to cross-reactive immune responses that target the cardiac tissue. The EBV has been implicated in systemic ADs such as SLE and MS. Epidemiological data revealed a higher prevalence of EBV antibodies in patients with these diseases than in healthy controls, suggesting a causal relationship.

The Role of Microbial Dysbiosis

Microbial dysbiosis, or an imbalance in the gut microbiota, is increasingly being recognized as a key factor in the pathogenesis of AD. Epidemiological studies have linked microbial alterations to diseases, such as RA, IBD, and T1D. Scher et al. (2013) conducted a longitudinal study on patients with RA and found that disease onset was correlated with an expansion of *P. copri* in the gut microbiota. Similarly, reduced levels of *F. prausnitzii*, a beneficial microbe, have been observed in patients with Crohn's. These findings suggest that microbial diversity and composition significantly influence immune regulation.

Climate Change and Autoimmune Disease Risks

Climate change indirectly influences the incidence of AD through its impact on environmental exposure and infectious disease patterns. Wallin et al. (2019) highlighted that rising temperatures and altered ecosystems increase exposure to infectious agents, potentially triggering autoimmune responses. Lyme disease, caused by *Borrelia burgdorferi*, has shown an increasing incidence in areas where warm climates extend the range of tick populations. Epidemiological studies have provided compelling evidence linking environmental changes to the rising incidence of ADs. Urbanization, pollution, dietary shifts, infections, and microbial dysbiosis collectively shape immune responses and influence the patterns of autoimmune diseases. These findings underscore the importance of multidisciplinary approaches to public health, integrating environmental policies, dietary guidelines, and infection control to mitigate the growing burden of ADs.

5.4.8 *Cutting-Edge Research and Future Directions*

5.4.8.1 Current Research Efforts Aimed at Identifying Environmental Triggers and Understanding Their Mechanisms

ADs represent a significant challenge to global health, and their etiology is often rooted in complex interactions between genetic predispositions and environmental exposure. Although the role of environmental factors in autoimmune disease pathogenesis has long been acknowledged, current research is increasingly focusing on elucidating the specific mechanisms by which these triggers influence immune functions. The following sections summarize recent advancements in the understanding of environmental triggers, including infections, dietary factors, pollutants, and microbiome alterations, as well as innovative approaches in the field.

Role of Infections in ADs

Infections are one of the most studied environmental triggers of ADs. Pathogens, such as viruses, bacteria, and parasites, disrupt immune homeostasis through molecular mimicry, bystander activation, and epitope spreading. Recent research has focused on characterizing these mechanisms and identifying specific pathogens associated with the onset of AD. Single-cell RNA sequencing techniques have been employed to analyze immune cell populations in EBV-infected individuals, revealing dysregulated gene expression patterns that correlate with disease severity. Similarly, the Environmental Determinants of Diabetes in the Young (TEDDY) Study identified enteroviruses as key environmental triggers of T1D. This longitudinal study demonstrated that early enterovirus infections increase the production of autoantibodies against pancreatic beta cells, initiating autoimmune destruction. Future studies should focus on developing vaccines and antiviral therapies to mitigate these risks.

Dietary Influences and Gut-Immune Axis

The gut-immune axis is a critical area of research in ADs and diet plays a central role in modulating immune responses. Advances in metabolomics and microbiology have enabled researchers to investigate the effects of dietary components on gut health and systemic immunity. Gluten is a well-established dietary trigger for celiac disease. Sollid and Jabri (2013) described how gluten peptides activate $CD4^{+}$ T cells in individuals carrying the HLA-DQ2 or HLA-DQ8 alleles, leading to intestinal inflammation. Recent studies using advanced imaging techniques have provided insights into how gluten-induced inflammation disrupts epithelial barrier integrity, allowing microbial antigens to enter systemic circulation. In IBD, high-fat and high-sugar diets associated with Western dietary patterns reduce gut microbial diversity and impair SCFA production. These changes promote intestinal permeability and systemic inflammation in mice. Ananthakrishnan (2015) highlighted how dietary interventions aimed at restoring the microbial balance can reduce IBD severity and prevent relapse.

Pollution and Autoimmune Disease Risks

Environmental pollutants, including air pollutants, heavy metals, and industrial chemicals, are significant contributors to ADs. Emerging research has focused on characterizing the molecular pathways affected by these factors. Jung et al. (2019) demonstrated that exposure to fine particulate matter (PM2.5) and nitrogen dioxide (NO2) correlated with an increased risk of SLE. Using in vitro models, researchers have identified that PM2.5 induces oxidative stress, mitochondrial damage, and immune activation through the production of reactive oxygen species. These findings have been corroborated by animal models in which pollutant exposure exacerbated autoimmune phenotypes in genetically susceptible mice. Heavy metals, such as mercury and cadmium, disrupt immune tolerance by altering antigen presentation and cytokine production. Recent mass cytometry studies have revealed that these metals selectively affect dendritic cell function, leading to the activation of autoreactive T cells. Future research aims to develop biomarkers for early detection of pollutant-induced ADs.

Microbiome Dysbiosis and ADs

The gut microbiome plays a pivotal role in the maintenance of immune homeostasis. Dysbiosis or microbial imbalance has been implicated in numerous ADs, including RA, MS, and T1D. Shotgun metagenomics techniques were used to profile microbial communities in patients with *AD*, revealing disease-specific patterns of dysbiosis. Scher et al. (2013) found that patients with RA exhibit an overrepresentation of *P. copri* and a reduction in beneficial microbes, such as *F. prausnitzii.* These microbial shifts are associated with altered immune signaling and increased systemic inflammation. Microbiota-targeted therapies, including the use of probiotics, prebiotics, and FMT, are currently being explored as interventions. Clinical trials have shown that FMT modulates immune responses in patients with IBD, leading to sustained remission in some cases. Future research should aim to identify specific microbial metabolites that regulate immune tolerance and provide novel therapeutic targets.

Epigenetic Modifications as Mediators of Environmental Influence

Epigenetic modifications, such as DNA methylation, histone acetylation, and noncoding RNA expression, mediate the impact of environmental factors on gene regulation. Advances in epigenomics have enabled researchers to investigate the contribution of these modifications to the pathogenesis of *AD*. Smoking-induced epigenetic changes in patients with RA have been characterized using whole-genome bisulfite sequencing. These studies revealed that smoking alters DNA methylation patterns in genes that regulate immune cell activation, leading to heightened autoimmune responses (Ringh et al. 2019). Similarly, noncoding RNAs, including microRNAs (miRNAs), have been implicated in the regulation of autoimmune genes. Recent studies using RNA-seq have identified dysregulated miRNA expression profiles in patients with lupus, providing potential biomarkers for disease activity and treatment responses.

5.4.8.2 Future Directions in Autoimmune Disease Research

The future of autoimmune disease research lies in leveraging multidisciplinary approaches and integrating cutting-edge technologies to address the multifactorial nature of autoimmune diseases. Environmental triggers, in particular, remain a critical focus because their interplay with genetic and immunological factors drives disease onset and progression. Below, we explore the key areas shaping future research and therapeutic advancements.

Precision Medicine and Predictive Modeling

Precision medicine aims to tailor interventions to individual risk profiles by considering genetic predispositions, environmental exposures, and immune system dynamics.

Advances in Omics Technologies

- Genomics: Whole-genome sequencing and genome-wide association studies (GWAS) continue to identify genetic variants associated with ADs, such as PTPN22 in RA and HLA-DQ8 in T1D.
- Proteomics: Analyses of protein expression and posttranslational modifications provide insights into disease-specific pathways and biomarkers. For example, protein citrullination in RA serves as both a diagnostic and therapeutic target.
- Microbiomics: Microbial profiling technologies, such as *16S* rRNA sequencing and metagenomics, uncover microbial signatures linked to diseases, such as IBD and MS, enabling targeted interventions.

AI in Risk Stratification

AI-driven predictive models have been developed to stratify patients based on genetic and environmental risk factors. Machine-learning (ML) algorithms analyze complex datasets, including genomic, environmental, and clinical information, to predict disease onset, progression, and therapeutic responses. AI models have been employed to identify individuals at high risk of SLE and T1D, facilitating early intervention.

Vaccination Strategies Against Infection-Driven Autoimmunity

Infections, such as those caused by the EBV and enteroviruses, are significant environmental triggers of ADs. Vaccination efforts targeting these pathogens hold promise in reducing the incidence of AD.

EBV Vaccination

EBV has been implicated in the pathogenesis of SLE and MS through molecular mimicry and B-cell activation. Vaccine candidates targeting the EBV envelope glycoproteins have been used in preclinical and clinical trials. Early evidence suggests that reducing EBV infection rates could significantly lower the incidence of MS.

Enterovirus Vaccination for T1D

Enteroviruses, particularly coxsackieviruses, are thought to initiate pancreatic beta-cell autoimmunity through molecular mimicry and bystander activation. Efforts to develop enteroviral vaccines, such as those targeting coxsackievirus B strains, are underway, with animal models demonstrating reduced beta-cell destruction.

Environmental Policy and Public Health Initiatives

Addressing environmental triggers requires coordinated public health strategies and regulatory measures to minimize exposure to harmful agents.

Reducing Pollutant Exposure

Policies targeting air pollution, water contamination, and occupational hazards are essential for mitigating autoimmune risks. For example, reducing silica exposure in the mining and construction industries lowers the incidence of SLE.

Promoting Sustainable Agricultural Practices

Restricting the use of organophosphate pesticides and adopting organic farming practices can reduce the risk of developing AD associated with pesticide exposure. Public awareness campaigns and subsidies for safer alternatives are critical for widespread adoption.

Global Collaboration

Initiatives, such as the Global Burden of ADs project, aim to standardize data collection and foster international collaboration to address the environmental determinants of autoimmunity.

Microbiome-Based Therapies

The pivotal role of the gut microbiome in immune regulation has spurred interest in microbiota-targeted interventions as a therapeutic option for ADs.

Personalized Probiotics

Advances in microbiome research have enabled the development of personalized probiotics tailored to microbial composition. For example, *Lactobacillus* and *Bifidobacterium* strains have been investigated for their potential to enhance Treg activity and suppress inflammation in IBD and RA.

FMT

FMT, already established for *Clostridium difficile* infections, is being explored for the treatment of ADs such as MS and IBD. Early trials indicated that restoring microbial diversity through FMT can reduce disease activity and improve patient outcomes.

Microbial Metabolite Supplementation

SCFAs such as butyrate have been studied for their ability to promote gut barrier integrity and regulate T-cell function. Preclinical studies using T1D and MS models have demonstrated the therapeutic potential of SCFA supplementation.

Epigenetic Therapeutics

Epigenetic modifications, including DNA methylation and histone acetylation, play crucial roles in the pathogenesis of AD. Drugs targeting these pathways offer novel therapeutic opportunities.

Histone Deacetylase (HDAC) Inhibitors

HDAC inhibitors, such as vorinostat and belinostat, have been evaluated for their ability to modulate proinflammatory gene expression. Preclinical studies using SLE and RA models have demonstrated reduced inflammation and improved disease outcomes.

DNA Methylation Modulators

Agents targeting DNA methyltransferases are being investigated for the treatment of diseases, such as SLE and MS. By reversing aberrant DNA methylation patterns, these therapies aim to restore immune tolerance.

Noncoding RNAs as Therapeutic Targets

MicroRNAs (miRNAs) and long noncoding RNAs (lncRNAs) regulate immune pathways and are currently being explored as biomarkers and therapeutic targets. miR-155 inhibition has shown promise for reducing inflammation in RA models. Autoimmune disease research is poised to transform our understanding and management of these complex conditions. By integrating precision medicine, microbiome-based therapies, vaccination strategies, and epigenetic approaches, researchers have developed targeted interventions to address the root causes of autoimmunity. Public health initiatives and environmental policies are essential for mitigating the risks posed by environmental triggers. These multidisciplinary efforts have the potential to reduce the burden of ADs and improve patient outcomes worldwide.

5.4.8.3 Emerging Technologies and Methodologies to Pinpoint Environmental Influences on ADs

The intricate interplay between environmental factors and genetic predispositions in ADs poses significant challenges in understanding disease mechanisms. However, the advent of cutting-edge technologies and methodologies has revolutionized the field, enabling researchers to reveal the complex pathways that link environmental exposure to autoimmune pathogenesis. These advancements span multiple disciplines, including genomics, microbiomics, epigenomics, and bioinformatics, and are crucial for identifying biomarkers, elucidating their mechanisms, and developing targeted interventions.

Multiomics Approaches in Autoimmune Disease Research

Multiomics methodologies integrate genomics, transcriptomics, proteomics, and metabolomics data to provide a holistic understanding of the influence of environmental factors on ADs. These technologies enable researchers to identify molecular signatures associated with environmental exposure and disease outcomes.

Genomics and Transcriptomics

Next-generation sequencing is pivotal in identifying genetic variants associated with ADs. GWAS have revealed numerous single nucleotide polymorphisms linked to diseases, such as RA, SLE, and T1D. While GWAS predominantly focus on inherited genetic factors, transcriptomics provides insights into gene expression changes driven by environmental exposure. RNA-seq studies have demonstrated how pollutants, such as PM2.5, alter the expression of genes involved in immune regulation, contributing to autoimmune responses (Jung et al. 2019).

Proteomics and Metabolomics

Proteomic analyses using mass spectrometry have identified posttranslational modifications, such as citrullination and oxidation, that create neoantigens that trigger autoimmunity. These findings are particularly relevant for smoking-induced RA. Metabolomics has highlighted the role of microbial metabolites, such as SCFAs, in maintaining immune tolerance. Dysbiosis-induced alterations in metabolite profiles have been linked to IBD and MS.

Integration of Multiomics Data

The integration of multiomics data provides a comprehensive view of the impact of environmental factors on multiple molecular layers. The combination of epigenomic and transcriptomic data has revealed how smoking-induced DNA methylation alters cytokine expression, promoting RA development (Liu et al. 2013). Such integrative approaches are increasingly used to identify novel biomarkers and therapeutic targets.

Single-Cell Technologies in Autoimmune Research

Single-cell technologies, including single-cell RNA sequencing (scRNA-seq) and mass spectrometry (CyTOF), have transformed our understanding of cellular heterogeneity in ADs. These methods allow researchers to profile individual cells and reveal how environmental exposures influence specific immune cell populations.

ScRNA-seq

scRNA-seq provides high-resolution insights into changes in gene expression at the single-cell level. This technology has been used to characterize autoreactive T and B cells, Tregs, and antigen-presenting cells. Kuret et al. (2022) utilized scRNA-seq to analyze Tregs in patients with MS, revealing how environmental triggers, such as viral infections, disrupt Treg function and promote neuroinflammation.

Mass Cytometry (CyTOF)

CyTOF enables simultaneous analysis of dozens of cellular markers, offering unparalleled insights into immune cell phenotypes and functions. This technology has been applied to study how pollutants, such as heavy metals, affect dendritic cells and macrophages, which are key players in antigen presentation and immune activation (Selmi et al. 2012). By mapping immune cell changes in exposed individuals, CyTOF can be used to identify early indicators of autoimmune disease risk.

Advanced Imaging and Spatial Transcriptomics

Visualizing interactions between immune cells and environmental factors at the tissue level is critical for understanding autoimmune pathogenesis. Advanced imaging techniques, such as multiplex immunofluorescence and spatial transcriptomics, provide spatial and molecular data in the same tissue context.

Multiplex Immunofluorescence

This technique allows simultaneous detection of multiple proteins in a single tissue section. It has been used to study how environmental exposures, such as UV radiation, alter skin immune cell populations in patients with SLE. UV-induced apoptosis of keratinocytes releases nuclear antigens that drive autoantibody production. Multiplex immunofluorescence provides spatial data on how immune cells respond to these antigens, thereby advancing our understanding of SLE pathogenesis.

Spatial Transcriptomics

Spatial transcriptomics integrates gene expression data with tissue morphology, enabling researchers to map transcriptional changes in specific tissue regions. This technology has been used to study gut biopsies from patients with IBD to reveal how dietary changes alter gut immune cell gene expression and microbial interactions.

High-Throughput Screening (HTS) and Organoid Models

HTS and organoid models are powerful tools for studying environmental triggers of ADs. These methodologies enable testing of multiple environmental factors and therapeutic candidates in controlled settings.

HTS

HTS platforms use automated systems to screen thousands of compounds or environmental agents for their effects on the immune cell function. HTS was used to identify environmental chemicals, such as pesticides and industrial byproducts, that disrupt immune tolerance. These screens provide valuable data for prioritizing regulatory actions and public health interventions.

Organoid Models

Organoid models, derived from patient stem cells, recapitulate the structure and function of specific tissues, such as the gut or pancreas. These models are particularly useful for studying the effects of environmental exposure on tissue-specific ADs. Pancreatic organoids have been used to investigate how viral infections trigger T1D through beta-cell destruction, whereas gut organoids help to elucidate the role of dietary components in IBD.

AI and ML

AI and ML are revolutionizing the analysis of complex datasets generated in autoimmune research. These technologies can be used to identify patterns and relationships that may not be apparent when traditional analytical methods are used.

Predictive Modeling

ML algorithms have been used to develop predictive models for autoimmune disease risk based on genetic, environmental, and clinical data. AI tools have successfully stratified patients into high-risk groups for T1D by integrating enterovirus exposure data with genetic profiles (Janani et al. 2015).

Natural Language Processing (NLP)

NLP technologies are increasingly used to mine scientific literature and electronic health records to gain insights into ADs. By analyzing vast datasets, NLP can identify previously unrecognized associations between environmental factors and disease outcomes.

Environmental Monitoring Technologies

Environmental monitoring technologies play a crucial role in identifying and quantifying exposures that contribute to ADs. Advances in wearable sensors, remote sensing, and environmental sampling have enabled the real-time collection of data.

Wearable Sensors

Wearable devices equipped with sensors for pollutants, UV radiation, and temperature provide personalized exposure data. This information can be integrated with health outcomes to investigate the roles of specific environmental triggers in autoimmune disease flares.

Remote Sensing

Satellite-based remote sensing technologies measure air quality, UV radiation, and other environmental variables at the population level. These data are increasingly being used in epidemiological studies to link environmental exposure to the incidence of AD.

Emerging technologies and methodologies are transforming our ability to identify environmental influences on ADs. By integrating multiomics data, leveraging single-cell and imaging technologies, and utilizing AI-driven analytics, researchers are attempting to unravel the complex interactions between genetic predispositions and environmental exposure. These advances hold immense promise for identifying novel biomarkers, elucidating disease mechanisms, and developing targeted prevention and intervention strategies.

5.5 Part IV: Practical Implications and Conclusion

5.5.1 Practical Steps for Patients and Healthcare Providers

The management of ADs requires a comprehensive approach that addresses both genetic predispositions and environmental triggers. Although genetic factors are nonmodifiable, patients and healthcare providers can significantly mitigate the impact of environmental exposure through informed decision-making and proactive strategies. This section provides practical advice for patients on minimizing environmental triggers and discusses the pivotal role of healthcare providers in educating patients and advocating environmental health policies.

5.5.1.1 Practical Advice for Patients on Minimizing Environmental Triggers

Dietary Changes

Diet plays a critical role in modulating immune responses and influencing the activity of AD. Evidence-based dietary interventions can help minimize the impact of environmental triggers.

- Antiinflammatory Diets: Incorporating an antiinflammatory diet rich in fruits, vegetables, whole grains, and omega-3 fatty acids reduces systemic inflammation. Studies, such as those by Abdulrazaq et al. (2017), have demonstrated the benefits of omega-3 supplementation in reducing disease activity in RA and lupus.
- Gluten-Free Diets: For patients with celiac disease or gluten sensitivity, strict adherence to gluten-free diets is essential. The elimination of gluten prevents intestinal damage and reduces systemic inflammation.
  Probiotic-rich foods, such as yogurt, kefir, and fermented vegetables, can help restore the gut microbial balance, which is often disrupted in autoimmune conditions such as IBD.
- Avoiding Processed Foods: Processed foods high in sugar, trans fats, and additives can exacerbate inflammation and disrupt the gut microbiome. Replacing these with whole-nutrient dense options is recommended.

Reducing Chemical Exposures

Chemical exposure to pollutants, pesticides, and industrial toxins has been linked to the onset and progression of AD. Patients can perform the following steps to minimize exposure:

- Indoor Air Quality: Using air purifiers and avoiding indoor smoking can reduce exposure to airborne pollutants. A study by Hart et al. (2013) found a strong correlation between particulate matter exposure and increased risk of SLE.
- Safe Cleaning Practices: Opting for natural cleaning agents and avoiding harsh chemicals reduces exposure to industrial toxins.
- Pesticide-Free Foods: Choosing organic produce and thoroughly washing fruits and vegetables can lower pesticide residues, which are associated with ADs, such as RA and SLE (Selmi et al. 2012).

Managing Stress

Stress has a profound impact on the immune system and is known to exacerbate ADs. Practical strategies for stress management include the following.

- Mindfulness and Meditation: Regular mindfulness practices reduce proinflammatory cytokine levels, thereby mitigating autoimmune disease activity.
- Regular Physical Activity: Moderate exercise improves physical health and reduces stress-related hormonal imbalances that affect immune regulation.
- Professional Counseling: Patients experiencing chronic stress or emotional distress may benefit from therapy or counseling to develop coping mechanisms.

5.5.1.2 Role of Healthcare Providers in Education and Advocacy

Healthcare providers play a crucial role in empowering patients to effectively manage their autoimmune conditions and advocate for broader environmental health measures. Their responsibilities included the following:

Patient Education

Healthcare providers must equip patients with knowledge and tools to make informed decisions about their health. This involves:

- Personalized Guidance: Developing individualized care plans that address each patient's unique environmental exposures and lifestyle factors. Dietary advice should be tailored based on the patient's autoimmune disease and genetic predispositions.
- Workshops and Resources: Conduct workshops or provide educational materials on topics, such as safe household practices, stress management techniques, and the importance of regular screenings for environmental toxin exposure.
- Behavioral Support: Assisting patients in making sustainable lifestyle changes, such as adopting an antiinflammatory diet or quitting smoking.

Advocacy for Environmental Health Policies

Healthcare providers play a critical role in shaping public health policies to reduce environmental risks associated with ADs. The key areas for advocacy include the following:

- Stricter Regulations on Pollutants: Supporting legislation aimed at reducing industrial emissions, improving air quality, and limiting pesticide use. Jung et al. (2019) emphasized the need for stringent air pollution regulations to decrease SLE incidence.
- Promoting Healthy Work Environments: Advocating for occupational safety standards that minimize exposure to hazardous substances, such as silica, which is linked to increased SLE risk.
- Community Outreach: Collaborating with local organizations to increase awareness of environmental triggers and their impact on ADs.

Collaboration and Research

Healthcare providers should engage in interdisciplinary collaborations to advance research and improve clinical practices:

- Participating in Research: Contributing to studies on the effects of environmental triggers on ADs and the efficacy of interventions. Participating in longitudinal studies, such as the Environmental Determinants of Diabetes in the Young (TEDDY) study.
- Integrating Technology: Leveraging tools, such as wearable sensors and remote monitoring, to track patient exposure to environmental triggers and correlate these with disease activity.

5.5.1.3 Practical Case Study

A 45-year-old female patient with a 10-year history of RA presented experienced frequent flares despite pharmacological management. After reviewing her environmental exposures, her healthcare provider identified chronic stress and high intake of processed foods as potential triggers.

Interventions

1. Dietary Modifications: The patient was introduced to an antiinflammatory diet rich in omega-3 fatty acids, whole grains, and probiotics.
2. Stress Management: She was enrolled in a mindfulness-based stress reduction program and encouraged to practice yoga.
3. Reducing Chemical Exposure: Switching to natural cleaning products and installing an air purifier in her home.

Outcome
Within six months, the patient reported a significant reduction in RA flares and an improved quality of life. This case highlights the importance of addressing environmental factors in the management of AD.

5.5.2 Conclusion: Toward a Holistic Approach

The study of ADs underscores the intricate relationship between genetic predispositions and environmental triggers during the development and progression of these complex conditions. We explored how environmental factors, ranging from chemical exposure and dietary patterns to infections and stress, interact with genetic susceptibility to disrupt immune tolerance and lead to chronic autoimmune responses. This chapter synthesizes these insights and advocates a comprehensive and holistic approach to manage ADs that integrates both prevention and personalized treatment strategies.

5.5.2.1 Recap of Key Points on Environmental Triggers and Autoimmune Responses

ADs arise due to the convergence of genetic susceptibility and environmental exposure. While genetic factors, such as specific HLA alleles, provide a foundational predisposition, environmental triggers act as catalysts for disease onset or exacerbation. The following key themes emerged from this study:

Diverse Environmental Triggers

- Chemical Exposures: Pollutants, industrial chemicals, and pesticides contribute to ADs, such as SLE and RA, via mechanisms, such as oxidative stress and immune activation.
- Dietary Factors: Gluten triggers celiac disease in genetically predisposed individuals, whereas Western-style diets exacerbate conditions, such as IBD, by disrupting the gut microbiota.
- Infections: Pathogens, including the EBV and enteroviruses, initiate autoimmune responses through molecular mimicry and bystander activation, contributing to diseases, such as MS and T1D.
- Stress: Chronic psychological stress disrupts immune regulation and aggravates autoimmune conditions through proinflammatory pathways.

Mechanisms Linking Triggers to Disease

- Environmental factors disrupt immune homeostasis by altering gene expression, promoting neoantigen formation, and impairing regulatory mechanisms. Smoking-induced citrullination generates antigens that drive RA, while UV-induced apoptosis releases nuclear antigens in SLE.
- Dysbiosis of the gut microbiota, which is influenced by diet and environmental exposure, further exacerbates immune dysregulation in ADs.

Emerging Insights from Cutting-Edge Research

- Advances in multiomics, single-cell technologies, and artificial intelligence have deepened our understanding of how environmental factors influence autoimmune pathogenesis. These methodologies have identified novel biomarkers and therapeutic targets, paving the way for personalized interventions.

- Longitudinal studies, such as the TEDDY study, emphasize the role of early-life exposure in shaping autoimmune risk, highlighting the importance of preventive strategies.

5.5.2.2 The Importance of a Holistic Approach

Management of ADs requires a holistic strategy that considers the interplay between genetic and environmental factors. This approach emphasizes prevention, early detection, and personalized care to address the multifaceted nature of these diseases.

Integrating Prevention into Healthcare

- Reducing environmental exposure: Public health policies should prioritize reducing pollutants, regulating occupational hazards, and promoting pesticide-free agricultural practices. Limiting silica exposure in industrial workers can significantly reduce the risk of SLE, as demonstrated in occupational studies.
- Promoting a healthy lifestyle: Dietary interventions, such as adopting antiinflammatory diets and avoiding processed foods, can help mitigate inflammation and support immune balance. Stress management through mindfulness and physical activity complements these efforts.

Empowering Patients and Healthcare Providers**

- Patients should be equipped with the knowledge and resources to identify and minimize their exposure to environmental triggers. Healthcare providers play a pivotal role in offering personalized guidance and fostering behavioral changes.
- Advocating environmental health policies is critical for creating safer living and working environments. Collaborative efforts between health care systems and policymakers can lead to meaningful changes.

Advancing Personalized Medicine

- Precision medicine approaches that leverage genetic, epigenetic, and environmental data enable tailored interventions to address individual risk profiles. Microbiota-targeted therapies, such as probiotics and dietary modifications, are promising tools for restoring immune homeostasis in IBD and RA.
- Vaccines targeting infection-driven autoimmunity, such as EBV vaccines for MS, represent a transformative avenue for disease prevention.

Fostering Collaborative Research

– Interdisciplinary research efforts are essential to unravel the complexity of ADs. Studies integrating multiomics, environmental monitoring, and clinical data will further enhance our understanding of disease mechanisms and therapeutic strategies.

5.5.2.3 Conclusion: A Path Forward

ADs are emblematic of the intricate interplay between genetics and environment, underscoring the need for a multifaceted and holistic approach for their management. By addressing environmental triggers alongside genetic predispositions, we can not only improve outcomes for those living with ADs but also work toward reducing their global burden. This journey begins with awareness, empowering patients to make informed decisions, equipping healthcare providers with tools to educate and advocate, and fostering a collaborative research ecosystem. Despite advances in precision medicine and environmental health, the potential to transform autoimmune disease management has not increased. Through continued research, education, and policy reform, we can move toward a future in which ADs are better understood, effectively managed, and ultimately preventable. This holistic vision serves as both a challenge and an opportunity for medical and scientific communities to embrace innovation and collaboration in the fight against ADs.

References

Abdulrazaq M, Innes JK, Calder PC (2017) Effect of ω-3 polyunsaturated fatty acids on arthritic pain: A systematic review. Nutrition 39-40:57–66

Ahmed SS, Lambert P-H (2014) Chapter 20—autoimmune diseases: the role for vaccines. In: Rose NR, Mackay IR (eds) The autoimmune diseases, 5th edn. Academic Press, Boston, pp 275–282

Albenberg LG, Wu GD (n.d.) Diet and the intestinal microbiome: associations, functions, and implications for health and disease. (1528-0012 (Electronic)). (In eng)

Ananthakrishnan AN (2015) Epidemiology and risk factors for IBD. Nat Rev Gastroenterol Hepatol 12(4):205–217.

Arpaia N, Campbell C, Fan X et al (2013) Metabolites produced by commensal bacteria promote peripheral regulatory T-cell generation. Nature 504(7480):451–455. https://doi.org/10.1038/nature12726

Bach J-F (2018) The hygiene hypothesis in autoimmunity: the role of pathogens and commensals. Nat Rev Immunol 18(2):105–120. https://doi.org/10.1038/nri.2017.111

Bach JF (2002) The effect of infections on susceptibility to autoimmune and allergic diseases. (1533–4406 (Electronic)) (In Eng)

Belkaid Y, Hand TW (n.d.) Role of the microbiota in immunity and inflammation. (1097-4172 (Electronic)). (In eng)

Berer K, Gerdes LA, Cekanaviciute E, et al. (n.d.) Gut microbiota from multiple sclerosis patients enables spontaneous autoimmune encephalomyelitis in mice. (1091-6490 (Electronic)). (In eng)

Bjornevik K, Cortese M, Healy BC et al (2022) Longitudinal analysis reveals high prevalence of Epstein-Barr virus associated with multiple sclerosis. Science 375(6578):296–301. https://doi.org/10.1126/science.abj8222

Bluestone JA, Herold K, Eisenbarth G (2010) Genetics, pathogenesis and clinical interventions in type 1 diabetes. Nature 464(7293):1293–1300. https://doi.org/10.1038/nature08933

Bottini N, Peterson EJ (n.d.) Tyrosine phosphatase PTPN22: multifunctional regulator of immune signaling, development, and disease. (1545-3278 (Electronic)). (In eng)

Brown K, DeCoffe D, Molcan E, Gibson DL (n.d.) Diet-induced dysbiosis of the intestinal microbiota and the effects on immunity and disease. (2072-6643 (Electronic)). (In eng)

Chittrakul J, Sapbamrer RA, Sirikul WA (2022) Pesticide exposure and risk of rheumatoid arthritis: a systematic review and meta-analysis. Toxics 10:207. https://doi.org/10.3390/toxics10050207. (2305-6304 (Electronic)) (In eng)

Cho I, Blaser MJ (2012) The human microbiome: at the interface of health and disease. Nat Rev Genet 13(4):260–270. https://doi.org/10.1038/nrg3182

D'Cruz DP, Khamashta MA, Hughes GRV (n.d.) Systemic lupus erythematosus. (1474-547X (Electronic)). (In eng)

Feinberg AP (2018) The key role of epigenetics in human disease prevention and mitigation. N Engl J Med 378(14):1323–1334. https://doi.org/10.1056/NEJMra1402513

Fife BT, Bluestone JA (n.d.) Control of peripheral T-cell tolerance and autoimmunity via the CTLA-4 and PD-1 pathways. (1600-065X (Electronic)). (In eng)

Filippi M, Bar-Or A, Piehl F, et al (n.d.) Multiple sclerosis. (2056-676X (Electronic)). (In eng)

Gregersen PK, Silver J, Winchester RJ (n.d.) The shared epitope hypothesis. An approach to understanding the molecular genetics of susceptibility to rheumatoid arthritis. (0004-3591 (Print)). (In eng)

Hart JE, Källberg H, Laden F et al (2013) Ambient air pollution exposures and risk of rheumatoid arthritis. Arthritis Care Res (Hoboken). 65(7):1190–6.

Hohlfeld R, Wekerle H (n.d.) Autoimmune concepts of multiple sclerosis as a basis for selective immunotherapy: from pipe dreams to (therapeutic) pipelines. (0027-8424 (Print)). (In eng)

Hutchinson D, Shepstone L Moots R, Moots R, Lear JT, Lear JT, Lynch MP, Lynch MP (2001) Heavy cigarette smoking is strongly associated with rheumatoid arthritis (RA), particularly in patients without a family history of RA. (0003-4967 (Print)). (In eng)

James JA, Robertson JM (2021) Lupus and Epstein-Barr. (1531–6963 (Electronic)). (In eng)

Janani C, Ranjitha Kumari BD (2015) PPAR gamma gene-a review. Diabetes Metab Syndr. 9(1):46–50.

Jung CR, Chung WT, Chen WT et al (2019) Long-term exposure to traffic-related air pollution and systemic lupus erythematosus in Taiwan: A cohort study. Sci Total Environ. 668:342–349.

Khan MA (n.d.) HLA-B27 and its subtypes in world populations. (1040–8711 (Print)). (In eng)

Klareskog L, Malmström V Lundberg K, Padyukov L, Alfredsson L (n.d.) Smoking, citrullination and genetic variability in the immunopathogenesis of rheumatoid arthritis. (1096-3618 (Electronic)). (In eng)

Klein L, Kyewski B, Allen PM, Hogquist KA (2014) Positive and negative selection of the T cell repertoire: what thymocytes see (and don't see). Nat Rev Immunol 14(6):377–391. https://doi.org/10.1038/nri3667

Kuret T, Sodin-Šemrl S, Leskošek B et al (2022) Single Cell RNA Sequencing in Autoimmune Inflammatory Rheumatic Diseases: Current Applications, Challenges and a Step Toward Precision Medicine. Front Med (Lausanne). 8:822804.

Kusters JG, van Vliet Ah Kuipers EJ (n.d.) Pathogenesis of Helicobacter pylori infection. (0893-8512 (Print)). (In eng)

Lerner A, Matthias T (n.d.) Changes in intestinal tight junction permeability associated with industrial food additives explain the rising incidence of autoimmune disease. (1873–0183 (Electronic)). (In eng)

Lessard CJ, Ice JA, Adrianto I, Adrianto I, Wiley GB, et al. (n.d.) The genomics of autoimmune disease in the era of genome-wide association studies and beyond. (1873–0183 (Electronic)). (In eng)

Liu Y, Aryee MJ, Padyukov L et al (2013) Epigenome-wide association data implicate DNA methylation as an intermediary of genetic risk in rheumatoid arthritis. Nat Biotechnol 31(2):142–147. https://doi.org/10.1038/nbt.2487

Marrack P, Kappler J Kotzin BL (n.d.) Autoimmune disease: why and where it occurs. (1078-8956 (Print)). (In eng)

Marshall GD Jr. (n.d.) The adverse effects of psychological stress on immunoregulatory balance: applications to human inflammatory diseases. (1557-8607 (Electronic)). (In eng)

McInnes IB, Schett G (n.d.) The pathogenesis of rheumatoid arthritis. (1533-4406 (Electronic)). (In eng)

Moayyedi P, Surette MG, Kim PT, et al (n.d.) Fecal microbiota transplantation induces remission in patients with active ulcerative colitis in a randomized controlled trial. (1528-0012 (Electronic)). (In eng)

Njoo MD, Westerhof W (n.d.) Vitiligo. Pathogenesis and treatment. (1175-0561 (Print)). (In eng)

Noble JA, Erlich HA (n.d.) Genetics of type 1 diabetes. (2157-1422 (Electronic)). (In eng)

Parks CG, Cooper GS, Nylander-French LA, Nylander-French LA, Sanderson WT, et al (2016) Occupational exposure to crystalline silica and risk of systemic lupus erythematosus: a population-based, case-control study in the southeastern United States. (0004-3591 (Print)). (In eng)

Patterson CC, Dahlquist GG, Gyürüs E, Green A, Soltész G (2009) Incidence trends for childhood type 1 diabetes in Europe during 1989–2003 and predicted new cases 2005–2020: a multicentre prospective registration study. (1474-547X (Electronic)) (In Eng)

Quan N, Banks WA (n.d.) Brain-immune communication pathways. (0889-1591 (Print)). (In eng)

Rojas M, Restrepo-Jiménez P, Monsalve DM, et al. (n.d.) Molecular mimicry and autoimmunity. (1095-9157 (Electronic)). (In eng)

Rook GA (n.d.) Hygiene hypothesis and autoimmune diseases. (1559–0267 (Electronic)). (In eng)

Ringh MV, Hagemann-Jensen M, Needhamsen M et al (2019) Tobacco smoking induces changes in true DNA methylation, hydroxymethylation and gene expression in bronchoalveolar lavage cells. EBioMedicine. 46:290–304

Sakaguchi S, Yamaguchi T, Nomura T, Ono M (n.d.) Regulatory T cells and immune tolerance. (1097-4172 (Electronic)). (In eng)

Saxelin M, Tynkkynen S, Mattila-Sandholm T, de Vos WM (n.d.) Probiotic and other functional microbes: from markets to mechanisms. (0958-1669 (Print)). (In eng)

Schallreuter KU, Bahadoran P, Picardo M, et al (2008) Vitiligo pathogenesis: autoimmune disease, genetic defect, excessive reactive oxygen species, calcium imbalance, or what else?. Exp Dermatol 17(2):139–40.

Scher JU, Sczesnak A, Longman RS, et al. Expansion of intestinal Prevotella copri correlates with enhanced susceptibility to arthritis. eLife 2013;2:e01202. https://doi.org/10.7554/eLife.01202

Selmi C, Leung PS, Sherr DH et al (2012) Mechanisms of environmental influence on human autoimmunity: a National Institute of Environmental Health Sciences expert panel workshop. J Autoimmun 39(4):272–284

Sharp RC, Abdulrahim M, Naser ES et al (2015) Genetic Variations of PTPN2 and PTPN22: Role in the Pathogenesis of Type 1 Diabetes and Crohn's Disease. Front Cell Infect Microbiol 24:5:95.

Simmons K, Michels AW (n.d.) Lessons from type 1 diabetes for understanding natural history and prevention of autoimmune disease. (1558–3163 (Electronic)). (In eng)

Sollid LM, Jabri B (2013) Triggers and drivers of autoimmunity: lessons from coeliac disease. (1474-1741 (Electronic)) (In Eng)

TEDDY Study Group. (2008) The Environmental Determinants of Diabetes in the Young (TEDDY) Study. (1749-6632 (Electronic)) (In eng)

Topol EJ (n.d.) Individualized medicine from prewomb to tomb. (1097-4172 (Electronic)). (In eng)

Trowsdale J, Knight JC (2013) Major histocompatibility complex genomics and human disease. (1545-293X (Electronic)) (In eng). DOI: D—NLM: EMS63323 EDAT- 2013/07/24 06:00 MHDA- 2013/12/16 06:00 PMCR- 2015/05/11 CRDT- 2013/07/24 06:00 PHST- 2013/07/24 06:00 [entrez] PHST- 2013/07/24 06:00 [pubmed] PHST- 2013/12/16 06:00 [medline]

PHST- 2015/05/11 00:00 [pmc-release] AID—10.1146/annurev-genom-091212-153455 [doi] PST—ppublish

Tsokos GC (n.d.) Systemic lupus erythematosus. (1533-4406 (Electronic)). (In eng)

Ueda H, Howson JMM, Esposito L et al (2003) Association of the T-cell regulatory gene CTLA4 with susceptibility to autoimmune disease. Nature 423(6939):506–511. https://doi.org/10.1038/nature01621

Vaarala O (2008) Leaky gut in type 1 diabetes. Current Opinion in Gastroenterology 24(6):701–706.

Wallin MT, Culpepper WJ, Campbell JD et al (2019) The prevalence of MS in the United States: A populationbased estimate using health claims data. Neurology 92(10):e1020–e1027.

Whitacre CC, Cummings SO, Griffin AC (1994) The effects of stress on autoimmune disease. In: Glaser R, Kiecolt-Glaser JK (eds) Handbook of human stress and immunity. Academic Press, San Diego, pp 77–100

Zhang X, Chen B-d, Zhao L-d, Li H (2020) The gut microbiota: emerging evidence in autoimmune diseases. Trends Mol Med 26(9):862–873. https://doi.org/10.1016/j.molmed.2020.04.001

Zhernakova A, van Diemen CC, Wijmenga C (n.d.) Detecting shared pathogenesis from the shared genetics of immune-related diseases. (1471-0064 (Electronic)) (In eng)

Chapter 6
Spectrum of Autoimmune Disorder: From Systemic to Organ-Specific Manifestation

Dahyoon Heo and Siyoung Yang

6.1 Introduction to Autoimmune Disorders

6.1.1 Definition and Prevalence of Autoimmune Disorders

Autoimmune disorders are conditions in which the immune system mistakenly identifies and attacks healthy cells, tissues, or organs, instead of protecting the body. This misdirected immune response leads to chronic inflammation, tissue damage, and functional impairment of affected organs. These disorders are highly variable, ranging from systemic diseases, such as systemic lupus erythematosus (SLE), which affects multiple organ systems, to organ-specific conditions, such as type 1 diabetes (T1D), which primarily targets the pancreas.

The prevalence of autoimmune disorders has steadily increased over the past few decades. According to the American Autoimmune Related Diseases Association, over 100 distinct autoimmune diseases have been identified, collectively affecting 4–7% of the global population. However, regional differences in prevalence exist owing to variations in genetic, environmental, and cultural factors. For example, diseases, such as multiple sclerosis (MS), are more common in populations living farther from the equator, likely because of differences in vitamin D levels and other environmental factors.

Women are disproportionately affected by autoimmune diseases, which account for nearly 80% of all cases. Conditions, such as SLE and Hashimoto's thyroiditis, are particularly prevalent among females, and hormonal differences are thought to play a key role. For instance, estrogen modulates immune responses, potentially increasing the risk of autoimmune conditions. Additionally, genetic factors, such as

D. Heo · S. Yang (✉)
Department of Biological Sciences, Sungkyunkwan University, Suwon, Republic of Korea
e-mail: hdy0523@g.skku.edu; yangsy@skku.edu

Y.-O. Son, M. Ghosh (eds.), *Cutting-Edge Strategies in Drug Delivery and Immunotherapy for Autoimmune Disorders*,
https://doi.org/10.1007/978-981-95-0520-3_6

X-chromosome-linked immune regulatory genes, may contribute to this disparity. The societal and economic impacts of autoimmune diseases are profound, encompassing healthcare costs, lost productivity, and reduced quality of life for patients and their families.

Autoimmune diseases are often comorbid; therefore, individuals diagnosed with one autoimmune condition are at a higher risk of developing another. Patients with SLE may develop Sjögren's syndrome or rheumatoid arthritis (RA). This clustering of diseases underscores the shared genetic and immunological mechanisms underlying autoimmunity and highlights the need for comprehensive diagnostic and therapeutic approaches.

6.1.2 How Autoimmune Disorders Occur

Autoimmune disease development is multifactorial, involving the complex interplay of genetic, environmental, and immunological factors. The core of these conditions is the failure of immune tolerance, a fundamental mechanism that prevents the immune system from attacking the body's own tissues. Immune tolerance can be categorized into two main types: central and peripheral tolerance.

Central tolerance occurs during the development of immune cells in primary lymphoid organs, such as the thymus and bone marrow. During this process, immature T and B lymphocytes, which strongly recognize self-antigens, are either deleted or rendered nonfunctional through a process called anergy. This ensures that the majority of self-reactive immune cells are eliminated before entering the bloodstream.

Peripheral tolerance occurs outside the primary lymphoid organs in peripheral tissues. This mechanism suppresses self-reactive immune cells that escape central tolerance. Regulatory T cells (Tregs) play a critical role in maintaining peripheral tolerance by secreting inhibitory cytokines, such as interleukin (IL)-10 and transforming growth factor-beta (TGF-β), which suppress aberrant immune responses.

Failure of these tolerance mechanisms is central to the pathogenesis of autoimmune diseases. Genetic predisposition is the primary factor that contributes to the breakdown of immune tolerance (Ramos et al. 2015). Specific genes, particularly those in the human leukocyte antigen (HLA) complex, are strongly associated with increased susceptibility to autoimmunity (Rosenblum et al. 2015). HLA-DR3 and HLA-DR4 are linked with T1D, whereas HLA-B27 is associated with ankylosing spondylitis (Matzaraki et al. 2017). These genes influence the presentation of self-antigens to immune cells, thereby increasing the likelihood of autoimmune reactions.

Environmental factors play a significant role in triggering autoimmune diseases in genetically predisposed individuals. For example, infections can lead to autoimmunity through molecular mimicry, epitope spreading, and bystander activation. Molecular mimicry occurs when pathogens express antigens that are structurally similar to their self-antigens, leading to cross-reactive immune responses. Rheumatic fever following streptococcal infection is a classic example of this phenomenon.

Epigenetic modifications, such as changes in DNA methylation and histone acetylation, contribute to the dysregulation of immune responses. These modifications can alter the expression of immune regulatory genes, thereby balancing autoimmunity. The hypomethylation of specific genes has been observed in patients with SLE, leading to increased activation of autoreactive T and B cells (Surace and Hedrich 2019).

The dysregulation of cytokine networks and immune signaling pathways is another key factor in the development of autoimmunity. Cytokines, such as tumor necrosis factor-alpha (TNF-α), IL-6, and IL-17, are often elevated in autoimmune diseases, perpetuating chronic inflammation and tissue damage. Immune signaling pathways, such as NF-κB, JAK-STAT, and PI3K-Akt, are crucial for regulating immune cell activation and function. The dysregulation of these pathways can lead to uncontrolled immune responses and exacerbate disease severity.

The progression of autoimmune diseases typically involves a preclinical phase, during which immune dysregulation silently occurs without noticeable symptoms. Over time, this dysregulation results in the accumulation of autoantibodies, autoreactive T cells, and inflammatory cytokines, eventually leading to clinically overt disease. Understanding these early mechanisms is critical for developing preventive strategies and targeted therapies.

6.2 Part I: Understanding Autoimmune Disorders

6.2.1 Mechanisms of Autoimmunity

6.2.1.1 Genetic Predispositions

Autoimmune disorders result from a complex interplay between genetic predisposition and environmental factors. Genetic predisposition is one of the most significant factors that contribute to the onset of these conditions (Ramos et al. 2015). The role of genetics is often highlighted by the familial clustering of autoimmune diseases, in which close relatives of affected individuals are more likely to develop similar or related conditions.

The HLA complex located on chromosome 6 is one of the most well-studied genetic contributors to autoimmunity. HLA molecules are responsible for presenting antigens to T cells, initiating immune responses (Rosenblum et al. 2015). Variations in specific HLA alleles are strongly associated with autoimmune disease (Dendrou et al. 2018):.

- *HLA-DR3 and HLA-DR4*: associated with T1D and alter the presence of pancreatic antigens in the immune system.
- *HLA-B27*: linked to ankylosing spondylitis and reactive arthritis, leading to a pro-inflammatory immune response in the joints and spine.

In addition to the HLA genes, other loci contribute to autoimmune susceptibility. The *PTPN22* gene, which encodes a tyrosine phosphatase involved in T-cell signaling, is associated with RA and T1D. Variants in the *IL2RA* gene, which regulates T cell activity, have been implicated in multiple sclerosis and autoimmune thyroid diseases. These studies highlighted the influence of genetic factors on immune cell activation and disease progression.

Recent advancements in *genome-wide association studies (GWAS)* have expanded our understanding of the genetic landscape of autoimmunity (Farh et al. 2015). GWAS have identified hundreds of risk loci that are often shared across multiple diseases, suggesting common pathways underlying different autoimmune conditions. Defects in apoptotic cell clearance mediated by genes, such as *C1Q*, have been linked to SLE.

Epigenetics plays a pivotal role in this process. Modifications, such as DNA methylation, histone acetylation, and regulation of noncoding RNAs, can alter gene expression without changing the DNA sequence (Farh et al. 2015). These modifications are often influenced by environmental exposures, such as infections or toxins, which can exacerbate genetic susceptibility. Hypomethylation of immune regulatory genes in patients with SLE has been associated with increased T-cell activation and autoantibody production (Hedrich 2017).

Understanding these genetic and epigenetic factors not only enhances our knowledge of autoimmune pathogenesis but also paves the way for precision medicine. Genetic testing and predictive models guide clinical decisions, allowing early intervention and tailored therapies.

6.2.1.2 Environmental Triggers

While genetic predispositions establish a baseline risk of autoimmune diseases, environmental triggers often act as catalysts, transforming latent risks into active diseases. These triggers include infections, dietary components, pollutants, and lifestyle factors, which together shape the immune response (Vojdani et al. 2014).

6.2.1.3 Role of Infections

Pathogens are among the most significant contributors of autoimmunity. A well-known mechanism involves *molecular mimicry*, in which microbial antigens resemble self-antigens, leading to cross-reactive immune responses:

- *Streptococcus pyogenes*: causes rheumatic fever, where antibodies targeting bacterial antigens attack the heart tissues.
- *Epstein-Barr virus*: implicated in diseases, such as multiple sclerosis and SLE, by modulating B-cell activity and increasing autoantibody production.

Another mechanism is *bystander activation*, in which nonspecific immune responses inadvertently activate autoreactive T cells during infections. *Epitope spreading*, in which tissue damage exposes new antigens to the immune system, exacerbates autoimmune responses.

6.2.1.4 Dietary and Lifestyle Factors

Diet plays a crucial role in modulating immune responses (Manzel et al. 2014). Gluten triggers celiac disease in genetically predisposed individuals, particularly those with HLA-DQ2 or HLA-DQ8 alleles. In addition to gluten, a high-sodium diet promotes inflammatory T helper 17 (Th17) cell responses, worsening autoimmune conditions, such as multiple sclerosis.

Lifestyle factors, such as smoking and stress, also significantly affect autoimmunity. Smoking is a well-established risk factor for RA that not only induces epigenetic changes but also promotes the production of autoantibodies, such as anti-citrullinated protein antibodies (ACPAs). Chronic stress through sustained cortisol elevation can dysregulate immune responses and trigger or worsen autoimmune conditions.

6.2.1.5 Environmental Pollutants

Exposure to environmental toxins, such as heavy metals (e.g., mercury) and industrial chemicals (e.g., bisphenol A), disrupts immune homeostasis. These pollutants can induce oxidative stress, promote pro-inflammatory cytokine production, and alter immune cell function.

6.2.1.6 Microbiome Dysbiosis

The gut microbiome, a critical regulator of immune system balance, is increasingly recognized for its role in autoimmunity. Dysbiosis, or an imbalance in the gut microbial composition, has been associated with diseases, such as inflammatory bowel disease and multiple sclerosis. A reduction in short-chain fatty acid-producing bacteria, such as *Lactobacillus*, can impair regulatory T cell (Treg) function, tipping the immune system toward autoimmunity.

Environmental factors offer unique opportunities for preventive intervention. Dietary modifications, lifestyle adjustments, and probiotic therapies targeting gut microbiota are promising strategies for reducing autoimmune risk and improving patient outcomes.

6.2.1.7 Immune Tolerance

Immune tolerance is the process by which the immune system distinguishes between self and nonself, thereby preventing autoreactive responses while effectively combating pathogens. This critical balance is maintained through two key mechanisms: central and peripheral tolerance.

6.2.1.8 Central Tolerance

Central tolerance occurs during the development of T and B lymphocytes in primary lymphoid organs, thymus, and bone marrow, respectively. Immature lymphocytes that strongly recognize self-antigens are eliminated through negative selection, a process that prevents autoreactive cells from entering circulation. Defects in genes, such as *autoimmune regulator (AIRE)*, can disrupt this process, resulting in autoimmune polyendocrinopathy-candidiasis-ectodermal dystrophy.

6.2.1.9 Peripheral Tolerance

Peripheral tolerance regulates mature lymphocytes in peripheral tissues. Key mechanisms include:

- *Tregs*: these cells suppress autoreactive immune cells through the secretion of inhibitory cytokines, such as IL-10 and TGF-β.
- *Anergy*: a state of functional inactivation induced in autoreactive cells that encounter self-antigens in the absence of costimulatory signals.
- *Clonal deletion*: apoptosis of autoreactive cells triggered in peripheral tissues.

Failures in these tolerance mechanisms can arise from:

1. *Genetic mutations*: variants in *FOXP3*, critical for Treg development, are associated with immune dysregulation, polyendocrinopathy, enteropathy, and X-linked syndrome.
2. *Epigenetic alterations*: environmental factors, such as UV radiation or pollutants, can induce epigenetic changes that impair immune regulation.
3. *Inflammation*: chronic inflammatory states overwhelm tolerance mechanisms, allowing autoreactive cells to proliferate.

The breakdown of immune tolerance is central to autoimmune pathogenesis. Therapeutic strategies aimed at restoring tolerance, such as Treg-based therapies and immune checkpoint inhibitors, represent promising frontiers in autoimmune disease management.

6.2.2 *Classification of Autoimmune Disorders*

Autoimmune disorders are broadly classified as systemic and organ-specific disorders (Table 6.1). This classification was based on the range of tissues and organs affected by the disease. Systemic autoimmune disorders involve multiple organs, whereas organ-specific autoimmune disorders involve specific tissues and organs. This distinction helps understand the pathophysiology, clinical manifestations, and management of these diseases.

Table 6.1 Comparison of systemic and organ-specific autoimmune disorders

	Systemic autoimmune disorders	Organ-specific autoimmune disorders
Examples	SLE, RA, Sjögren's syndrome	T1D, Hashimoto's thyroiditis, Celiac diseases
Target organs	Multiple organs	Specific organs (e.g., pancreas, thyroid)
Autoantibodies	ANA, RF, anti-dsDNA	Anti-TPO, anti-tTG
Common symptoms	Fatigue, fever, multiorgan involvement	Symptoms related to specific organ failure
Treatment approach	Immunosuppressants, biologics	Targeted therapy, lifestyle modification

Systemic autoimmune disorders are characterized by their ability to affect multiple organ systems, either simultaneously or sequentially. These diseases arise when the immune system mounts a generalized attack on connective tissues or proteins that are widely distributed throughout the body. A hallmark of systemic autoimmune disorders is chronic inflammation, which is often accompanied by the production of autoantibodies that drive disease progression. A well-known example is SLE, a disease affecting nearly every organ system. In SLE, autoantibodies, such as anti-double-stranded DNA (anti-dsDNA) and anti-Smith (anti-Sm) antibodies, form immune complexes that are deposited in tissues, leading to widespread inflammation and damage. Similarly, RA predominantly targets the synovial joints but has systemic manifestations, including cardiovascular and pulmonary complications. Other examples include systemic sclerosis (SSc), which is marked by excessive fibrosis of the skin and internal organs, and Sjögren's syndrome, which primarily affects the exocrine glands but with systemic involvement in severe cases.

The pathogenesis of systemic autoimmune disorders often involves a common mechanism of action. Genetic predisposition plays a critical role in certain HLA alleles such as HLA-DR2 and HLA-DR3, which are common across multiple diseases. Dysregulated cytokine networks, including elevated levels of TNF-α and IL-6, drive chronic inflammation. Autoantibodies exacerbate tissue damage by forming immune complexes that activate the complement system. Clinically, these diseases present with diverse symptoms, such as fatigue, fever, polyarthritis, and organ-specific manifestations, such as lupus nephritis or pulmonary fibrosis. Diagnosis typically involves a combination of clinical evaluation, serological markers, and imaging studies, whereas treatment focuses on immunosuppressants, biologics, and symptomatic therapies tailored to the affected organ systems.

Understanding systemic autoimmune disorders requires a deeper exploration of their effects on diverse organ systems. In SSc, excessive fibrosis leads to pulmonary arterial hypertension (PAH) and interstitial lung disease (ILD), significantly reducing the life expectancy. Similarly, RA can result in systemic complications, such as rheumatoid lung disease, chronic anemia, and vasculitis, emphasizing the importance of early and accurate diagnosis to mitigate disease progression.

In addition to these prominent examples, other systemic conditions, such as polymyositis and dermatomyositis, demonstrate how autoimmune processes can target muscles and connective tissues, leading to significant functional impairment. Mixed connective tissue disease (MCTD) further illustrates the complexity of systemic autoimmune disorders by presenting with overlapping features of SLE, SSc, and polymyositis. Anti-U1-ribonucleoprotein antibodies are a hallmark of MCTD and help distinguish it from other systemic conditions.

Systemic autoimmune disorders highlight the interconnected nature of immune-mediated diseases. The shared pathways of inflammation and immune dysregulation often result in comorbidities. Patients with SLE may develop antiphospholipid syndrome, which increases the risk of thrombosis and pregnancy-related complications. Similarly, Sjögren syndrome frequently coexists with RA, underscoring the overlapping features of systemic autoimmunity.

Organ-specific autoimmune disorders are confined to specific tissues or organs and lead to localized inflammation and dysfunction. These conditions often result from immune responses directed against unique antigens expressed in specific organs. In T1D, mellitus is a prime example, where autoreactive T cells target insulin-producing beta cells in the pancreas, leading to hyperglycemia. Another common organ-specific autoimmune disorder, Hashimoto's thyroiditis, is characterized by chronic lymphocytic infiltration of the thyroid gland, resulting in hypothyroidism. Graves' disease is an autoimmune hyperthyroidism that occurs when autoantibodies stimulate thyroid-stimulating hormone (TSH) receptors, causing increased thyroid hormone production.

Multiple sclerosis (MS) is another notable example of the central nervous system (CNS). In MS, autoreactive T cells target myelin proteins, leading to demyelination and neurological deficits, such as vision problems, muscle weakness, and sensory disturbances. Celiac disease, an autoimmune condition triggered by gluten, leads to inflammation and villous atrophy in the small intestine. It is strongly associated with the HLA-DQ2 and HLA-DQ8 alleles. The pathogenesis of organ-specific autoimmune disorders involves genetic predisposition—with specific HLA alleles predisposing individuals to diseases—and environmental triggers, such as infections or dietary antigens. Autoantibodies and autoreactive T cells play pivotal roles in the attack on specific cell types or tissues.

The clinical spectrum of organ-specific autoimmune diseases extends to conditions, such as autoimmune hepatitis, in which the liver is the primary target, resulting in chronic inflammation and potential cirrhosis. Similarly, primary biliary and sclerosing cholangitis are characterized by immune-mediated damage to the bile ducts, leading to progressive liver dysfunction. These disorders highlight the diversity of organ-specific autoimmunities.

Organ-specific autoimmune diseases manifest symptoms directly related to the loss of function in the targeted organ. T1D leads to hyperglycemia, polyuria, and weight loss owing to insulin deficiency. Similarly, Hashimoto's thyroiditis results in fatigue, weight gain, and cold intolerance due to hypothyroidism. Diagnosis often relies on the detection of organ-specific autoantibodies, imaging studies, and functional tests, such as glucose tolerance tests for T1D or thyroid function tests for Hashimoto's thyroiditis. Treatment typically includes the replacement of lost function—such as insulin therapy for T1D or thyroid hormone replacement for Hashimoto's thyroiditis—and immunomodulatory therapies for conditions such as MS.

Further examination of organ-specific autoimmune disorders reveals their profound impact on the quality of life and long-term health. In T1D, complications, such as diabetic nephropathy and retinopathy, arise from poor glycemic control. Similarly, untreated celiac disease increases the risk of intestinal lymphoma, highlighting the need for strict dietary adherence. The targeted nature of these disorders underscores the importance of precision medicine approaches in their management.

The management of organ-specific autoimmune disorders requires a multidisciplinary approach. For example, in MS, neurologists, physiotherapists, and occupational therapists collaborate to address neurological deficits and maintain functionality. Similarly, endocrinologists play a critical role in managing thyroid-related autoimmune conditions, ensuring optimal hormonal balance, and mitigating systemic effects.

Understanding the classification of autoimmune disorders provides a framework for studying the underlying mechanisms, identifying diagnostic markers, and developing targeted therapeutic approaches. By distinguishing between systemic and organ-specific conditions, researchers and clinicians can address the diverse challenges posed by these complex diseases. This distinction not only aids in accurate diagnosis and effective management but also guides research efforts aimed at unraveling the complex interplay between genetic predisposition, environmental triggers, and immune dysregulation in autoimmunity. This understanding is essential for advancing personalized medicine and improving patient outcomes in the treatment of autoimmune disorders.

6.3 Part II: Systemic Autoimmune Disorders

6.3.1 Common Systemic Autoimmune Disorders

6.3.1.1 Systemic Lupus Erythematosus (SLE)

SLE is a chronic autoimmune disease that affects multiple organs and leads to inflammation and tissue damage (Kaul et al. 2016). It predominantly affects women of childbearing age, with a female-to-male ratio of approximately 9:1 (Tsokos 2011). The disease arises from immune dysregulation, which is influenced by genetic predisposition and environmental triggers. Autoantibodies, particularly antinuclear antibodies (ANAs), are central to SLE pathogenesis and form immune complexes that are deposited in various tissues, causing widespread inflammatory responses. Environmental factors, such as ultraviolet radiation, infections, and certain medications, can exacerbate disease activity. Hormonal influences, particularly estrogen, may contribute to the higher prevalence in women.

Clinical Manifestations

SLE presents a diverse range of clinical manifestations that reflect its multisystem involvement (Fanouriakis et al. 2021). Dermatologic symptoms, including the characteristic malar rash, or "butterfly rash," and photosensitivity, are among the most recognizable features. Arthralgia and arthritis occur in most patients, typically manifesting as nonerosive but often painful joint inflammation. Renal involvement, particularly lupus nephritis, is a severe complication that can present with proteinuria and hematuria, often progressing to nephrotic syndrome or renal failure if left untreated. Neuropsychiatric manifestations, such as seizures, psychosis, and cognitive dysfunction, represent another severe spectrum of SLE. Cardiovascular involvement includes pericarditis and an increased risk of atherosclerosis, whereas pulmonary symptoms may range from pleuritis to ILD. Hematological abnormalities, including anemia, leukopenia, and thrombocytopenia, are frequent, and antiphospholipid syndrome increases the risk of thrombotic events (Fig. 6.1).

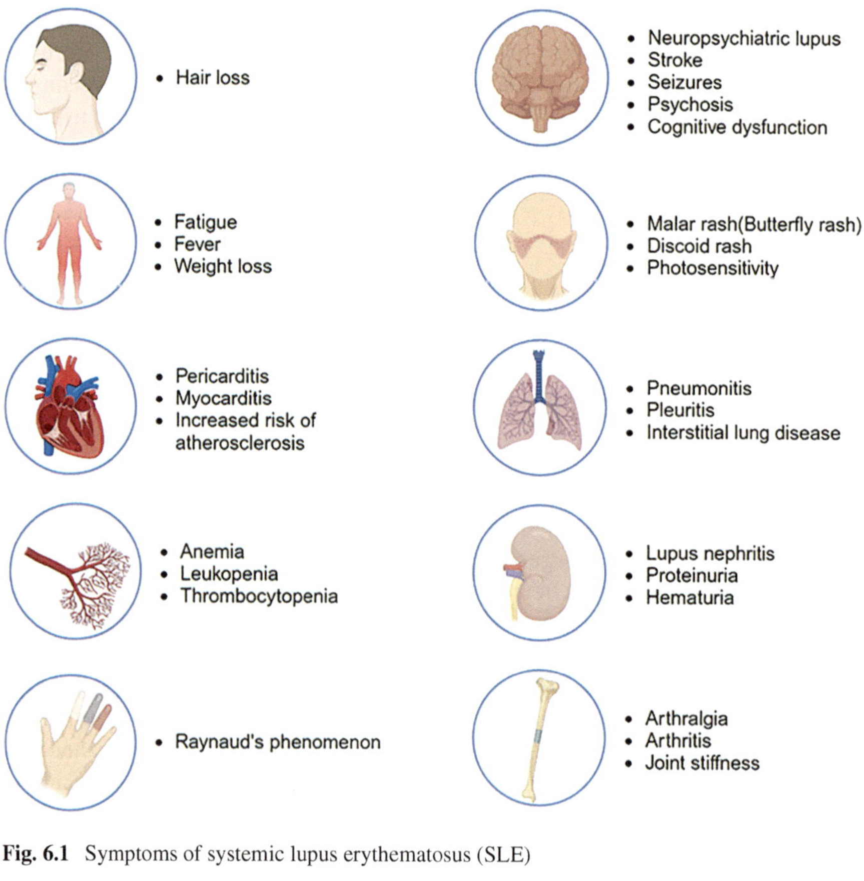

Fig. 6.1 Symptoms of systemic lupus erythematosus (SLE)

Diagnosis

The diagnosis of SLE is challenging because of its heterogeneous presentation and overlap with that of other autoimmune diseases. The diagnostic criteria established by the American College of Rheumatology (ACR) and the European League Against Rheumatism (EULAR) emphasize the integration of clinical and laboratory findings. Autoantibody testing is pivotal, with ANA positivity found in >95% of patients, although it lacks specificity. Anti-double-stranded DNA (anti-dsDNA) and anti-Smith antibodies are specific to SLE and correlate with disease activity. Hypocomplementemia, indicated by low C3 and C4 levels, reflects active disease. Renal biopsy is often required in cases of lupus nephritis to determine the extent of renal damage and guide treatment strategies. Imaging studies, such as echocardiography and high-resolution CT, are used to evaluate cardiopulmonary involvement, whereas serological and inflammatory markers assist in monitoring disease progression.

Treatment

SLE treatment is highly individualized, focusing on reducing inflammation, preventing flares, and minimizing organ damage. In mild cases, hydroxychloroquine remains the cornerstone therapy, demonstrating efficacy in reducing disease flares and improving survival. Nonsteroidal anti-inflammatory drugs (NSAIDs) are used to manage musculoskeletal symptoms and serositis. For moderate-to-severe disease, corticosteroids are used to rapidly control inflammation, while immunosuppressive agents, such as mycophenolate mofetil, azathioprine, or methotrexate, are added for long-term management. Biological therapies, including belimumab, which targets B-cell activating factor (BAFF), are effective in refractory cases. Rituximab, although not universally approved for SLE, is used for certain severe manifestations, such as neuropsychiatric lupus or vasculitis. Lupus nephritis requires aggressive immunosuppressive therapy, with cyclophosphamide or mycophenolate mofetil as first-line agents, followed by maintenance therapy to sustain remission. Patients with antiphospholipid syndrome often require lifelong anticoagulation therapy to mitigate the thrombosis risk.

Emerging Approaches

Emerging treatments targeting specific immune pathways, such as Janus kinase (JAK) inhibitors and type I interferon blockers, are under investigation, reflecting a move toward precision medicine for SLE management (Dörner and Furie 2019). In addition to pharmacological interventions, non-pharmacological strategies, including patient education, sun protection, and cardiovascular risk management, are critical for optimizing outcomes. A multidisciplinary approach involving rheumatologists, nephrologists, and other specialists ensures comprehensive care of this complex and multifaceted disease.

6.3.1.2 Rheumatoid Arthritis (RA)

RA is a chronic systemic autoimmune disorder that primarily affects the synovial joints, leading to inflammation, pain, and progressive joint damage (McInnes and Schett 2011) (Fig. 6.2). The disease has a global prevalence of approximately 0.5% to 1% and is more common in women, with a female-to-male ratio of approximately 3:1. RA typically manifests between 30 and 60 years of age, although it can occur at any age. Genetic predisposition, particularly the presence of HLA-DR4 and HLA-DR1 alleles, combined with environmental triggers, such as smoking and microbial infections, plays a substantial role in disease onset. The disease process involves an aberrant immune response in which autoantibodies, such as rheumatoid factor (RF) and ACPAs, drive chronic synovial inflammation and joint destruction.

Clinical Manifestations

RA is persistent symmetrical polyarthritis involving the small joints of the hands and feet, although the larger joints may also be affected as the disease progresses. Patients often report morning stiffness lasting for more than an hour, which improves

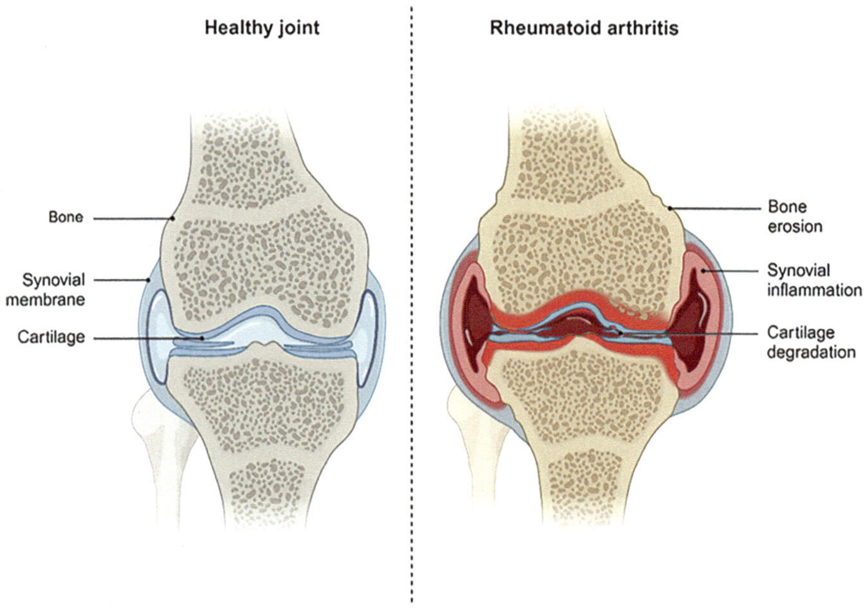

Fig. 6.2 Comparison of healthy and rheumatoid arthritis affected joints

with activity and is a key diagnostic feature. As the disease advances, untreated inflammation can lead to joint deformities, such as ulnar deviation, swan-neck deformities, and boutonnière deformities.

Extra-articular manifestations are common in RA and reflect the systemic nature of the disease. Pulmonary involvement includes ILD and pleuritis, whereas cardiovascular risks are elevated due to chronic inflammation, leading to accelerated atherosclerosis. Other systemic features include anemia due to chronic disease, vasculitis, scleritis, and keratoconjunctivitis sicca. Advanced RA can also affect the nervous system, resulting in peripheral neuropathy or cervical spine instability, particularly of the atlantoaxial joints.

Diagnosis

Diagnosis of RA involves the integration of clinical evaluation with laboratory and imaging findings. The ACR/EULAR classification criteria are widely used and focus on joint involvement, serological markers, and symptom duration.

Laboratory testing is central to diagnosis. RF is positive in approximately 70–80% of patients but lacks specificity. ACPAs, detected in up to 70% of RA cases, are highly specific for RA and often correlated with disease severity. Elevated levels of acute-phase reactants, such as erythrocyte sedimentation rate and C-reactive protein, indicate active inflammation. Imaging studies, including radiography, ultrasonography, and magnetic resonance imaging (MRI), are used to assess joint damage and synovitis. Early imaging findings include soft tissue swelling and periarticular osteopenia, which progress to erosions and joint space narrowing in advanced disease.

Treatment

The primary goals of RA treatment are to control inflammation, prevent joint damage, and improve overall quality of life. Early and aggressive interventions are critical to achieve these objectives.

First-line therapy often includes conventional disease-modifying antirheumatic drugs (DMARDs), with methotrexate as the cornerstone, owing to its efficacy and tolerability. Other conventional DMARDs, such as leflunomide, sulfasalazine, and hydroxychloroquine, are used either as monotherapy or in combination. Biologic DMARDs targeting specific immune pathways have been introduced for patients with moderate-to-severe disease or an inadequate response to conventional DMARDs. These include TNF-α inhibitors (e.g., adalimumab and etanercept), IL-6 inhibitors (e.g., tocilizumab), and B cell-depleting agents (e.g., rituximab). JAK inhibitors, such as tofacitinib, represent an emerging class of targeted synthetic DMARDs that can be orally administered.

Corticosteroids are used for short-term symptom control during flares, whereas NSAIDs provide symptomatic relief but do not alter disease progression. Physical and occupational therapies play integral roles in maintaining joint function and improving mobili (Sharif et al. 2018). Lifestyle modifications, such as smoking cessation, regular exercise, and a balanced diet, are essential adjuncts to pharmacological therapy.

Emerging Approaches

Advances in RA research have shifted focus to personalized medicine. Biomarkers that predict treatment response are being investigated to tailor therapies more effectively. Combination regimens involving biological agents and targeted synthetic DMARDs are under study to optimize outcomes in refractory cases. Additionally, novel therapies targeting pathways, such as the granulocyte-macrophage colony-stimulating factor, are being developed (Cornish et al. 2009).

Preventive strategies, particularly for individuals with a genetic predisposition or preclinical markers of RA, are gaining attention. Early screening and intervention during the "window of opportunity," a period when disease progression may be halted or reversed, hold promise for transforming RA management. A multidisciplinary approach involving rheumatologists, physiotherapists, and primary care providers is essential for holistic patient care.

6.3.1.3 Sjögren's Syndrome

Sjögren's syndrome is a systemic autoimmune disorder that primarily affects the exocrine glands, resulting in dry mouth (xerostomia) and dry eyes (keratoconjunctivitis sicca) (Brito-Zerón et al. 2016). This disease can occur as a primary or secondary condition in other autoimmune disorders, such as RA or SLE. It predominantly affects women, with a female-to-male ratio of approximately 9:1, and typically manifests between 40 and 60 years of age (Mariette and Criswell 2018). Its underlying pathogenesis involves lymphocytic infiltration of the salivary and lacrimal glands, leading to glandular dysfunction and subsequent systemic manifestations. Autoantibodies, such as anti-Ro/SSA and anti-La/SSB, play a significant role in disease diagnosis and are implicated in its pathophysiology.

Clinical Manifestations

The hallmark symptoms of Sjögren's syndrome are xerostomia and keratoconjunctivitis sicca, often referred to as "sicca symptoms." Patients may report difficulty in swallowing, a burning sensation in the mouth, and frequent dental caries due to reduced saliva production. Eye symptoms include irritation, redness, and a gritty sensation, which may progress to corneal damage if left untreated. In addition to these glandular symptoms, systemic features are common and include fatigue, arthralgia, and myalgia. Some patients develop extraglandular manifestations, such as ILD, peripheral neuropathy, or vasculitis. The risk of developing non-Hodgkin lymphoma, particularly mucosa-associated lymphoid tissue lymphoma, is significantly increased in patients with Sjögren syndrome.

Diagnosis

Diagnosis of Sjögren's is based on a combination of clinical evaluation, serological testing, and histopathological examination. ACR and EULAR classification criteria are commonly used.

Clinical assessments focused on the presence of sicca symptoms and signs of systemic involvement. Objective tests for glandular function include Schirmer's test, which measures tear production and unstimulated whole saliva flow. Serologic testing for anti-Ro/SSA and anti-La/SSB antibodies is essential for confirming the diagnosis, as these autoantibodies are present in approximately 60–70% of cases. Additional tests, such as ANA and RF tests, may also show positive results.

Histopathological examination of the minor salivary glands obtained via lip biopsy is a diagnostic cornerstone. The presence of focal lymphocytic sialadenitis, with a focus score ≥ 1 (indicating the number of lymphocytic foci per 4 mm^2 of glandular tissue), confirms the diagnosis. Imaging studies, such as salivary gland ultrasonography and sialography, are increasingly used to noninvasively evaluate glandular abnormalities.

Treatment

Management of Sjögren's syndrome focuses on alleviating sicca symptoms, preventing complications, and addressing systemic manifestations. Treatment is often multidisciplinary and involves rheumatologists, ophthalmologists, and dentists.

Artificial tears and lubricating ointments are the first-line therapies for dry eyes. In severe cases, cyclosporine ophthalmic drops or lifitegrasts are prescribed to reduce inflammation and improve tear production. Punctal plugs can be used to prevent tear drainage and maintain ocular moisture.

For xerostomia, salivary substitutes and sialogogues, such as pilocarpine and cevimeline, are commonly used to stimulate salivary flow. Patients are advised to maintain meticulous oral hygiene and undergo regular dental checkups to prevent dental caries and oral infections.

Systemic manifestations require immunosuppressive therapy tailored for specific organ involvement. Hydroxychloroquine is frequently used to treat arthralgia and fatigue. In cases of severe systemic involvement, corticosteroids and immunosuppressive agents, such as methotrexate, azathioprine, or mycophenolate mofetil, may be necessary. Biological therapies targeting B cells, such as rituximab, have shown efficacy in refractory cases and reducing the risk of lymphoma.

Emerging Approaches

Recent advances in the understanding of the pathophysiology of Sjögren's syndrome have led to the development of novel therapeutic approaches. Targeted therapies targeting cytokines, such as IL-6 and BAFF, are under investigation (Nocturne and Mariette 2013). Additionally, the use of advanced imaging techniques, such as high-resolution ultrasonography and positron emission tomography (PET) enhances the early detection and monitoring of glandular and systemic involvement.

Nonpharmacological strategies, including lifestyle modifications and dietary adjustments, play an adjunctive role in management. Patients were encouraged to stay hydrated, avoid environmental triggers, such as dry air, and use humidifiers to maintain mucosal moisture. A comprehensive, patient-centered approach remains critical for improving quality of life and reducing disease-related complications.

6.3.1.4 Systemic Sclerosis (SSc)

SSc, also known as scleroderma, is a rare systemic autoimmune disorder characterized by widespread fibrosis of the skin and internal organs, along with microvascular dysfunction and immune system abnormalities (Denton and Khanna 2017). SSc is classified into two main subtypes: limited cutaneous SSc (lcSSc) and diffuse cutaneous SSc (dcSSc) (Walker et al. 2007). The disease predominantly affects women, with a female-to-male ratio of approximately 4:1, and typically manifests between 30 and 50 years of age (Allanore et al. 2015). Although its exact etiology remains unclear, genetic predisposition; environmental triggers, such as silica exposure, and immune dysregulation have been implicated in its pathogenesis. The hallmark of SSc is excessive collagen deposition driven by fibroblast activation and vascular injury.

Clinical Manifestations

The clinical features of SSc vary depending on the subtype and the extent of organ involvement. In limited cutaneous SSc, the skin thickening is generally confined to the hands, forearms, face, and neck. Patients often present with Raynaud's phenomenon, which precedes other symptoms by years, and may also develop telangiectasia, calcinosis, and esophageal dysmotility. Diffuse cutaneous SSc involves extensive skin thickening, affecting the trunk and proximal limbs, and is associated with early and severe visceral involvement.

Pulmonary complications are the leading cause of morbidity and mortality in patients with SSc. ILD and PAH are common manifestations of progressive dyspnea, reduced exercise tolerance, and severe respiratory failure. Gastrointestinal involvement, which is present in up to 90% of the patients, includes esophageal dysmotility, gastroesophageal reflux disease (GERD), and intestinal pseudo-obstruction. Renal complications, such as scleroderma renal crisis, characterized by acute hypertension and renal failure, are life-threatening but less frequent in the era of angiotensin-converting enzyme (ACE) inhibitor therapy. Cardiovascular complications, including arrhythmias and pericardial effusion, further contribute to the systemic disease burden.

Diagnosis

Diagnosis of SSc is based on clinical features, serological testing, and imaging studies. The 2013 ACR/EULAR classification criteria provide a framework for diagnosis, emphasizing skin thickening, nail fold capillaroscopy abnormalities, and serological markers.

Autoantibody testing is central to the diagnosis. ANA are present in more than 90% of patients. Subtype-specific autoantibodies include anti-centromere (associated with lcSSc) and anti-topoisomerase I (Scl-70) antibodies (associated with dcSSc and ILD). Imaging studies, such as high-resolution computed tomography (HRCT) are essential for detecting ILD, whereas echocardiography and right heart catheterization are used to diagnose and monitor PAH. Nailfold capillaroscopy is a noninvasive tool for identifying microvascular changes characteristic of SSc.

Treatment

The management of SSc is challenging because of its heterogeneous nature and the lack of a universal cure. Treatment strategies focus on managing symptoms, preventing complications, and improving the quality of life. Immunosuppressive therapies, including mycophenolate mofetil and cyclophosphamide, are commonly used to treat ILD and other severe conditions (Volkmann and Varga 2019). The efficacy of antifibrotic agents, such as nintedanib, slows down the progression of ILD.

Targeted therapies, such as endothelin receptor antagonists (e.g., bosentan), phosphodiesterase-5 inhibitors (e.g., sildenafil), and prostacyclin analogs (e.g., epoprostenol) are used for PAH (Naranjo and Hassoun 2021). Gastrointestinal symptoms are managed using proton pump inhibitors for GERD and prokinetic agents for dysmotility. In cases of scleroderma renal crisis, prompt treatment with ACE inhibitors is critical to reduce mortality.

Emerging therapies, including autologous hematopoietic stem cell transplantation (HSCT), offer hope for patients with severe refractory diseases (Volkmann and Varga 2019). Clinical trials are exploring the potential of novel antifibrotic agents, such as pirfenidone, and immune-modulating therapies targeting specific cytokines and pathways.

Emerging Approaches

Advances in the understanding of the molecular mechanisms of SSc have led to the development of targeted treatments aimed at fibrosis and vascular dysfunction. Research on biomarkers for early detection and disease progression is ongoing, with the goal of personalizing treatment approaches. Multidisciplinary care, involving rheumatologists, pulmonologists, cardiologists, and other specialists, is essential for comprehensive management. Non-pharmacological strategies, including physical therapy, skin care, and psychosocial support, play critical roles in improving patient outcomes and quality of life (Sharif et al. 2018). With continued research and innovation, better therapies and improved prognoses in patients with SSc will be available in the future.

6.3.2 *Overlap Syndromes and Complex Cases*

6.3.2.1 Overlap Syndromes and Their Features

Overlapping syndromes are conditions in which the features of multiple autoimmune diseases coexist in a single patient, such as MCTD, which combines the clinical characteristics of SLE, RA, and SSc, and anti-synthetase syndrome, characterized by myositis, ILD, and Raynaud's phenomenon (Ramos et al. 2015). These syndromes often have specific serological markers, such as anti-U1 RNP antibodies in MCTD or anti-Jo-1 antibodies in antisynthetase syndrome. Patients with overlapping syndromes may present with a wide range of clinical features, such as severe muscle weakness, joint inflammation, skin changes, or pulmonary complications.

6.3.2.2 Diagnostic Challenges and Considerations

Diagnosing overlapping syndromes can be particularly challenging because of the simultaneous presence of the symptoms of different autoimmune diseases. A comprehensive clinical evaluation, including a detailed patient history and physical examination, is essential. Laboratory tests for autoantibodies play a critical role in identifying specific overlapping syndromes, with markers such as anti-U1 RNP, anti-Jo-1, and anti-Ro/SSA aiding in the diagnosis. Imaging studies, such as HRCT for lung involvement and echocardiography for cardiac assessment, are often required to evaluate organ-specific manifestations. Tissue biopsies may also be necessary to confirm the extent of damage to the affected organs, such as the skin, lungs, or muscles.

6.3.2.3 Management Approaches

Treatment strategies for overlapping syndromes focus on managing the dominant disease features while addressing the associated systemic complications. Immunosuppressive therapies, including corticosteroids, methotrexate, and mycophenolate mofetil, are commonly used to reduce inflammation and prevent organ damage. Biological agents, such as rituximab or tocilizumab, may be employed in cases with refractory symptoms or severe systemic involvement. In patients with pulmonary arterial hypertension or ILD, vasodilators such as bosentan and antifibrotic agents, such as nintedanib are often integrated into the treatment plan.

Regular follow-ups and multidisciplinary care are essential for managing these complex conditions. Pulmonologists, rheumatologists, and other specialists collaborate to provide comprehensive care that addresses both autoimmune and organ-specific manifestations. Lifestyle modifications, including smoking cessation and tailored exercise programs, play an adjunctive role in improving the overall outcomes.

6.3.2.4 Emerging Therapies and Future Directions

Research into the pathogenesis of overlapping syndromes has led to the development of targeted therapies aimed at modulating specific immune pathways. Inhibitors of the JAK pathway and therapies targeting IL-12/23 are under investigation for their potential to manage overlapping inflammatory and autoimmune features. Advances in precision medicine, including the use of genetic and biomarker profiling, are promising for early detection and personalized treatment of overlapping syndromes. Ongoing clinical trials are investigating novel biological agents and small molecules that can improve patient outcomes.

Overall, the management of overlapping syndromes requires a nuanced approach that considers the complexity of coexisting autoimmune conditions. Early diagnosis, targeted therapies, and multidisciplinary teams are key to optimizing care and minimizing disease-related morbidity and mortality.

6.4 Part III: Organ-Specific Autoimmune Disorders

6.4.1 Major Organ-Specific Autoimmune Disorders

6.4.1.1 Type 1 Diabetes (T1D)

Pathophysiology

T1D is a chronic autoimmune disorder characterized by the destruction of insulin-producing beta cells in the pancreas (Bluestone et al. 2010). This autoimmune attack is primarily mediated by T lymphocytes, particularly CD8+ cytotoxic T cells, which target beta cells through a combination of genetic susceptibility and environmental triggers. Genetic predisposition is strongly associated with HLA class II alleles, such as HLA-DR3 and HLA-DR4. Environmental factors, including viral infections and dietary components, are thought to initiate or accelerate beta cell destruction in genetically predisposed individuals. As beta-cell mass declines, the pancreas becomes unable to produce sufficient insulin to regulate blood glucose levels, leading to hyperglycemia and the clinical manifestations of diabetes.

Autoantibodies, such as glutamic acid decarboxylase (GAD), insulin autoantibodies (IAA), and zinc transporter 8 (ZnT8), are often detectable before the onset of hyperglycemia, marking an extended preclinical phase of T1D (Kawasaki 2023). The progression from autoimmunity to overt diabetes varies among individuals, and factors such as immune regulation, beta cell resilience, and environmental influences play critical roles.

Clinical Manifestations

T1D typically presents with acute-onset symptoms related to hyperglycemia and metabolic dysregulation. Classic symptoms include polyuria (frequent urination), polydipsia (excessive thirst), and polyphagia (increased hunger), accompanied by unintentional weight loss. Fatigue and irritability are common early symptoms. In severe cases, the initial presentation may include diabetic ketoacidosis (DKA), a life-threatening condition characterized by hyperglycemia, ketonemia, and metabolic acidosis. DKA often manifests as abdominal pain, nausea, vomiting, rapid breathing, and altered mental status.

Long-term complications of T1D result from chronic hyperglycemia and include microvascular and macrovascular damage. Microvascular complications include diabetic retinopathy, nephropathy, and neuropathy, whereas macrovascular complications include accelerated atherosclerosis, which increases the risk of cardiovascular diseases. Early and intensive glycemic control is essential to mitigate these risks and improve long-term outcomes.

Diagnostic Criteria

The diagnosis of T1D is based on clinical presentation, laboratory testing, and the exclusion of alternative causes of hyperglycemia. The diagnostic criteria were as follows:

- *Fasting plasma glucose*: ≥126 mg/dL (≥7.0 mmol/L).
- *Random plasma glucose*: ≥200 mg/dL (≥11.1 mmol/L) in the presence of classic symptoms.
- *Oral glucose tolerance test*: two-hour plasma glucose ≥200 mg/dL (≥11.1 mmol/L) after a 75-g glucose load.
- *Hemoglobin A1c*: ≥6.5%, reflecting chronic hyperglycemia.

Autoantibody testing, including GAD, IAA, and ZnT8 antibodies, helps confirm the autoimmune nature of diabetes and distinguishes T1D from type 2 and other forms of diabetes. C-peptide levels were also assessed to evaluate endogenous insulin production, with low or undetectable levels supporting the diagnosis of T1D.

Treatment Strategies

The cornerstone of T1D management is insulin replacement therapy, which is designed to mimic physiological insulin secretion. Modern insulin regimens include multiple daily injections of basal and prandial insulin or continuous subcutaneous insulin infusion using an insulin pump. Advances in insulin formulations, such as ultra-rapid-acting and long-acting analogs, have improved glycemic control and reduced the risk of hypoglycemia.

Continuous glucose monitoring systems and insulin pumps with automated insulin delivery, also known as artificial pancreatic systems, have revolutionized T1D management by providing real-time glucose monitoring and precise insulin delivery. These technologies improve glycemic variability, reduce the burden of diabetes self-management, and enhance quality of life.

Adjunctive therapies, including pramlintide, a synthetic analog of amylin, can modulate postprandial glucose spikes. Additionally, ongoing research on beta cell replacement therapies, such as islet cell transplantation and stem cell-derived beta cell therapy, holds promise for restoring endogenous insulin production.

Non-pharmacological interventions, including dietary counseling, regular physical activity, and psychosocial support, are integral components of comprehensive diabetes care. Patient education on carbohydrate counting, insulin dose adjustment, and hypoglycemia management empowers individuals to manage their conditions effectively and prevent complications.

Emerging Therapies and Future Directions

Advances in immunotherapy aim to preserve residual beta cell function and delay or prevent the onset of T1D in at-risk individuals. Agents, such as anti-CD3 monoclonal antibodies and regulatory T cell therapies, are under investigation for their potential to modulate autoimmune responses. Beta cell regeneration strategies that leverage gene editing and cellular reprogramming represent a cutting-edge frontier in T1D research. Precision medicine approaches, which integrate genetic, immunological, and metabolic data, promise to further individualize care and improve outcomes in individuals living with T1D.

6.4.1.2 Hashimoto's Thyroiditis

Pathophysiology

Hashimoto's thyroiditis, also known as chronic lymphocytic thyroiditis, is an autoimmune disorder characterized by the gradual destruction of the thyroid gland, leading to hypothyroidism (Ralli et al. 2020). This condition results from an aberrant immune response in which autoreactive T cells and autoantibodies target thyroid antigens, such as thyroid peroxidase (TPO) and thyroglobulin (Tg).

Immune-mediated destruction of thyroid tissue involves cytotoxic T lymphocytes, cytokine-mediated inflammation, and antibody-dependent cellular cytotoxicity. Genetic predisposition to Hashimoto's thyroiditis is associated with HLA-DR3 and HLA-DR5 alleles, along with other immune regulatory genes. Environmental factors, such as iodine excess, selenium deficiency, infections, and stress, contribute to disease onset in genetically susceptible individuals.

Clinical Manifestations

Hashimoto's thyroiditis progresses insidiously, and patients often remain asymptomatic during the early stages. The initial phase may involve transient thyrotoxicosis, known as hashitoxicosis, owing to the release of preformed thyroid hormones from damaged follicles. As the disease advances, patients develop symptoms of hypothyroidism, including fatigue, weight gain, cold intolerance, constipation, dry skin, hair thinning, and depression. Other features included bradycardia, hoarseness, and myxedema in severe cases. A hallmark physical finding is a diffusely enlarged, firm, and nontender thyroid gland (goiter), although the gland may atrophy in advanced stages.

Subclinical hypothyroidism, characterized by elevated thyroid-stimulating hormone (TSH) levels and normal free thyroxine (T4) levels, is common in early disease and may progress to overt hypothyroidism if left untreated. Hashimoto's thyroiditis is also associated with an increased risk of other autoimmune disorders, such as T1D, pernicious anemia, and Addison's disease.

Diagnostic Criteria

Hashimoto thyroiditis was diagnosed based on a combination of clinical, laboratory, and imaging findings. The key diagnostic features included the following:

- *Serological markers*: Elevated levels of anti-thyroid peroxidase (anti-TPO) and anti-thyroglobulin (anti-Tg) antibodies are hallmarks of Hashimoto's thyroiditis. These autoantibodies are present in most cases and are crucial for confirming the autoimmune nature of the disease.
- *Thyroid function tests*: Hypothyroidism is reflected by elevated TSH and low free T4 levels. In subclinical hypothyroidism, the TSH levels are elevated, whereas the free T4 levels remain within the normal range.
- *Ultrasound imaging*: Thyroid ultrasonography may reveal a diffusely hypoechoic and heterogeneous gland with reduced vascularity. These findings are consistent with those of chronic thyroiditis and help distinguish Hashimoto's thyroiditis from other thyroid disorders.

Treatment Strategies

The cornerstone of Hashimoto's thyroiditis management is thyroid hormone replacement therapy to correct hypothyroidism and alleviate the symptoms. Currently, levothyroxine, a synthetic thyroxine (T4), is the treatment of choice. The dosage was tailored to achieve euthyroidism, as indicated by normalization of TSH levels. Regular monitoring of TSH and free T4 levels is essential to adjust the dosage and ensure optimal therapy.

Patients with subclinical hypothyroidism may not require immediate treatment unless TSH levels are significantly elevated (≥10 mIU/L) or if symptoms are present. In pregnant women or those planning a pregnancy, early treatment is critical to prevent adverse maternal and fetal outcomes.

Surgical intervention, such as thyroidectomy, is rarely indicated but may be considered in cases of large goiters causing compressive symptoms or when there is suspicion of malignancy. Selenium supplementation has been investigated for its potential to reduce thyroid antibody levels and slow disease progression; however, the evidence remains inconclusive.

Emerging Insights and Future Directions

Ongoing research on the immunopathogenesis of Hashimoto's thyroiditis is exploring novel therapeutic approaches aimed at modulating the immune response and preserving thyroid function. Experimental therapies targeting cytokines, such as IL-17 and TNF-α, are under investigation. Advances in genetic and biomarker research have enhanced the ability to identify at-risk individuals and tailor preventive strategies.

Lifestyle modifications, including a balanced diet, stress management, and the avoidance of excessive iodine intake, play an adjunctive role in disease management. Patient education and regular follow-ups are critical to ensure adherence to therapy and early detection of complications, such as myxedema or associated autoimmune conditions. With continued advancements in understanding and treatment, the long-term prognosis of patients with Hashimoto's thyroiditis remains favorable.

6.4.1.3 Celiac Disease

Pathophysiology

Celiac disease is a chronic autoimmune disorder triggered by the ingestion of gluten, a protein found in wheat, barley, and rye (Lindfors et al. 2019). It primarily affects genetically predisposed individuals carrying HLA-DQ2 or HLA-DQ8 alleles. In susceptible individuals, gluten peptides, particularly gliadins, are not completely digested by the small intestine. These peptides are deamidated by tissue transglutaminases (tTG), which increase their immunogenicity. Deamidated peptides are presented by antigen-presenting cells to CD4+ T cells in the lamina propria and trigger a cascade of immune responses. This leads to the production of proinflammatory cytokines, anti-tTG autoantibodies, and intestinal damage. Villous atrophy, crypt hyperplasia, and intraepithelial lymphocytosis are the hallmark histological features of celiac disease, resulting in malabsorption and a variety of systemic manifestations (Fig. 6.3).

Clinical Manifestations

Celiac disease exhibits a broad spectrum of clinical presentations, ranging from classic gastrointestinal symptoms to non-classical and asymptomatic forms. The classical presentation includes chronic diarrhea, steatorrhea, abdominal pain,

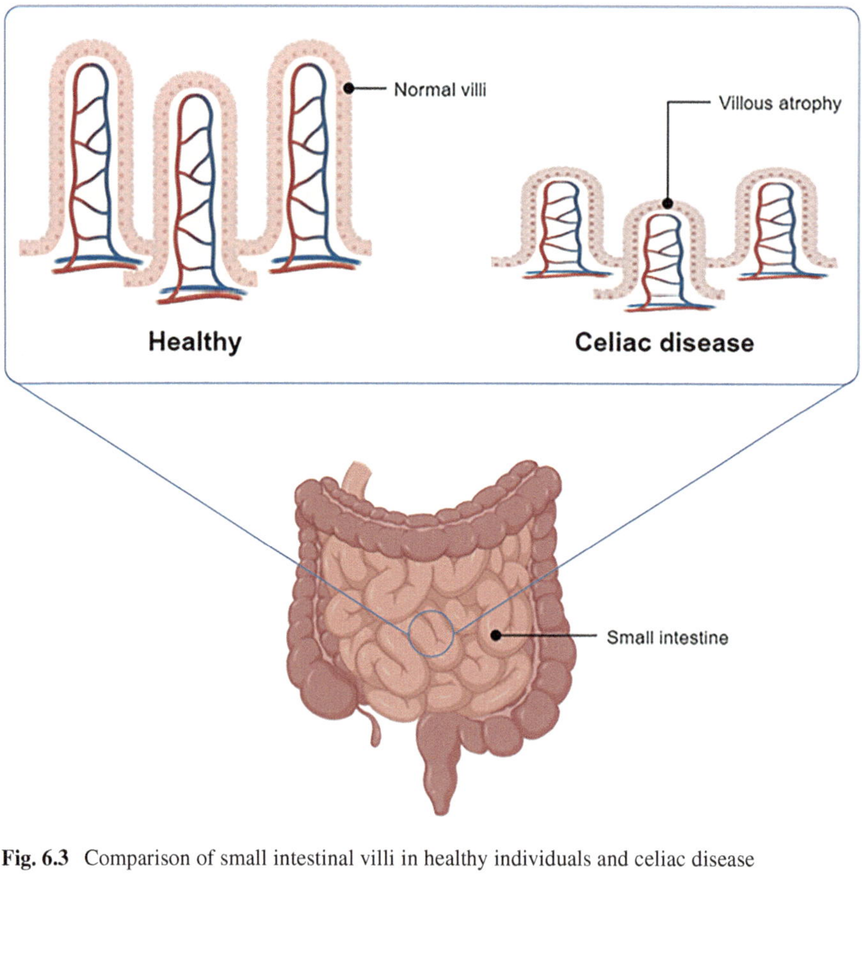

Fig. 6.3 Comparison of small intestinal villi in healthy individuals and celiac disease

bloating, and weight loss, reflecting malabsorption. Children may experience growth or developmental delays.

Nonclassical symptoms are increasingly recognized and include iron-deficiency anemia, osteoporosis, chronic fatigue, and neurological manifestations, such as peripheral neuropathy and ataxia. Dermatitis herpetiformis, a pruritic blistering skin condition, is a well-known extraintestinal manifestation. Patients with untreated celiac disease are at risk of complications, such as small bowel adenocarcinoma, enteropathy-associated T cell lymphoma, and refractory celiac disease, where symptoms persist despite adherence to a gluten-free diet.

Diagnostic Criteria

Diagnosis of celiac disease requires a combination of serological testing, histological examination, and a gluten-free diet. The key diagnostic steps were as follows:

- *Serologic Testing*: Measurement of serum anti-tTG IgA levels is a primary screening test, with high sensitivity and specificity. Total serum IgA levels should also be measured to rule out IgA deficiency, which can lead to false negative results. Anti-endomysial antibodies and deamidated gliadin peptide antibodies may be used in additional tests.
- *Duodenal Biopsy:* Endoscopic biopsy of the duodenum is the gold standard for diagnosis. The histological findings included villous atrophy, crypt hyperplasia, and increased intraepithelial lymphocytes.
- *Genetic Testing*: HLA-DQ2 and HLA-DQ8 testing can help exclude celiac disease in uncertain cases, as their absence makes a diagnosis unlikely.

Patients should remain on a gluten-containing diet during testing to avoid false negative results. A confirmed diagnosis requires a positive serological test and characteristic histological changes, with symptom resolution on a gluten-free diet.

Treatment Strategies

The cornerstone of celiac disease management is strict lifelong adherence to a gluten-free diet. This eliminates immune triggers, thereby allowing intestinal healing and symptom resolution. Patients require dietary counseling to identify hidden sources of gluten and prevent cross-contamination. Adherence to diet is critical for preventing complications and improving the quality of life.

In cases of nutrient deficiency, supplementation with iron, calcium, vitamin D, and vitamin B is necessary to address the deficiencies caused by malabsorption. Patients with persistent symptoms despite a gluten-free diet should be evaluated for inadvertent gluten exposure, refractory celiac disease, and alternative diagnoses. For refractory celiac disease, immunosuppressive therapies, such as corticosteroids or biologics, may be considered under specialist guidance.

Emerging Therapies and Future Directions

Ongoing research on the pathogenesis of celiac disease has spurred the development of novel therapeutic approaches beyond dietary management. These include:

- *Enzyme therapy*: Glutenases are designed to degrade gluten peptides in the gastrointestinal tract before triggering an immune response.
- *Tight junction modulators*: Agents, such as larazotide acetate, reduce intestinal permeability and limit gluten peptide translocation.
- *Immunotherapies*: Vaccines and monoclonal antibodies targeting specific immune pathways that induce tolerance to gluten are currently under investigation.
- *Microbiome modulation*: Research on the role of the gut microbiota in celiac disease suggests potential benefits from probiotics or fecal microbiota transplantation.

Lifestyle modifications, including psychological support and engagement with patient advocacy groups, play vital roles in managing the psychosocial burden of celiac disease. Early diagnosis, adherence to a gluten-free diet, and emerging therapies hold promise for improving the outcomes and quality of life in patients with celiac disease.

Emerging Insights and Research Advances

Recent Advancements

Recently, marked progress has been made in understanding the pathogenesis and treatment of autoimmune disorders. Advances in genomics and proteomics have led to the identification of novel genetic variants and protein biomarkers associated with disease susceptibility, progression, and response to therapy. Next-generation sequencing technologies have facilitated the discovery of previously unknown genetic loci linked to autoimmune diseases, such as T1D, RA, and multiple sclerosis. These findings shed light on the intricate molecular networks driving autoimmunity and open new avenues for targeted therapeutic intervention.

Epigenetic research has elucidated the roles of DNA methylation, histone modifications, and noncoding RNAs in modulating gene expression in autoimmune diseases. MiRNAs have critical functions in regulating immune cell activation and cytokine production, providing potential targets for novel therapies. High-throughput single-cell technologies are transforming our understanding of the immune landscape, allowing researchers to map immune cell heterogeneity and interactions at an unprecedented resolution.

Advancements in imaging modalities have enhanced disease diagnosis and monitoring. High-resolution techniques, such as MRI, PET, and novel contrast agents, enable the earlier detection of organ-specific involvement, as seen in RA and SSc.

These tools are invaluable for assessing disease activity and treatment efficacy and can guide more precise and personalized management.

The integration of artificial intelligence (AI) and machine learning into autoimmune research is another groundbreaking development. AI-driven algorithms are used to analyze large datasets, predict disease outcomes, and identify patient subgroups likely to benefit from specific therapies. These innovations pave the way for precision medicine approaches that optimize patient care based on individual biological profiles.

Promising Therapies

The therapeutic landscape for autoimmune diseases is rapidly evolving, with numerous promising approaches. Biological therapies targeting specific immune pathways have revolutionized treatment paradigms; however, next-generation strategies aim to go beyond the suppression of immune responses to restore immune tolerance.

One promising avenue is cellular therapy, which includes Treg infusion. Tregs play a pivotal role in maintaining immune homeostasis, and their dysfunction has been implicated in many autoimmune diseases. Efforts to expand and enhance Treg function ex vivo have shown potential in preclinical and early clinical trials for conditions such as T1D and multiple sclerosis.

Immunotherapies targeting cytokines and chemokines are rapidly being developed. Inhibitors of the IL-17, IL-23, and JAK pathways have demonstrated efficacy in diseases such as psoriasis, ankylosing spondylitis, and ulcerative colitis. Dual cytokine blockade, which targets synergistic inflammatory pathways, is an emerging strategy with the potential to improve outcomes in refractory cases.

Another cutting-edge approach involves antigen-specific immunotherapy designed to desensitize the immune system to specific autoantigens without inducing global immunosuppression. This strategy includes peptide-based vaccines, tolerogenic dendritic cells, and nanoparticle-based delivery systems, which are under investigation for treating conditions such as multiple sclerosis and celiac disease. These therapies retrain the immune system and offer a paradigm shift in treatment.

Advances in gene-editing technologies, particularly CRISPR-Cas9, have opened new possibilities for correcting the genetic defects underlying autoimmune diseases. Preclinical studies have explored the use of CRISPR to edit immune cell populations or modulate gene expression to restore immune tolerance. Additionally, RNA-based therapeutics, such as small interfering RNAs and antisense oligonucleotides, have been developed to selectively inhibit disease-associated proteins or pathways.

Stem cell-based therapies, including autologous HSCT and mesenchymal stem cell therapy, are promising for the treatment of severe refractory autoimmune diseases. HSCT aims to reset the immune system, and stem cells exhibit immunomodulatory and regenerative properties. Clinical trials are ongoing to establish the safety and efficacy of these approaches in a range of autoimmune conditions.

As these promising therapies progress from the bench to bedside, the focus remains on improving their safety, efficacy, and accessibility. Continued collaboration among researchers, clinicians, and industry stakeholders is essential for translating these advancements into meaningful benefits for patients with autoimmune disorders.

6.4.1.4 Multiple Sclerosis (MS)

Pathophysiology

MS is a chronic immune-mediated neurological disorder characterized by inflammation, demyelination, and neurodegeneration of the CNS (Reich et al. 2018). This disease primarily affects young adults, with a higher prevalence in women and a female-to-male ratio of approximately 3:1. Its pathogenesis involves an aberrant immune response in which autoreactive T cells, B cells, and macrophages infiltrate the CNS, targeting the myelin sheath and oligodendrocytes. This immune-mediated attack leads to demyelination, axonal injury, and eventually, neurodegeneration, thereby disrupting neural transmission. Genetic susceptibility, particularly HLA-DRB1*1501, and environmental factors, such as vitamin D deficiency, Epstein-Barr virus infection, and smoking, contribute to disease development. Lesions or plaques observed on MRI are the pathological hallmarks of MS and reflect areas of inflammation, demyelination, and gliosis.

Clinical Manifestations

MS is a heterogeneous disease with a wide spectrum of clinical presentations, reflecting the multifocal nature of CNS involvement. Symptoms vary depending on the location of the lesions but often include motor, sensory, and autonomic disturbances.

Patients frequently present with fatigue (a hallmark of MS), motor weakness, spasticity, and coordination difficulties. Sensory abnormalities such as numbness, tingling, and Lhermitte's sign (an electric shock-like sensation triggered by neck flexion) are common. Optic neuritis, which presents as unilateral vision loss and pain, is often an early manifestation. Brain and cerebellar involvement can lead to diplopia, vertigo, ataxia, and dysarthria. Autonomic dysfunction includes bladder and bowel disturbances, as well as sexual dysfunction.

The disease course was classified into distinct phenotypes. Relapsing-remitting MS is the most common form, characterized by episodes of neurological dysfunction, followed by partial or complete recovery. Secondary progressive MS involves the gradual worsening of symptoms over time, whereas primary progressive MS is characterized by steady disease progression without distinct relapses.

Diagnostic Criteria

The diagnosis of MS is based on clinical evaluation, imaging, and laboratory findings following the McDonald criteria. A key principle is the demonstration of dissemination in space and dissemination in time, which indicates lesions in multiple CNS regions and evidence of disease progression over time.

MRI is the cornerstone of MS diagnosis and reveals hyperintense lesions on T2-weighted and FLAIR sequences, particularly in the periventricular, juxtacortical, infratentorial, and spinal cord regions. Gadolinium-enhancing lesions indicated active inflammation, whereas T1-hypointense "black holes" indicated chronic damage. Cerebrospinal fluid analysis may reveal oligoclonal bands and an elevated immunoglobulin G index, supporting an immune-mediated etiology.

Potential studies, such as visual-evoked potentials, can detect subclinical demyelination, particularly in the optic pathway. Blood tests were performed to exclude other conditions, such as neuromyelitis optica spectrum disorder or systemic autoimmune diseases.

Treatment Strategies

Management of MS aims to reduce disease activity, prevent relapse, and slow the progression of disability. Treatment is tailored based on disease phenotype, severity, and patient-specific factors.

Acute relapses are typically managed with high-dose intravenous corticosteroids, such as methylprednisolone, to reduce inflammation and hasten recovery. Disease-modifying therapies (DMTs) are the cornerstone of long-term management, targeting the underlying immune dysregulation. First-line DMTs include injectable agents, such as interferon-beta and glatiramer acetate, which modulate immune activity.

Oral therapies, including dimethyl fumarate, teriflunomide, and fingolimod, offer convenient alternatives with established efficacy. For more aggressive disease or inadequate response to first-line agents, monoclonal antibodies, such as natalizumab, alemtuzumab, and ocrelizumab, are employed. These therapies target specific immune components, such as lymphocyte trafficking or B cell depletion, to suppress disease activity. Cladribine, a selective immune reconstitution therapy, is another option for the treatment of highly active MS.

Symptomatic management is critical for improving quality of life. Physical and occupational therapies address motor and functional impairments, whereas medications, such as baclofen or tizanidine, alleviate spasticity. Fatigue is managed with energy-conservation strategies and pharmacological agents, such as amantadine or modafinil. Cognitive rehabilitation and psychosocial support are essential for addressing cognitive dysfunction and mental health challenges, such as depression and anxiety.

Emerging Approaches and Future Directions

Advances in MS research have led to the development of novel therapeutic approaches aimed at halting the disease progression and promoting neural repair. Remyelination strategies, including the use of agents targeting oligodendrocyte precursor cells, have been investigated to restore myelin integrity. Neuroprotective therapies, such as sodium channel blockers, aim to preserve axonal function and prevent neurodegeneration.

Biomarker research is advancing personalized medicine for MS, enabling earlier diagnosis and more precise treatment selection. High-resolution imaging techniques, such as ultra-high-field MRI, provide deeper insights into lesion pathology and brain atrophy. Additionally, emerging DMTs with dual anti-inflammatory and neuroprotective effects are promising for altering disease course.

Lifestyle modifications, including vitamin D supplementation, smoking cessation, and regular exercise, play supportive roles in disease management. A multidisciplinary approach involving neurologists, physiotherapists, and mental health professionals is essential to deliver comprehensive care and improve outcomes in patients with MS.

6.5 Part IV: Holistic Management and Future Directions

6.5.1 Holistic Approach to Autoimmune Disorders

6.5.1.1 Multidisciplinary Care and Lifestyle Factors

The management of autoimmune disorders requires a multidisciplinary approach that reflects the complexity and systemic nature of these diseases. Autoimmune conditions often affect multiple organs and systems, necessitating collaboration between rheumatologists, endocrinologists, dermatologists, nephrologists, and other specialists. This integrated care model ensures that all aspects of the disease, including primary and secondary symptoms, are comprehensively addressed.

A key component of multidisciplinary care is an early and accurate diagnosis, which allows for timely interventions to prevent irreversible organ damage. Regular monitoring of disease activity and treatment response through laboratory tests and imaging studies is crucial for tailoring therapies to individual needs. Patients with SLE may benefit from a combination of rheumatological care for joint inflammation and nephrological evaluation for lupus nephritis. Similarly, individuals with MS require close coordination between neurologists and rehabilitation specialists to manage their neurological symptoms and maintain their functional independence.

Lifestyle modifications play a pivotal role in improving the outcomes and enhancing the quality of life of patients with autoimmune disorders. Regular physical activity, tailored to the patient's abilities and disease stage, helps reduce inflammation, improve cardiovascular health, and alleviate fatigue. Yoga and low-impact exercises, such as swimming and walking, are particularly beneficial for individuals with joint or mobility limitations.

Dietary interventions also have significant potential. An anti-inflammatory diet rich in fruits, vegetables, whole grains, lean proteins, and omega-3 fatty acids may modulate immune activity. Specific dietary approaches, such as gluten-free diets for celiac disease or iodine restriction for Hashimoto's thyroiditis, are essential for the management of disease-specific triggers. Maintaining a healthy weight reduces stress on the joints and lowers the risk of cardiovascular complications, particularly in conditions such as RA.

Stress management is a critical aspect of holistic care. Chronic stress exacerbates inflammation and triggers disease flares in autoimmune conditions. Mindfulness-based stress reduction, meditation, and cognitive-behavioral therapy improve psychological well-being and reduce disease activity. Ensuring adequate sleep is equally important, as sleep deprivation negatively affects immune regulation and overall health. Education on sleep hygiene and the management of sleep disorders, such as insomnia or restless leg syndrome, is integral to comprehensive care.

Preventive care measures, including vaccination and regular screening for comorbid conditions, are vital components of autoimmune disease management. Immunosuppressive therapies increase susceptibility to infections, making vaccination a priority. Patients should also undergo regular monitoring for osteoporosis, cardiovascular diseases, and malignancies, as these complications are common in long-standing autoimmune disorders.

Complementary Therapies

Complementary therapies, when used alongside conventional medical treatments, can enhance symptom management and improve quality of life in patients with autoimmune disorders. Although these therapies should not replace evidence-based treatments, they can provide additional benefits in alleviating pain, reducing stress, and supporting overall well-being.

Acupuncture has gained recognition for its potential to relieve chronic pain associated with autoimmune diseases, such as RA and fibromyalgia. By stimulating specific points in the body, acupuncture modulates pain pathways and promotes the release of endorphins, thereby reducing the need for pharmacological interventions.

Herbal and nutritional supplements are frequently used by patients seeking alternative approaches for symptom relief. Turmeric, which contains the active compound curcumin, has anti-inflammatory properties and may be beneficial in conditions such as inflammatory bowel disease and psoriasis. Omega-3 fatty acids, commonly found in fish oil supplements, are associated with reduced inflammation and improved cardiovascular health. However, the use of supplements should be carefully monitored by healthcare providers to avoid potential interactions with the prescribed medications.

Massage and physical therapies are valuable adjuncts in the management of musculoskeletal symptoms. Massage alleviates muscle tension and improves circulation, whereas physical therapy focuses on restoring muscle mobility, strength, and function. Tailored rehabilitation programs are particularly beneficial for patients recovering from disease flares and surgical interventions.

Mind-body practices, such as tai chi and qi gong, combine gentle physical movements with mindfulness and deep-breathing techniques. These practices improve balance, flexibility, and mental clarity, while reducing stress. Similarly, art and music therapy offer creative outlets for emotional expression and coping, helping patients navigate the psychological challenges of living with chronic conditions.

Aromatherapy, using essential oils such as lavender and eucalyptus, has been reported to alleviate anxiety and promote relaxation. Although its effects are primarily anecdotal, many patients find aromatherapy a soothing addition to their care routines.

It is essential that patients discuss complementary therapies with their healthcare teams to ensure their safety and integration into conventional treatments. A holistic approach that combines medical expertise with lifestyle interventions and complementary practices empowers patients to manage their condition actively and achieve optimal health outcomes.

6.5.2 Patient Perspectives and Support

6.5.2.1 Patient Stories and Challenges

Living with an autoimmune disorder is often a complex and arduous journey, marked by substantial physical, emotional, and social challenges. Patient experiences provide critical insights into the multifaceted impact of these diseases. For many, the road to diagnosis involves delays and misdiagnosis, as the symptoms of autoimmune conditions can mimic those of other illnesses. This diagnostic odyssey often leaves the patients frustrated and invalidated. A patient with SLE may experience years of unexplained joint pain, fatigue, and skin rashes before receiving a definitive diagnosis. Similarly, individuals with Hashimoto's thyroiditis may initially attribute their weight gain, depression, and lethargy to lifestyle factors rather than an underlying autoimmune process.

Once diagnosed, patients face the challenge of managing their lifelong treatment regimens. Adhering to strict dietary modifications, such as a gluten-free diet for celiac disease, or dealing with the side effects of immunosuppressive medications for conditions such as RA, adds layers of complexity to daily life. Patients frequently report difficulties balancing medical appointments, medication schedules, and the financial burden of treatment with personal and professional responsibilities. A young adult with MS might struggle to maintain employment owing to fatigue and cognitive impairments while simultaneously managing the high costs of disease-modifying therapies.

Therefore, the emotional toll of autoimmune diseases cannot be overestimated. Feelings of isolation are common, especially in those with conditions that involve invisible symptoms, such as chronic pain or fatigue. Adolescents and young adults with autoimmune disorders may face unique social challenges, including difficulty participating in peer activities or explaining their conditions to friends and colleagues. Older adults may experience the compounded effects of aging and autoimmune diseases, such as increased comorbidities and reduced mobility.

Thus, family dynamics are profoundly affected. Caregivers often assume substantial responsibilities, ranging from managing medications to providing emotional support, which can lead to caregiver fatigue. Patients themselves may feel guilt or frustration over the burden that their illness places on their loved ones. Despite these challenges, many individuals find ways to adapt and thrive by leveraging available resources and support systems.

6.5.2.2 Resources and Support Networks

Support networks are indispensable for patients in navigating the complexities of autoimmune disorders. These networks provide not only practical assistance but also a sense of community and understanding that can significantly improve quality of life. Patient advocacy organizations, such as the Lupus Foundation of America, the Multiple Sclerosis Society, and the Celiac Disease Foundation, are invaluable resources. These organizations offer educational materials, research updates, and access to clinical trials. They also host events, such as webinars, workshops, and support groups, where patients can connect with others facing similar challenges.

Online platforms have become increasingly popular in fostering patient communities. Social media groups, forums, and mobile apps designed for chronic disease management allow individuals to share their experiences, seek advice, and celebrate milestones in virtual spaces. These platforms are particularly beneficial for patients in rural or underserved areas who may lack access to in-person support groups. Peer-to-peer mentorship programs, in which experienced patients guide newly diagnosed individuals, are another effective way of providing emotional and practical support.

Healthcare providers play a critical role in connecting patients with these resources. Multidisciplinary clinics that bring together specialists, psychologists, dietitians, and social workers under one roof exemplify a patient-centered approach to care. These clinics not only streamline medical management but also address the holistic needs of patients, from mental health support to nutritional counseling.

Financial assistance programs are essential for many patients. Organizations, such as the Patient Access Network Foundation and co-pay assistance programs provided by pharmaceutical companies, help alleviate the economic burden of treatment. For patients with limited financial resources, nonprofit grants and community fundraising initiatives can provide much-needed relief.

Educational initiatives targeting patients and their families empower individuals to play active roles in managing their health. Workshops on disease management, stress reduction techniques, and navigating the healthcare system equip patients with the tools they need to live well, despite their diagnosis. Public awareness campaigns aim to reduce the stigma surrounding autoimmune diseases, fostering greater understanding and acceptance in a broader society.

By leveraging these resources and support networks, patients with autoimmune disorders can better navigate the challenges of their conditions, build resilience, and improve their overall well-being. The collective efforts of advocacy organizations, healthcare providers, and community support systems are essential to create a more inclusive and supportive environment for those living with autoimmune diseases.

6.6 Conclusion: Looking Ahead

6.6.1 *Key Insights*

Understanding the diverse spectrum of autoimmune disorders is essential for advancing clinical care and scientific research. These conditions, which collectively affect millions of people worldwide, highlight the intricate interplay between genetic predisposition, environmental triggers, and immune dysregulation. Each autoimmune disease presents unique challenges, ranging from diagnostic delays to the management of chronic symptoms and the prevention of complications. Through a deeper appreciation of these complexities, healthcare providers, researchers, and policymakers can develop effective strategies to improve patient outcomes.

The importance of multidisciplinary approaches for addressing autoimmune disorders cannot be overstated. Collaboration among specialists enables comprehensive care tailored to the individual needs of the patients, addressing both primary disease manifestations and secondary complications. Equally critical is the integration of lifestyle modifications, patient education, and psychosocial support, which empowers individuals to play an active role in their health management. As illustrated throughout this text, autoimmune disorders demand a holistic approach that encompasses medical, emotional, and social dimensions.

6.6.2 *Future Directions*

The future of autoimmune disorder management lies in precision medicine, which seeks to personalize treatment based on a patient's unique genetic, immunological, and environmental profiles. Advances in genomics and proteomics are paving the way for identifying biomarkers that can predict disease onset, progression, and response to therapy. These biomarkers hold promise for tailoring interventions to individual patients, minimizing trial-and-error, and optimizing outcomes.

Emerging research on immune tolerance offers hope for the development of curative therapies. Strategies, such as antigen-specific immunotherapy, aim to reprogram the immune system to recognize and tolerate self-antigens, potentially halting disease progression without broad immunosuppression. Cellular therapies, including regulatory T-cell infusions and mesenchymal stem cell transplantation, are also promising in clinical trials, particularly for refractory autoimmune conditions.

Technological advances have transformed disease monitoring and management. AI and machine learning algorithms have been employed to analyze vast datasets, identify patterns in disease activity, and predict flares before they occur. Wearable devices and mobile health applications provide real-time insights into patient health, enabling more proactive and personalized care.

Preventive strategies are another critical focus area. Vaccination campaigns, public health initiatives targeting modifiable risk factors, such as smoking and obesity, and early screening programs for at-risk populations are gaining traction. For example, identifying individuals with preclinical markers of T1D or RA and intervening during the unknown "window of opportunity" may delay or even prevent disease onset.

The role of patient advocacy and support networks in shaping the future of autoimmune care should not be overlooked. These organizations play a vital role in raising awareness, funding research, and fostering a sense of community among patients. By amplifying the voices of those affected, they ensure that research priorities and healthcare policies align with patients' needs.

The integration of cutting-edge science, innovative technology, and patient-centered care holds the potential to transform the landscape of autoimmune disorders. By continuing to invest in research, embrace multidisciplinary approaches, and prioritize patient empowerment, we can move closer to achieving a world in which autoimmune diseases are not only manageable but also preventable.

References

Allanore Y, Simms R, Distler O et al (2015) Systemic sclerosis. Nat Rev Dis Prim 1(1):15002. https://doi.org/10.1038/nrdp.2015.2

Bluestone JA, Herold K, Eisenbarth G. Genetics, pathogenesis and clinical interventions in type 1 diabetes. Nature. 2010;464(7293):1293–300 (In eng). https://doi.org/10.1038/nature08933.

Brito-Zerón P, Baldini C, Bootsma H et al (2016) Sjögren syndrome. Nat Rev Dis Prim 2(1):16047. https://doi.org/10.1038/nrdp.2016.47

Cornish AL, Campbell IK, McKenzie BS, Chatfield S, Wicks IP (2009) G-CSF and GM-CSF as therapeutic targets in rheumatoid arthritis. Nat Rev Rheumatol 5(10):554–559. https://doi.org/10.1038/nrrheum.2009.178

Dendrou CA, Petersen J, Rossjohn J, Fugger L (2018) HLA variation and disease. Nat Rev Immunol 18(5):325–339. https://doi.org/10.1038/nri.2017.143

Denton CP, Khanna D. Systemic sclerosis. Lancet. 2017;390(10103):1685–99 (In eng). https://doi.org/10.1016/s0140-6736(17)30933-9.

Dörner T, Furie R (2019) Novel paradigms in systemic lupus erythematosus. Lancet 393(10188):2344–2358. https://doi.org/10.1016/S0140-6736(19)30546-X

Fanouriakis A, Tziolos N, Bertsias G, Boumpas DT (2021) Update on the diagnosis and management of systemic lupus erythematosus. Ann Rheum Dis 80(1):14–25. https://doi.org/10.1136/annrheumdis-2020-218272

Farh KK-H, Marson A, Zhu J et al (2015) Genetic and epigenetic fine mapping of causal autoimmune disease variants. Nature 518(7539):337–343. https://doi.org/10.1038/nature13835

Hedrich CM (2017) Epigenetics in SLE. Curr Rheumatol Rep 19(9):58. (In eng). https://doi.org/10.1007/s11926-017-0685-1

Kaul A, Gordon C, Crow MK et al (2016) Systemic lupus erythematosus. Nat Rev Dis Prim 2(1):16039. https://doi.org/10.1038/nrdp.2016.39

Kawasaki E. Anti-islet autoantibodies in type 1 diabetes. Int J Mol Sci. 2023;24(12) (In eng). https://doi.org/10.3390/ijms241210012.

Lindfors K, Ciacci C, Kurppa K et al (2019) Coeliac disease. Nat Rev Dis Prim 5(1):3. https://doi.org/10.1038/s41572-018-0054-z

Manzel A, Muller DN, Hafler DA, Erdman SE, Linker RA, Kleinewietfeld M. Role of "Western diet" in inflammatory autoimmune diseases. Curr Allergy Asthma Rep. 2014;14(1):404. (In eng). https://doi.org/10.1007/s11882-013-0404-6.

Mariette X, Criswell LA (2018) Primary Sjögren's syndrome. N Engl J Med 378(10):931–939. https://doi.org/10.1056/NEJMcp1702514

Matzaraki V, Kumar V, Wijmenga C, Zhernakova A (2017) The MHC locus and genetic susceptibility to autoimmune and infectious diseases. Genome Biol 18(1):76. https://doi.org/10.1186/s13059-017-1207-1

McInnes IB, Schett G (2011) The pathogenesis of rheumatoid arthritis. N Engl J Med 365(23):2205–2219. https://doi.org/10.1056/NEJMra1004965

Naranjo M, Hassoun PM. Systemic sclerosis-associated pulmonary hypertension: spectrum and impact. Diagnostics (Basel). 2021;11(5) (In eng). https://doi.org/10.3390/diagnostics11050911.

Nocturne G, Mariette X (2013) Advances in understanding the pathogenesis of primary Sjögren's syndrome. Nat Rev Rheumatol 9(9):544–556. https://doi.org/10.1038/nrrheum.2013.110

Ralli M, Angeletti D, Fiore M et al (2020) Hashimoto's thyroiditis: an update on pathogenic mechanisms, diagnostic protocols, therapeutic strategies, and potential malignant transformation. Autoimmun Rev 19(10):102649. https://doi.org/10.1016/j.autrev.2020.102649

Ramos PS, Shedlock AM, Langefeld CD (2015) Genetics of autoimmune diseases: insights from population genetics. J Hum Genet 60(11):657–664. https://doi.org/10.1038/jhg.2015.94

Reich DS, Lucchinetti CF, Calabresi PA (2018) Multiple Sclerosis. N Engl J Med 378(2):169–180. https://doi.org/10.1056/NEJMra1401483

Rosenblum MD, Remedios KA, Abbas AK (2015) Mechanisms of human autoimmunity. J Clin Invest 125(6):2228–2233. https://doi.org/10.1172/JCI78088

Sharif K, Watad A, Bragazzi NL, Lichtbroun M, Amital H, Shoenfeld Y (2018) Physical activity and autoimmune diseases: get moving and manage the disease. Autoimmun Rev 17(1):53–72. (In eng). https://doi.org/10.1016/j.autrev.2017.11.010

Surace AEA, Hedrich CM (2019) The role of epigenetics in autoimmune/inflammatory disease. Front Immunol 10:1525. (In eng). https://doi.org/10.3389/fimmu.2019.01525

Tsokos GC (2011) Systemic lupus erythematosus. N Engl J Med 365(22):2110–2121. https://doi.org/10.1056/NEJMra1100359

Vojdani A, Pollard KM, Campbell AW (2014) Environmental triggers and autoimmunity. Autoimmune Dis 2014:798029. (In eng). https://doi.org/10.1155/2014/798029

Volkmann ER, Varga J (2019) Emerging targets of disease-modifying therapy for systemic sclerosis. Nat Rev Rheumatol 15(4):208–224. https://doi.org/10.1038/s41584-019-0184-z

Walker UA, Tyndall A, Czirják L et al (2007) Clinical risk assessment of organ manifestations in systemic sclerosis: a report from the EULAR Scleroderma Trials and Research group database. Ann Rheum Dis 66(6):754–763. (In eng). https://doi.org/10.1136/ard.2006.062901

Chapter 7
Therapeutic Modalities for Autoimmune Diseases: Current Challenges and Future Directions

Khushwant Nandal, Priyanka Dahiya, Manjeet Kaur, and Surender Jangra

7.1 Introduction

Autoimmune diseases (ADs), affecting approximately 5–8% of the global population, are complex conditions in which the immune system mistakenly attacks the body's own tissues, leading to a range of health challenges and complications, including ocular manifestations and systemic inflammation. The prevalence of these diseases is on the rise, contributing to significant healthcare burdens worldwide. Factors such as genetic predisposition, environmental triggers, and epigenetic modifications, particularly DNA methylation, play crucial roles in their pathogenesis (Albarbar and Aga 2024). Current treatment strategies, including those targeting innate immune sensors (also known as pattern recognition receptors), aim to manage symptoms but often fall short of providing long-term solutions, leading to relapses (Shukla et al. 2024). Moreover, the emergence of systemic autoinflammatory diseases (AIDs) highlights the overlapping nature of immune dysfunctions, necessitating precise diagnostic and therapeutic approaches to improve patient outcomes (Rajalingam and Ganjiwale 2024). The complexity and heterogeneity of ADs demand ongoing research and innovative treatment strategies.

Current therapeutic strategies for ADs face several limitations that hinder their long term effectiveness. Traditional treatments, such as broad-acting immunosuppressants like cyclophosphamide, often cause significant toxicity and fail to induce complete remission, necessitating prolonged drug treatment for established diseases

K. Nandal · P. Dahiya · M. Kaur
Department of Biotechnology, University Institute of Engineering and Technology, Maharshi Dayanand University, Rohtak, Haryana, India

S. Jangra (✉)
Department of Chemistry and Biochemistry, Sharda School of Engineering and Sciences, Sharda University, Greater Noida, Uttar Pradesh, India

Y.-O. Son, M. Ghosh (eds.), *Cutting-Edge Strategies in Drug Delivery and Immunotherapy for Autoimmune Disorders*,
https://doi.org/10.1007/978-981-95-0520-3_7

(Tyndall 2012). While newer biologic therapies targeting specific cytokines and immune pathways have shown promise, they still require continuous administration and often fail to induce long-term immune tolerance, which remains a long standing goal in ADs therapy (Chasov et al. 2024b; Rosato et al. 2010). The variability in ADs, influenced by genetic, environmental, and immunological factors, complicates treatment protocols, making it challenging to apply a one-size-fits-all approach (Celichowska et al. 2024; Chandrashekara 2012). Additionally, nonpharmacological interventions, although beneficial, lack robust evidence-based recommendations, limiting their integration into standard care (Celichowska et al. 2024).

The increasing prevalence of ADs highlights the urgent need for novel therapeutic strategies as current treatments primarily focus on nonspecific immunosuppression, failing to restore immune tolerance or provide a cure (Park and Kwok 2022; Sármay 2021). Traditional approaches, including disease-modifying antirheumatic drugs (DMARDs) and immunosuppressants, often cause significant side effects and fail to adequately address the underlying dysregulation of the immune system (Park and Kwok 2022). Innovative therapies, such as Bi-specific AutoAntigen-T cell Engagers (BiAATEs) and autologous stem cell treatments, have shown promise in selectively targeting autoreactive B-cells and promoting tissue regeneration, respectively (Kaliterna and Barac 2023; Perico et al. 2023). The exploration of antigen-specific therapies and the use of tolerogenic cells aim to restore self-tolerance without compromising the immune response to pathogens (Park and Kwok 2022; Sármay 2021). The development of these novel therapies is important to improve patient outcomes and address the limitations of existing treatments.

7.2 Current Therapeutic Approaches

The management of ADs presents significant challenges regarding the efficacy of current therapies and their associated adverse effects. While biologic agents have transformed treatment paradigms in conditions like rheumatoid arthritis (RA) and systemic lupus erythematosus (SLE), they can induce immunosuppression, increasing the risk of infections and paradoxical autoimmune responses (Her and Kavanaugh 2016). The limited efficacy of existing treatments in certain patients necessitates the exploration of novel approaches, such as low-dose interleukin-2 (LD-IL-2), which has shown promise in enhancing regulatory T cell (Treg) populations and reducing disease activity without serious adverse events (Bharti et al. 2024; Su et al. 2024b). However, the complexity of ADs, characterized by diverse symptoms and treatment resistance, complicates effective management (Cicuttini 2017; Pandey et al. 2024). Thus, while advancements in therapeutic options exist, the balance between efficacy and safety remains a critical concern in the treatment of autoimmune disorders.

Conventional treatments for major ADs have evolved significantly, yet they still rely heavily on established therapies. For instance, in SLE, hydroxychloroquine is considered the cornerstone of treatment, while corticosteroids, methotrexate, azathioprine, and cyclophosphamide are commonly used immunosuppressive agents (Athanassiou and Athanassiou 2023). These therapies aim to manage symptoms and reduce disease activity, although they can lead to treatment-related complications and side effects (Quattrocchio et al. 2016). In addition, traditional therapies like glucocorticoids and other DMARDs continue to play a crucial role in managing various autoimmune conditions by targeting inflammation (Hahn and King 2014). Despite the introduction of newer biologic therapies, these conventional treatments remain essential for many patients, serving as a benchmark for evaluating the safety and efficacy of emerging options (Tyndall 2012).

7.3 Novel Therapeutic Modalities

7.3.1 Cell Therapies

Cell therapies are emerging as a promising approach to treat ADs, addressing the limitations of conventional immunosuppressive treatments. Strategies include engineering antigen-presenting cells (APCs) to induce antigen-specific tolerance and utilizing chimeric antigen receptor (CAR) T-cell therapy, which targets autoreactive immune cells with greater specificity and fewer side effects (Puccetti et al. 2024; Yu et al. 2024b). Recent advancements have shown that CAR T-cells can effectively mitigate autoimmune responses in conditions like SLE and multiple sclerosis (MS), offering a tailored therapeutic option (Liu et al. 2024a). Mesenchymal stromal cells (MSCs) and Tregs have been successfully integrated into treatment protocols, enhancing immune regulation and tissue repair (Greco et al. 2024; Mehta et al. 2024). These innovative cellular therapies not only aim to restore immune tolerance but also to regenerate damaged tissues, representing a significant shift in the management of ADs and improving patient quality of life (Mehta et al. 2024).

Tregs play a pivotal role in the pathogenesis and treatment of ADs, as they are essential for maintaining immune homeostasis and self-tolerance. In autoimmune conditions, dysfunction or deficiency of Tregs can disrupt the balance between protective and pathogenic immune responses, thereby exacerbating disease symptoms (Honing et al. 2024). Recent studies have highlighted the therapeutic potential of Treg-based interventions, demonstrating their safety and feasibility in clinical trials for various ADs (Yu et al. 2024a). Moreover, advancements in targeting Tregs migration to inflammatory sites using small molecules have shown promise in enhancing their therapeutic efficacy (Martin et al. 2024). Understanding the distinct subpopulations of Tregs, such as thymically derived Tregs (tTregs) and peripherally induced Tregs (pTregs), is crucial for developing effective treatments, as their functional characteristics and mechanisms of action differ (Hardtke-Wolenski and Landwehr-Kenzel 2024). Tregs represent a promising avenue for innovative therapies aimed at restoring immune balance in ADs like RA (Su et al. 2024a).

CAR T-cell therapy has emerged as a promising treatment for various ADs, including SLE, idiopathic inflammatory myositis, and systemic sclerosis. This innovative approach addresses the limitations of conventional therapies, which often lack specificity and can lead to significant side effects due to indiscriminate immune suppression (Liu et al. 2024a; Yu et al. 2024b). Clinical studies have demonstrated that autologous CD19 CAR T-cells can induce remission or substantial improvement in patients with these conditions, with long-term safety and efficacy observed over follow-up periods of up to 3 years (Taubmann et al. 2024). The therapy not only reduces serum autoantibody levels but also promotes sustained remissions, showcasing its potential to reset aberrant autoimmunity (Alcaraz-Serna et al. 2024; Ohno and Nakamura 2024). Despite challenges such as manufacturing complexities and potential CAR T-cell exhaustion, ongoing research continues to explore and refine this therapeutic strategy for ADs.

Myeloid-derived suppressor cells (MDSCs) are a heterogeneous population of immature myeloid cells that play crucial roles in the immune regulation of ADs, including multiple sclerosis (MS) and others. MDSCs can be categorized into granulocytic (G-MDSCs) and monocytic (M-MDSCs) subsets, each exhibiting distinct immunosuppressive functions and mechanisms of action, which can either promote or inhibit T-cell activity, thereby influencing disease progression or remission (Bekić and Tomić 2023; Zanghì et al. 2024). In autoimmune conditions, MDSCs have been shown to promote the differentiation of T-helper 17 (Th 17) cells and release proinflammatory cytokines, contributing to tissue damage (Xiong et al. 2024). Studies indicate that MDSC populations increase during disease exacerbation, particularly in MS, where their activity correlates with disease severity (Jiang et al. 2024; Radygina et al. 2023). While MDSCs have potential therapeutic implications, their dual roles necessitate careful consideration in developing effective treatments for ADs (Bekić and Tomić 2023; Jiang et al. 2024).

MSCs play a complex role in ADs, exhibiting both therapeutic potential and pathological implications. In conditions like RA and SLE, MSCs demonstrate immunomodulatory capabilities, which can help regulate the immune response and promote tissue repair (Pignatti et al. 2024; Swain et al. 2024). However, their efficacy varies; for instance, while allogeneic MSCs from healthy donors can lead to clinical remission in SLE patients, autologous MSCs often show compromised functionality, exacerbating the disease (Barbado 2024; Hosseini et al. 2024). MSCs can inhibit autoreactive T-cells and modulate B-cell activity, potentially inducing self-tolerance and reducing inflammation (Barbado 2024). In RA, MSCs have shown promise in preclinical and clinical studies, but challenges remain regarding their long-term efficacy and the need for standardized treatment protocols (Dehnavi et al. 2023; Pignatti et al. 2024).

Targeting pathogenic immune responses in ADs, particularly SLE, involves several mechanisms of action. Key strategies include modulating immune cell function, such as enhancing regulatory T-cell activity and inhibiting effector T- and B-cell responses, which are often dysregulated in SLE (Fang et al. 2024). Recent advancements in understanding the role of type I interferon signaling and B lineage abnormalities have led to targeted therapies like belimumab and anifrolumab, which aim to correct these imbalances (Gatto et al. 2023). Also, gene modulation techniques, such as using siRNA to inhibit proinflammatory cytokines, show promise in tailoring treatments (Lee et al. 2020). Metabolic regulation of immune cells also presents a novel therapeutic avenue, as metabolic abnormalities in autoimmune conditions can be targeted to restore immune homeostasis (Teng et al. 2019), which highlights the potential for precision medicine in managing ADs by addressing specific pathogenic mechanisms (Dörner 2023). Different advanced therapeutic strategies for autoimmune diseases are shown in Table 7.1.

The clinical application of novel therapeutic strategies for ADs faces several challenges. One significant issue is the heterogeneity of ADs, which leads to varied patient responses to treatments, including conventional DMARDs and targeted therapies (Balogh et al. 2024; Felten et al. 2023). The complexity of autoimmune pathogenesis complicates the development of effective therapies, as many current treatments do not restore immune tolerance and may result in adverse effects due to their broad immunosuppressive actions (Park and Kwok 2022; Rezaei and Yazdanpanah 2022). Furthermore, translational studies are essential to bridge the gap between laboratory findings and clinical application, yet differences in human and animal physiology can hinder this process (Rezaei and Yazdanpanah 2022). Lastly, the emergence of new therapeutic modalities, such as cell therapies and combination treatments, requires careful patient selection to maximize efficacy and minimize risks (Balogh et al. 2024; Felten et al. 2023).

7.3.2 *Nucleotide-Based Therapies*

mRNA therapeutics represent a groundbreaking approach in the treatment of ADs, utilizing the body's own cellular machinery to produce therapeutic proteins. These therapies can induce immune tolerance toward autoantigens, thereby mitigating pathological immune responses (Krienke et al. 2021). The versatility of mRNA allows for the development of targeted treatments that can modulate immune pathways, as seen with antisense oligonucleotides (ASOs) designed to reduce T-cell activation in conditions like RA (Iyer 2022). Furthermore, advancements in non-coding RNA (ncRNA) research highlight the role of aberrant ncRNA expression in autoimmune disorders, suggesting that modifying these molecules could offer new therapeutic avenues (Zhang et al. 2018). The rapid evolution of mRNA technology, particularly in response to the COVID-19 pandemic, has opened doors for innovative treatments for various autoimmune conditions, including diabetes and neurodegenerative diseases (Sarfraz et al. 2024). mRNA therapeutics can be personalized medicine in AD management (Deyhimfar et al. 2024).

Gene editing technologies, particularly CRISPR/Cas9 and emerging base editing techniques, are revolutionizing the management of ADs by enabling precise modifications of genetic material. CRISPR/Cas9 allows for targeted editing of genes associated with ADs, facilitating the modulation of immune responses and the restoration of immune tolerance, which is crucial in conditions like RA (Chasov et al.

Table 7.1 Different advanced therapeutic strategies for autoimmune diseases

Modality	Description	Targeted condition	References
Antigen-presenting cells (APCs)	Engineered to induce antigen-specific immune tolerance.	Autoimmune diseases (general)	Puccetti et al. (2024)
CAR-T-cell therapy	Targets autoreactive immune cells with high specificity, reducing side effects. CAR-T cells induce remission and promote immune reset.	SLE, IIM, SSc	Liu et al. (2024a), Yu et al. (2024b), Taubmann et al. (2024)
Mesenchymal stromal cells (MSCs)	Exhibit immunomodulatory properties, enhance tissue repair, and modulate immune responses by inhibiting autoreactive T- and B-cells.	SLE, RA	Pignatti et al. (2024), Swain et al. (2024), Barbado (2024)
Regulatory T-cells (Tregs)	Maintain immune homeostasis and self-tolerance. Treg-based therapies aim to restore immune balance and are effective in clinical trials.	RA, other autoimmune diseases	Honing et al. (2024), Yu et al. (2024a), Su et al. (2024a)
Myeloid-derived suppressor cells (MDSCs)	Modulate T-cell activity, promote or inhibit immune responses, and play dual roles depending on the disease stage.	MS, other autoimmune conditions	Bekić and Tomić (2023), Zanghì et al. (2024), Jiang et al. (2024)
Small molecule Treg targeting	Enhances Treg migration to inflamed tissues, improving therapeutic outcomes.	Autoimmune diseases (general)	Martin et al. (2024)
Type I IFN and B-cell modulation	Inhibits type I IFN signaling and addresses B-cell abnormalities using targeted therapies like belimumab and anifrolumab.	SLE	Fang et al. (2024), Gatto et al. (2023)
Gene modulation (siRNA)	siRNA targets proinflammatory cytokines to suppress aberrant immune responses.	SLE	Lee et al. (2020)
Metabolic regulation	Targets immune cell metabolic abnormalities to restore homeostasis.	Autoimmune diseases (general)	Teng et al. (2019)
Combination and personalized therapy	Combines new modalities with existing treatments, addressing disease heterogeneity and enhancing therapeutic outcomes with patient-specific approaches.	Autoimmune diseases (general)	Balogh et al. (2024), Felten et al. (2023)

2024a; Lee et al. 2022). Recent advancements also highlight the potential of CRISPR-based epigenome editors to address epigenetic dysregulation in T-cells, a significant factor in autoimmune pathogenesis (Yahsi et al. 2024). The versatility of CRISPR technology extends to understanding the underlying mechanisms of ADs, offering insights into nonviral triggers and enhancing therapeutic strategies (Lee et al. 2022; Singh 2024).

Precision targeting of immune cells and inflammatory pathways in ADs involves a multifaceted approach that leverages advancements in understanding disease pathophysiology. Recent research highlights the significance of inflammatory mediators and immune cell interactions in the development of ADs, suggesting that targeting specific inflammatory cytokines, such as TNF, IL-6, and IL-17, can effectively modulate immune responses and improve patient outcomes (Jung and Kim 2022; Xiang et al. 2023). The integration of computational models and multiomics profiling allows for the identification of distinct proinflammatory pathways, enabling the design of combination therapies tailored to individual patient profiles (Desvaux et al. 2022). Exploring immunometabolic pathways, particularly the mTOR pathway, offers new therapeutic targets by regulating immune cell fate and function [Shukla et al. 2024], and aims to enhance treatment efficacy while minimizing side effects, ultimately leading to more personalized care for patients with ADs (Velikova et al. 2024).

The translation of preclinical findings to clinical settings in ADs faces several limitations. One significant challenge is the physiological differences between animal models and humans, which can lead to discrepancies in treatment efficacy and safety profiles (Rezaei and Yazdanpanah 2022). The heterogeneity of the human immune system, influenced by genetic, environmental, and experiential factors, complicates the understanding of disease mechanisms and responses to therapies (Buckner 2023). This variability means that a treatment effective in one individual may not work for another, even with the same diagnosis (Buckner 2023); while preclinical models, such as humanized mice, are being optimized to better reflect human responses, they still often fall short in predicting clinical outcomes (Houeiss et al. 2022; Shandil et al. 2022). The complexity of ADs, characterized by multiple immune pathways and diverse manifestations, necessitates a more nuanced approach to treatment development and application.

7.3.3 Antibody-Based Therapies

Bispecific antibodies (bsAbs) are engineered proteins that can simultaneously bind to two different antigens or receptors, offering a novel approach to modulate immune responses in ADs. They can target specific immune cells, such as T- or B-cells, and inhibit their activation or differentiation, thereby reducing inflammation and promoting immune tolerance (Lamendour et al. 2024; Zhao 2020). For instance, certain bsAbs have been shown to induce regulatory T-cell differentiation and enhance the secretion of antiinflammatory cytokines like IL10, which is crucial in managing autoimmune conditions (Lamendour et al. 2024). bsAbs can alter signaling pathways, such as TLR4, to produce immunosuppressive effects, further aiding in the treatment of ADs (Nicolette et al. 2018). The development of dual-Fc bsAbs enhances their efficacy by increasing binding avidity and effector functions, making them promising candidates for future therapies (Zhou et al. 2024). bsAbs represent a significant advancement in the precise modulation of immune responses in autoimmune disorders.

Bispecific antibodies represent a promising advancement in the treatment of ADs by targeting multiple disease-related molecules, thereby enhancing therapeutic efficacy and reducing side effects compared with traditional monoclonal antibodies (Segués et al. 2022; Zhao 2020). Recent developments have shown bsAbs effectively deplete pathogenic T- or B-cells, inhibit T-cell activation, and neutralize proinflammatory cytokines, with ongoing clinical trials demonstrating both successes and failures in various autoimmune conditions (Zhao 2020). However, challenges remain, including the complexities of manufacturing these antibodies, ensuring high yields and purity, and addressing the need for reliable assembly and purification processes (Underwood et al. 2022). The specificity of bsAbs must be carefully managed to avoid unintended immune responses, as seen in the development of bispecific CD8 Treg modulators that selectively target pathogenic CD4 T-cells (Gardell et al. 2022).

Antibody-drug conjugates (ADCs) represent a promising targeted therapeutic approach for ADs by combining the specificity of monoclonal antibodies with potent cytotoxic agents. This innovative strategy allows for precise targeting of immune components, thereby minimizing off-target toxicity and enhancing therapeutic efficacy (Huang et al. 2024; Pandey et al. 2023). Recent advancements in ADC technology have shown potential in treating conditions like multiple sclerosis, where antigen-drug conjugates (AgDCs) have demonstrated improved efficacy by codelivering autoantigens with immunosuppressants, leading to better outcomes than traditional therapies (Pickens et al. 2019). The development of ADCs is bolstered by the success of biological drugs and small molecule inhibitors that target inflammatory cytokines and immune cells, which have transformed the treatment landscape for ADs (Jung and Kim 2022). ADCs indicate potential in addressing unmet medical needs in autoimmune therapy, offering a more targeted and effective treatment paradigm (Lordick et al. 2023). A comparison of different nucleotide and antibody-based therapies is shown in Table 7.2.

7.4 Drug Repurposing and Novel Drug Development

Drug repurposing involves identifying existing drugs that can be used for new therapeutic indications, particularly in ADs where traditional treatments are limited. By leveraging bioinformatics and transcriptomic analyses, researchers can uncover molecular pathways that are dysregulated in specific autoimmune conditions, such as SLE and scleroderma. For instance, studies have highlighted the PI3K/AKT pathway as a significant target for drug candidates identified through repositioning efforts, including KU-0063794 and YM-155, which show promise for treating systemic ADs (Oktem and Yazar 2022). Additionally, integrating gene regulatory networks and transcriptome data has led to the identification of novel compounds that can reverse aberrant gene expression in monocytes from SLE patients, targeting

Table 7.2 Nucleotide and antibody-based therapies

Therapeutic modality (nucleotide and antibody based)	Description	Targeted conditions	Key features	References
mRNA therapeutics	Utilizes mRNA to produce therapeutic proteins for immune modulation and tolerance induction.	Autoimmune diseases (e.g., RA, diabetes)	Personalized treatments, rapid production, potential to mitigate autoimmune responses.	Krienke et al. (2021), Iyer (2022), Sarfraz et al. (2024), Deyhimfar et al. (2024)
Gene editing (CRISPR/Cas9)	Precise genetic editing for immune modulation and correction of autoimmune-related gene defects.	RA, lupus, T1D	High precision, potential for durable immune tolerance, epigenome editing capabilities.	Chasov et al. (2024a), Lee et al. (2022), Yahsi et al. (2024), Singh (2024)
Bispecific antibodies (BsAbs)	Engineered antibodies that bind two antigens simultaneously for immune modulation.	RA, lupus, autoimmune uveitis	Enhances Treg differentiation, targets immune pathways, reduces inflammation.	Lamendour et al. (2024), Zhao (2020), Zhou et al. (2024)
Antibody-drug conjugates (ADCs)	Combines monoclonal antibodies with cytotoxic agents to precisely target immune components.	Multiple sclerosis, RA	Precise targeting, reduced off-target effects, codelivery of antigens and immunosuppressants.	Pickens et al. (2019), Lordick et al. (2023), Pandey et al. (2023)
Targeting immune pathways	Focuses on modulating inflammatory cytokines and immunometabolic pathways like mTOR.	RA, MS, SLE	Personalized therapies via multiomics profiling, reduced inflammation, immune cell fate control.	Jung and Kim (2022), Xiang et al. (2023), Velikova et al. (2024)
Preclinical translation	Optimization of preclinical models to enhance clinical applicability of autoimmune therapies.	Various autoimmune diseases	Improved humanized models, enhanced prediction of treatment outcomes, overcoming variability.	Rezaei and Yazdanpanah (2022), Buckner (2023), Houeiss et al. (2022)

pathways like NF-κB and HSP90 (Nikolakis et al. 2023). This systems biology approach enhances our understanding of disease mechanisms and facilitates the discovery of effective therapies for ADs, as shown in Fig. 7.1 (Devaprasad et al. 2021; Grammer and Lipsky 2017).

Recent studies have identified several promising candidates for drug repurposing in ADs. A combined transcriptome-reversal and network-based approach revealed

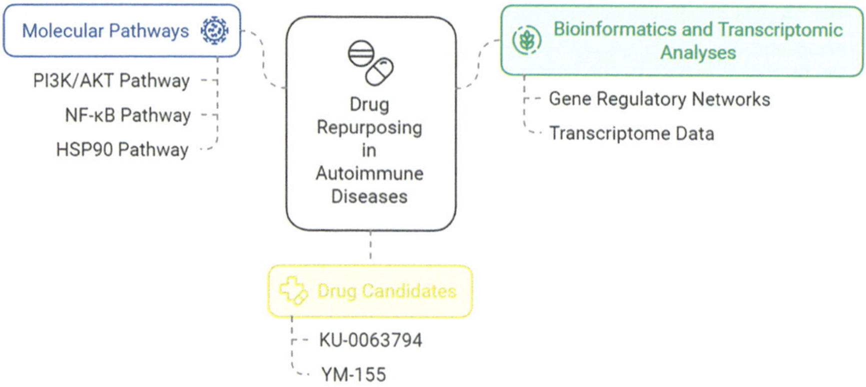

Fig. 7.1 Different tools/procedures used in drug repurposing

inhibitors targeting the NF-κB pathway, heat shock protein 90 (HSP90), and the Pim-1/NFATc1/NLRP3 signaling axis as potential therapies for SLE (Nikolakis et al. 2023). Six small molecule drug candidates, including KU-0063794 and YM-155, were identified through integrative analyses of transcriptomes from various systemic ADs, highlighting the PI3K/AKT pathway as a key target (Oktem and Yazar 2022). Natural products with antiinflammatory properties, such as those derived from fungi and bacteria, also show potential as new immunosuppressants, often acting through the inhibition of classical inflammatory signaling pathways (Saurin et al. 2022; Nandal et al. 2023). Artemisinin and its derivatives have demonstrated immunoregulatory properties, suggesting their viability in treating autoimmune conditions (Gao et al. 2024).

Recent advancements in understanding the molecular mechanisms underlying ADs have led to the identification of several promising drug targets. For instance, in RA, new therapeutic agents are being developed that target CD40, PD-1, and GM-CSF, alongside other molecules like IRAK1, IRAK4, and Tyk2, which have shown potential in preclinical studies (Wallace et al. 2024). In SLE, targeted treatments such as belimumab and anifrolumab have emerged, focusing on interferon signaling and B-cell pathways, with over 90 investigational drugs currently in development (Askanase et al. 2023; Gatto et al. 2023). A study identified 12 circulating proteins as potential therapeutic targets for various ADs, including IL12B for ankylosing spondylitis and SWAP70 for RA (Jin et al. 2024). This evolving landscape of targeted therapies aimed at improving patient outcomes in autoimmune disorders (Sandhu and Thelma 2022).

7.5 Personalized Medicine in Autoimmunity

Multiomics approaches are pivotal in characterizing ADs by integrating various biological data types, including genomics, transcriptomics, proteomics, and metabolomics. These methods enable a comprehensive understanding of the complex interactions within the immune system and the molecular underpinnings of diseases like SLE and immune thrombocytopenia (Kiumarsi 2024; Shinde et al. 2024). Multiomics analyses can reveal distinct molecular signatures associated with disease activity in SLE, highlighting the roles of the complement system and lipid metabolism (Huang et al. 2022). Advanced bioinformatic tools are essential for integrating these diverse datasets, facilitating insights into cellular interactions and therapeutic strategies (Borrego-Yaniz et al. 2024). The application of single-cell technologies and machine learning further enhances the potential for personalized treatment approaches, ultimately improving diagnostic and prognostic capabilities in autoimmune disorders (Shinde et al. 2024).

Patient-reported outcomes (PROs) are essential tools for optimizing treatment in ADs, as they provide insights into patients' health-related quality of life and treatment effectiveness from the patient's perspective. Research indicates that incorporating PROs can enhance the understanding of disease activity and treatment responses, particularly in conditions like SLE and psoriatic arthritis, where symptoms can be heterogeneous and complex (LaBere et al. 2023; Mease et al. 2022; Parodis and Studenic 2022). Patients initiating biologic treatments often report significant improvements in PROs, highlighting the importance of these measures in evaluating real-world clinical effectiveness (Beukelman et al. 2023). The development of disease-specific PRO tools, informed by patient input, ensures that the measures are relevant and capture the full patient experience, facilitating better communication between patients and healthcare providers (Mease et al. 2022). PROs not only aid in treatment optimization but also foster a collaborative approach to disease management.

The integration of personalized approaches in clinical practice for ADs is pivotal for enhancing treatment efficacy and patient outcomes. Personalized medicine utilizes individual patient characteristics, including genetic, phenotypic, and environmental factors, to tailor therapeutic strategies, moving away from the traditional "one-size-fits-all" model (Chakraborty et al. 2022). This approach is particularly beneficial in ADs, where diverse clinical manifestations and responses to treatment are common (Díaz-Peña 2021). Advances in genomic technologies, such as next-generation sequencing, have facilitated the identification of specific biomarkers that guide diagnosis, prognosis, and treatment response prediction (Ajayi et al. 2020). Pharmacogenetics plays a crucial role in determining how patients respond to various drugs, allowing for more effective and safer treatment regimens (Qamar and Mukherjee 2021). By leveraging these personalized strategies, clinicians can improve compliance, reduce adverse drug reactions, and ultimately provide more effective management of autoimmune conditions (Chakraborty et al. 2022).

Implementing personalized therapies in ADs faces several significant challenges. Firstly, the complexity and heterogeneity of diseases like multiple sclerosis complicate the identification of specific biomarkers and disease subtypes necessary for precision medicine (Pathak 2023). While advancements in machine learning and biological markers are promising, there remains a lack of consensus on standardized diagnostic criteria and treatment guidelines, which hampers the effective application of personalized approaches (Ramo-Tello 2022; Chakraborty et al. 2022). The acceptance of personalized medicine varies among patients, necessitating a deeper understanding of their experiences and attitudes toward such treatments to ensure they are beneficial rather than burdensome (Henschke et al. 2021). The development of targeted therapies must navigate ethical and legal considerations, as well as the economic implications of personalized treatment strategies, which can be substantial (Carballido et al. 2020; Pathak 2023).

7.6 Gut Microbiota and Autoimmune Diseases

Gut dysbiosis refers to an imbalance in the gut microbiota, which can disrupt immune homeostasis and contribute to the development of ADs such as SLE, type 1 diabetes, and RA. This dysbiosis can result from various factors, including diet, age, and antibiotic use, leading to altered immune responses and increased susceptibility to autoimmunity through mechanisms like molecular mimicry and translocation of microbes (Nandal et al. 2024; Sadeghpour Heravi 2024; Sai et al. 2024). For instance, in Graves' disease, dysbiosis has been linked to immune dysregulation, characterized by altered B-cell distribution and elevated proinflammatory cytokines (Liu et al. 2024b). Gut dysbiosis can promote islet autoimmunity by enhancing T-cell attraction to pancreatic islets via chemokines like CXCL10, which is induced by low-grade endotoxemia from bacterial lipopolysaccharides (Pöysti et al. 2023). The interplay between gut microbiota and immune function is crucial in understanding the pathogenesis of autoimmune conditions.

Microbiome-modulating therapies, including prebiotics, probiotics, and fecal microbiota transplants (FMT), have emerged as promising interventions for ADs. FMT involves transferring gut microbiota from healthy donors to patients, aiming to restore immune balance and improve clinical outcomes in conditions like MS and SLE (Nandal et al. 2024; Seida et al. 2024; Zhanel et al. 2023). FMT can enhance intestinal motility and motor abilities in MS patients, with effects lasting for years (Zhanel et al. 2023). Probiotics and dietary interventions have also shown positive impacts on autoimmune symptoms, such as reducing relapse rates and improving disability scores (Tsogka et al. 2023; Zeng et al. 2024). Gut microbiota-based therapies have been linked to improved inflammatory profiles and symptom relief across various autoimmune conditions, although their efficacy can vary (Liu et al. 2023; Zeng et al. 2024). These therapies represent a novel approach to managing ADs by targeting the gut microbiome's role in immune regulation. Figure 7.2 shows how gut microbiota leads to the progression of ADs.

Clinical evidence suggests that microbiome-based therapies hold significant promise for managing ADs. Research indicates that dysbiosis, or an imbalance in gut microbiota, contributes to the pathogenesis of conditions like RA and SLE by promoting immune dysregulation and increased gut permeability (Amarnani and Silverman 2023; Balasundaram et al. 2024; Mo et al. 2024). Probiotics and other microbiome-modulating interventions have shown potential in alleviating symptoms and improving disease outcomes, as they can restore microbial balance and produce beneficial metabolites (Balasundaram et al. 2024; Nayak and Orellana 2024). Future directions in this field include personalized treatment approaches that consider individual microbiome compositions, as well as the exploration of extracellular vesicles and their role in microbiome modulation (Amarnani and Silverman 2023; Peshkova et al. 2024). Ongoing research into the mechanisms of microbiome interactions with the immune system will be crucial for developing effective therapies tailored to specific autoimmune conditions.

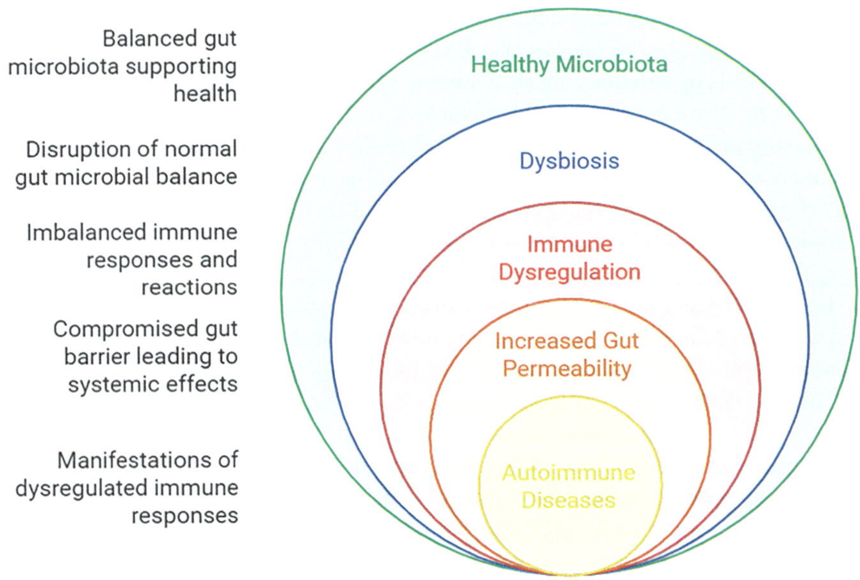

Fig. 7.2 Gut microbiota and progression of autoimmune diseases

7.7 Targeting Specific Cellular and Molecular Mechanisms

Dendritic cells (DCs) play a pivotal role in the modulation of immune responses in ADs by balancing immune activation and tolerance. In healthy individuals, tolerogenic DCs promote the expansion of Tregs, which are crucial for maintaining peripheral tolerance and preventing autoimmunity (Cai and Thomas 2024). However, in autoimmune conditions, DCs can become activated by pathogen-associated and damage-associated molecular patterns, leading to the expansion of self-reactive T-cells and chronic inflammation (El Jamal and Shibli 2024). This dysregulation is further complicated by the interactions between DCs and other immune cells, such as B-cells, which can exacerbate autoimmune responses (Ismail et al. 2021). Therapeutic strategies targeting DCs, including ex vivo manipulation and costimulatory molecule modulation, are being explored to restore immune balance and improve outcomes in ADs (El Jamal and Shibli 2024; Honing et al. 2024). Understanding these mechanisms is essential for developing effective immunotherapies aimed at reestablishing tolerance in autoimmune disorders.

Citrullinated neoantigens have emerged as significant therapeutic targets in ADs such as RA and SLE. In RA, citrullinated proteins trigger the production of anticitrullinated protein antibodies (ACPA), which are crucial in disease pathogenesis, as they bind to citrullinated autoantigens like thrombospondin-1 and β-actin, leading to inflammation and joint damage (Gómez-Bañuelos et al. 2022; Xu et al. 2023). The role of peptidylarginine deiminase (PAD) enzymes in citrullination presents a potential therapeutic avenue, as pan-PAD inhibitors have shown promise in preclinical models by ameliorating autoimmune symptoms (Bruggeman et al. 2021). In SLE, the treat-to-target approach emphasizes managing disease activity and minimizing organ damage, where targeting citrullinated neoantigens could enhance treatment efficacy (Nikolopoulos et al. 2024). The T-cell responses to citrullinated peptides can inform the development of targeted therapies, as these responses are complex and not solely dependent on binding affinity to HLA-DR alleles (Becart et al. 2021).

Epigenetic therapies are emerging as promising strategies to modulate immune responses in ADs by targeting the epigenetic mechanisms that underlie immune dysregulation. These therapies, which include DNA methyltransferase inhibitors and histone deacetylase inhibitors, aim to reverse aberrant epigenetic modifications that contribute to autoimmune pathogenesis, such as those observed in SLE and RA (Hui et al. 2023; Klein 2023). Epigenetic alterations, including histone modifications and noncoding RNAs, play a crucial role in the activation and differentiation of immune cells, particularly B-cells, which are pivotal in autoimmune responses (Xiao et al. 2022). The identification of epigenetic biomarkers can enhance our understanding of disease mechanisms and patient phenotypes, facilitating personalized treatment approaches (Ibáñez-Cabellos et al. 2022). The integration of epigenetic insights into therapeutic strategies holds significant potential for improving outcomes in autoimmune disorders.

7.8 Prevention and Early Intervention Strategies

Early diagnosis of ADs is crucial as it significantly reduces the risk of morbidity, disability, and mortality, allowing for timely therapeutic interventions that can lead to better patient outcomes (Patel et al. 2024). The complexity and variability of ADs, such as SLE and autoimmune gastritis, often result in delayed diagnoses, which can exacerbate the disease and complicate treatment (Sebastiani et al. 2016; Ozhiganova et al. 2023). Advanced diagnostic techniques, including multiplexed electrochemical bioplatforms for autoantibody detection, enhance the accuracy and speed of diagnosis, facilitating the differentiation between various autoimmune conditions (Arévalo et al. 2023). Genetic predispositions and immune mechanisms involved in these diseases can inform early intervention strategies, ultimately aiming for remission and preventing functional deterioration (Konforte et al. 2012), and integration of innovative diagnostic methods and a deeper understanding of autoimmune pathogenesis is essential for improving early diagnosis and treatment outcomes in ADs.

Preventing the progression of ADs involves a multifaceted approach that includes both pharmacological and nonpharmacological strategies. Immunotherapeutic interventions, such as LD-IL-2, have shown promise in reducing autoimmune responses and promoting Tregs, which can help mitigate disease progression in conditions like acute rheumatic fever (Rafeek et al. 2024). Antigen-specific immunotherapy (ASI) aims to induce immune tolerance through repeated exposure to specific antigens, which can effectively modulate the immune response without systemic suppression (Shakya and Nandakumar 2018). Nonpharmacological strategies, including dietary modifications and physical exercise, have also been recognized for their potential to improve quality of life and manage symptoms (Celichowska et al. 2024). Targeted delivery systems, such as microparticles, can enhance the efficacy of tolerogenic therapies by specifically modulating immune responses in diseases like RA and MS (Peine et al. 2019; Shakya and Nandakumar 2018). Collectively, these strategies represent a comprehensive approach to managing ADs and preventing their progression.

Immunological tolerance in ADs refers to the immune system's ability to remain unresponsive to self-antigens, thereby preventing harmful autoimmune responses. Achieving this state without chronic immunosuppression is crucial, as traditional therapies often lead to significant side effects, including increased infection risk and cancer (Cobbold 2009; Stanway and Isaacs 2020). Strategies to induce tolerance include enhancing the function of Tregs, which play a vital role in maintaining immune balance and preventing autoimmunity (Nasonov et al. 2014). Advances in nanomedicine also offer promising avenues for targeted delivery of immunomodulatory agents, potentially improving treatment efficacy while minimizing systemic adverse reactions (Gharagozloo et al. 2015). Specific interventions, such as antigen-specific therapies and stem cell transplantation, can promote drug-free remission and restore immune tolerance (Cobbold 2009; Stanway and Isaacs 2020). The focus is shifting toward personalized, targeted approaches that can achieve long-lasting tolerance without the need for ongoing immunosuppressive therapy.

7.9 Translating Preclinical Findings to Clinical Success

Overcoming challenges in animal models and clinical trials for ADs requires addressing several key issues. Firstly, while animal models, particularly murine models, have been instrumental in understanding autoimmune pathologies, they often fail to fully replicate the complexities of human diseases due to genetic and environmental differences (Chen et al. 2022; Polis et al. 2024). This necessitates the development of more refined models, such as humanized mice, which can better mimic human immune responses (Chen et al. 2022). The exploration of microbial pathomechanisms highlights the need for standardized techniques in studying the interplay between pathogens and the immune system, as inconsistent results hinder progress (Aloizou et al. 2023). In the context of autoimmune uveitis, the use of genetically modified animals and in vitro systems can provide deeper insights into disease mechanisms, yet translating these findings to human applications remains a challenge (Al-Hakam and Subramanian 2024). The approach that combines innovative model development with rigorous clinical trial designs is essential for advancing therapeutic strategies in ADs.

Translating novel therapies for ADs to human patients involves several critical considerations. First, understanding the heterogeneity of immune-mediated inflammatory diseases (IMIDs) is essential, as this variability can affect patient responses to targeted treatments, necessitating personalized therapeutic strategies (Felten et al. 2023). The safety and efficacy of new therapies, particularly those involving cell-based approaches, must be rigorously evaluated to avoid complications such as organ damage or infections associated with traditional immunosuppressants (Mizui and Kono 2024). The development of antigen-specific therapies, which aim to restore self-tolerance without broadly suppressing the immune system, represents a promising avenue, but requires careful design to ensure specificity and minimize adverse effects (Sármay 2021). The manipulation and expansion of tolerogenic cell populations, such as Tregs, must be optimized for clinical application (Park and Kwok 2022). Immunopathogenesis and patient-specific factors are crucial for successful translation.

Successful clinical transitions in ADs are exemplified through various case studies highlighting the importance of structured transition programs. For instance, a cohort study in France demonstrated that formal transition programs significantly improved the regularity of follow-up and disease control in patients with AIDs transitioning from pediatric to adult care, with a notable 66% achieving disease control at their first adult consultation (Elhani et al. 2024). A case study involving a 25-year-old woman with severe pulmonary arterial hypertension associated with SLE showcased a successful transition from intravenous epoprostenol to oral selexipag and inhaled iloprost, leading to sustained improvement without exacerbations (Saiki et al. 2022). The implementation of individualized transition policies in AIDs emphasizes equipping young adults with self-management skills, thereby enhancing their quality of life and engagement in their care (Pratsidou-Gertsi 2018). These examples underscore the critical role of structured transitions in managing chronic autoimmune conditions effectively.

7.10 Future Directions

The integration of new technologies in autoimmune therapy is revolutionizing treatment approaches for ADs. Biomarker-driven strategies, supported by AI-powered analytics, enable precision medicine by identifying clinically relevant biomarkers (BMKs) through extensive molecular profiling and multiomics technologies, which enhance patient stratification and predict treatment responses (Kumar et al. 2021; Laigle et al. 2023). Additionally, regenerative therapies utilizing autologous adult stem cells have shown promise, particularly in skin-related autoimmune conditions like cutaneous lupus erythematosus, demonstrating significant improvement in patient outcomes (Kaliterna and Barac 2023). Furthermore, advancements in cell therapeutics, including the use of immunoregulatory cell populations such as mesenchymal stromal cells and Tregs, are being explored to restore immune tolerance and mitigate autoimmunity (Kumar et al. 2021). Collectively, these technological innovations are paving the way for more effective, personalized treatment strategies in managing complex ADs.

Refining clinical trials for dynamic treatment approaches in ADs, particularly SLE, involves implementing innovative trial designs that enhance efficiency and patient stratification. Adaptive trial designs, such as those used in studies of efavaleukin alfa and rozibafusp alfa, allow for modifications based on interim data, improving the statistical power and reducing the need for multiple phase III trials (Garces et al. 2023). Mathematical modeling can identify effective dosage ranges and predict treatment outcomes based on patient-specific immune profiles, such as the Tcon/Treg ratio (Gao et al. 2022). The integration of systems pharmacology approaches, including Boolean network models, can help identify drug targets and patient subpopulations, addressing the heterogeneity of SLE (Ruiz-Cerdá et al. 2016). The application of personalized medicine through regenerative therapies, like autologous stem cells, shows promise in treating autoimmune conditions (Kaliterna and Barac 2023), which aim to optimize treatment efficacy while minimizing resource use in clinical trials.

Future research priorities in the management of ADs focus on several key areas to enhance treatment efficacy and patient outcomes. There is a pressing need for high-quality research to support the management of specific conditions, such as hidradenitis suppurativa, with an emphasis on innovative treatments like laser hair removal in conjunction with medical therapies (Leighton et al. 2023). Understanding the molecular pathways involved in diseases like SLE is important, as ongoing investigations into targeted therapies and new drug developments are promising

(Askanase et al. 2023). The exploration of microbiome-based therapies, including probiotics and FMT, presents a novel approach to personalized medicine in autoimmune treatment (Tsigalou et al. 2023). Optimizing monitoring tools and treatment strategies based on real-world data is essential for improving chronic disease management and minimizing disability (Macaron and Willis 2021). Such research directions refine therapeutic approaches and enhance patient care in ADs.

7.11 Conclusion

ADs present significant challenges, including the lack of effective diagnostic tools and validated markers for disease prediction, which complicates prevention and treatment efforts (Yazdanpanah and Rezaei 2023). Current therapies often target the immune system broadly, leading to adverse effects and highlighting the need for more precise treatment strategies. The physiological differences between humans and animal models further complicate the translation of research findings into clinical practice (Rezaei and Yazdanpanah 2022). However, emerging opportunities exist, such as the development of nanomedicine, which can specifically target dysregulated immune cells while minimizing systemic effects (Thatte et al. 2024). Additionally, Treg therapies offer promising avenues for restoring immune balance and tolerance, although they face challenges in clinical application (Bharti et al. 2024). Recent advances in understanding cytokine pathways in diseases like RA also provide a framework for targeted therapies, making disease remission a more attainable goal.

Innovative therapies are poised to transform the treatment landscape for ADs, which affect a significant portion of the population. Nanomedicine offers targeted delivery systems that can specifically address dysregulated immune responses while minimizing systemic side effects, thus enhancing therapeutic efficacy (Thatte et al. 2024). Additionally, phosphatidylserine (PS)-liposomes encapsulating autoantigens have shown promise in restoring self-tolerance across various autoimmune conditions, including RA and myasthenia gravis (Almenara-Fuentes et al. 2023). Furthermore, cell therapies present a dual approach by regenerating damaged tissues and modulating pathological immune responses (Mehta et al. 2024). The development of monoclonal antibodies and bsAbs, alongside gene editing technologies like CRISPR/Cas9, further expands the arsenal against ADs by targeting specific molecular pathways and correcting genetic defects (Felten et al. 2023; Lee et al. 2022). Collectively, these innovative strategies highlight a shift toward more personalized and effective treatments for autoimmune disorders.

References

Ajayi A, Adebayo O, Adebayo E (2020) Precision medicine of autoimmune diseases. In: Saxena SK, Prakash H (eds) Innate immunity in health and disease. IntechOpen. https://doi.org/10.5772/intechopen.95248

Albarbar B, Aga H (2024) A review on autoimmune diseases: recent advances and future perspectives. AlQalam J Med Appl Sci 7(3):718–729

Alcaraz-Serna A, Porret R, Trueb L, Ribi C, Seebach J, Muller YD (2024) CAR-T therapy for autoimmune diseases. Rev Med Suisse 20(868):688–693

Al-Hakam AM, Subramanian A (2024) Experimental models in autoimmune uveitis. In: Alenzi FQ (ed) Animal models in experimental medicine. Bentham Science Publisher, pp 246–260. https://doi.org/10.2174/9789815196382124010015

Almenara-Fuentes L, Rodriguez-Fernandez S, Rosell-Mases E, Kachler K, You A, Salvado M, Andreev D, Steffen U, Bang H, Bozec A (2023) A new platform for autoimmune diseases. Inducing tolerance with liposomes encapsulating autoantigens. Nanomedicine 48:102635

Aloizou A-M, Siokas V, Efthymiou G, Tsigalou C, Dardiotis E, Bogdanos DP (2023) Current challenges in research with exploring the microbial pathomechanisms of autoimmune diseases. In: Dwivedi MK, Sankaranarayanan A, Kemp EH, Shoenfeld Y (eds) Role of microorganisms in pathogenesis and management of autoimmune diseases, volume II: kidney, central nervous system, eye, blood, blood vessels & bowel. Springer, pp 469–488. https://doi.org/10.1007/978-981-19-4800-8_24

Amarnani A, Silverman GJ (2023) Understanding the roles of the microbiome in autoimmune rheumatic diseases. Rheumatol Immunol Resh 4(4):177–187

Arévalo B, Serafín V, Garranzo-Asensio M, Barderas R, Yáñez-Sedeño P, Campuzano S, Pingarrón JM (2023) Early and differential autoimmune diseases diagnosis by interrogating specific autoantibody signatures with multiplexed electrochemical bioplatforms. Biosens Bioelectron: X 13:100325

Askanase A, Khalili L, Tang W, Mertz P, Scherlinger M, Sebbag E, Chasset F, Felten R, Arnaud L (2023) New and future therapies: changes in the therapeutic armamentarium for SLE. Best Pract Res Clin Rheumatol 37(4):101865

Athanassiou P, Athanassiou L (2023) Current treatment approach, emerging therapies and new horizons in systemic lupus erythematosus. Life (Basel) 13(7):1496

Balasundaram D, Veerasamy V, Singarayar MS, Neethirajan V, Devanesan AA, Thilagar S (2024) Therapeutic potential of probiotics in gut microbial homeostasis and rheumatoid arthritis. Int Immunopharmacol 137:112501

Balogh L, Oláh K, Sánta S, Majerhoffer N, Németh T (2024) Novel and potential future therapeutic options in systemic autoimmune diseases. Front Immunol 15:1249500

Barbado J (2024) Mesenchymal stem cell transplantation may be able to induce immunological tolerance in systemic lupus erythematosus. Biom J 47(4):100724

Becart S, Whittington KB, Prislovsky A, Rao NL, Rosloniec EF (2021) The role of posttranslational modifications in generating neo-epitopes that bind to rheumatoid arthritis-associated HLA-DR alleles and promote autoimmune T cell responses. PLoS One 16(1):e0245541

Bekić M, Tomić S (2023) Myeloid-derived suppressor cells in the therapy of autoimmune diseases. Eur J Immunol 53(12):2250345

Beukelman T, Long MD, Rhee RL, Kappelman MD, Merkel PA, Nowell WB, Clinton C, Ringold S, Del Gaizo V, Price B (2023) Assessment of real-world patient-reported outcomes in patients initiating biologic agents for the treatment of autoimmune diseases: an observational study in four patient-powered research networks. Patient Relat Outcome Meas 14:171–180

Bharti S, Guralarasan G, Bharti M (2024) Current challenges and future prospects of regulatory T-cell based therapeutics in autoimmune diseases. In: Dwivedi MK, Fairweather D (eds) Regulatory T cells and autoimmune diseases (pp. 897–910). Academic Press. https://doi.org/10.1016/B978-0-443-13947-5.00031-2

Borrego-Yaniz G, Terrón-Camero LC, Kerick M, Andrés-León E, Martin J (2024) A holistic approach to understanding immune-mediated inflammatory diseases: bioinformatic tools to integrate omics data. Comput Struct Biotechnol J 23:96–105

Bruggeman Y, Sodré FMC, Buitinga M, Mathieu C, Overbergh L, Kracht MJL (2021) Targeting citrullination in autoimmunity: insights learned from preclinical mouse models. Expert Opin Ther Targets 25(4):269–281

Buckner JH (2023) Translational immunology: applying fundamental discoveries to human health and autoimmune diseases. Eur J Immunol 53(12):2250197

Cai B, Thomas R (2024) Dendritic cells and antigen-specific immunotherapy in autoimmune rheumatic diseases. Best Pract Res Clin Rheumatol 38(2):101940

Carballido JM, Regairaz C, Rauld C, Raad L, Picard D, Kammüller M (2020) The emerging jamboree of transformative therapies for autoimmune diseases. Front Immunol 11:472

Celichowska M, Bentkowska Z, Bogoń A, Szpyra J, Dziuba G, Serkis B, Dębińska J, Kałuża I, Golemo J, Miazga M (2024) Methods of prevention and mitigation of autoimmune diseases – a review of the literature. Quality Sport 17:53121

Chakraborty S, Wagh A, Goel P, Phatak S (2022) Personalized medicine in India: mirage or a viable goal? Indian J Rheumatol 17(1):57–64. https://doi.org/10.4103/injr.injr_152_21

Chandrashekara S (2012) The treatment strategies of autoimmune disease may need a different approach from conventional protocol: a review. Indian J Pharmacol 44(6):665–671

Chasov V, Ganeeva I, Zmievskaya E, Davletshin D, Gilyazova E, Valiullina A, Bulatov E (2024a) Cell-based therapy and genome editing as emerging therapeutic approaches to treat rheumatoid arthritis. Cells 13(15):1282. https://doi.org/10.3390/cells13151282

Chasov V, Zmievskaya E, Ganeeva I, Gilyazova E, Davletshin D, Khaliulin M, Kabwe E, Davidyuk YN, Valiullina A, Rizvanov A, Bulatov E (2024b) Immunotherapy strategy for systemic autoimmune diseases: betting on CAR-T cells and antibodies. Antibodies (Basel) 13(1):10. https://doi.org/10.3390/antib13010010

Chen J, Liao S, Zhou H, Yang L, Guo F, Chen S, Li A, Pan Q, Yang C, Liu H (2022) Humanized mouse models of systemic lupus erythematosus: opportunities and challenges. Front Immunol 12:816956. https://doi.org/10.3389/fimmu.2021.816956

Cicuttini FM (2017) Autoimmune rheumatic diseases: recent advances and current challenges. Med J Aust 206(5):201–202

Cobbold SP (2009) Future therapeutics for the induction of peripheral immune tolerance in autoimmune disease and organ transplantation. Immunotherapy 1(3):447–460

Dehnavi S, Sadeghi M, Afshari JT, Mohammadi M (2023) Interactions of mesenchymal stromal/stem cells and immune cells following MSC-based therapeutic approaches in rheumatoid arthritis. Cell Immunol 393-394:104771. https://doi.org/10.1016/j.cellimm.2023.104771

Desvaux E, Aussy A, Hubert S, Keime-Guibert F, Blesius A, Soret P, Guedj M, Pers J-O, Laigle L, Moingeon P (2022) Model-based computational precision medicine to develop combination therapies for autoimmune diseases. Expert Rev Clin Immunol 18(1):47–56

Devaprasad A, Radstake TRDJ, Pandit A (2021) Integration of immunome with disease-gene network reveals common cellular mechanisms between IMIDs and drug repurposing strategies. Front Immunol 12:669400

Deyhimfar R, Izady M, Shoghi M, Kazazi MH, Ghazvini ZF, Nazari H, Fekrirad Z, Arefian E (2024) The clinical impact of mRNA therapeutics in the treatment of cancers, infections, genetic disorders, and autoimmune diseases. Heliyon 10(5):e26971. https://doi.org/10.1016/j.heliyon.2024.e26971

Díaz-Peña R (2021) Personalized medicine in autoimmune diseases. J Pers Med 11(11):1181. https://doi.org/10.3390/jpm11111181

Dörner T (2023) 31 other targets: expectations from research on pathogenic mechanisms. Lupus Sci Med 10(2). https://doi.org/10.1136/lupus-2023-la.31

El Jamal M, Shibli F (2024) Dendritic cell and co-stimulatory molecule targeted therapy for autoimmune diseases: a review of the newly implemented strategies. Explor Immunol 4(2):189–210

Elhani I, Hentgen V, Quartier P, Melki I, Bader-Meunier B, Neven B, Koné-Paut I, Rossi-Semerano L, Meinzer U, Grateau G, Savey L, Georgin-Lavialle S (2024) *POS1184* the transition to adult care in autoinflammatory diseases: a cohort of 112 French patients. Ann Rheum Dis 83:552–553

Fang J, Lin L, Xiao X, Tian J, Zhang M, Saw PE (2024) Advances of immune cells in pathogenesis and targeted therapy of systemic lupus erythematosus. J Cell Immunol 6(2):98–112

Felten R, Mertz P, Sebbag E, Scherlinger M, Arnaud L (2023) Novel therapeutic strategies for autoimmune and inflammatory rheumatic diseases. Drug Discov Today 28(7):103612

Gao X, He J, Sun X, Li F (2022) Dynamically modeling the effective range of IL-2 dosage in the treatment of systemic lupus erythematosus. iScience 25(9):104911. https://doi.org/10.1016/j.isci.2022.104911

Gao X, Lin X, Wang Q, Chen J (2024) Artemisinins: promising drug candidates for the treatment of autoimmune diseases. Med Res Rev 44(2):867–891. https://doi.org/10.1002/med.22001

Garces S, Karis E, Merrill JT, Askanase AD, Kalunian K, Mo M, Milmont CE (2023) Improving resource utilisation in SLE drug development through innovative trial design. Lupus Sci Med 10(2):e000890

Gardell JL, Crane C, Bowser J, Fasnacht R, Julien S, Maurer ME, Templeton M, Therriault J, Yang SJ, Swiderek KM (2022) Bispecific CD8 Treg modulators regulate a novel regulatory CD8 T cell network and eliminate pathogenic CD4 T cells in live cell co-culture system. J Immunol 208(1_Supplement):174.16. https://doi.org/10.4049/jimmunol.208.Supp.174.16

Gatto M, Depascale R, Stefanski AL, Schrezenmeier E, Dörner T (2023) Translational implications of newly characterized pathogenic pathways in systemic lupus erythematosus. Best Pract Res Clin Rheumatol 37(4):101864

Gharagozloo M, Majewski S, Foldvari M (2015) Therapeutic applications of nanomedicine in autoimmune diseases: from immunosuppression to tolerance induction. Nanomedicine 11(4):1003–1018

Gómez-Bañuelos E, Konig MF, Andrade F (2022) Microbial pathways to subvert host immunity generate citrullinated neoantigens targeted in rheumatoid arthritis. Curr Opin Struct Biol 75:102423

Grammer AC, Lipsky PE (2017) Drug repositioning strategies for the identification of novel therapies for rheumatic autoimmune inflammatory diseases. Rheum Dis Clin North Am 43(3):467–480

Greco R, Alexander T, Del Papa N, Müller F, Saccardi R, Sanchez-Guijo F, Schett G, Sharrack B, Snowden JA, Tarte K et al (2024) Innovative cellular therapies for autoimmune diseases: expert-based position statement and clinical practice recommendations from the EBMT practice harmonization and guidelines committee. eClinicalMedicine 69:102476

Hahn BH, King JK (2014) Treatment of autoimmune disease: established therapies. In: Rose NR, Mackay IR (eds) The autoimmune diseases, 5th edn. Academic Press, pp 1209–1220. https://doi.org/10.1016/B978-0-12-384929-8.00080-0

Hardtke-Wolenski M, Landwehr-Kenzel S (2024) Tipping the balance in autoimmunity: are regulatory T cells the cause, the cure, or both? Molecular and Cellular Pediatrics 11(1):3. https://doi.org/10.1186/s40348-024-00176-8

Henschke A, Desborough J, Parkinson A, Brunoro C, Fanning V, Lueck C, Brew-Sam N, Brüstle A, Drew J, Chisholm K (2021) Personalizing medicine and technologies to address the experiences and needs of people with multiple sclerosis. Journal of Personalized Medicine 11(8):791. https://doi.org/10.3390/jpm11080791

Her M, Kavanaugh A (2016) Alterations in immune function with biologic therapies for autoimmune disease. J Allergy Clin Immunol 137(1):19–27

Honing DY, Luiten RM, Matos TR (2024) Regulatory T cell dysfunction in autoimmune diseases. Int J Mol Sci 25(13):7171. https://doi.org/10.3390/ijms25137171

Hosseini S, Mahmoudi M, Rezaieyazdi Z, Shapouri-Moghaddam A, Hosseinzadeh A, Arab FL, Tabasi NS, Esmaeili S-A (2024) Lupus mice derived mesenchymal stromal cells: beneficial or detrimental on SLE disease outcome. Int Immunopharmacol 126:111306. https://doi.org/10.1016/j.intimp.2023.111306

Houeiss P, Boitard C, Luce S (2022) Preclinical models to evaluate the human response to autoantigen and antigen-specific immunotherapy in human type 1 diabetes. Front Endocrinol 13:883000. https://doi.org/10.3389/fendo.2022.883000

Huang X, Luu LDW, Jia N, Zhu J, Fu J, Xiao F, Liu C, Li S, Shu G, Hou J (2022) Multi-platform omics analysis reveals molecular signatures for pathogenesis and activity of systemic lupus erythematosus. Front Immunol 13:833699

Huang Z, Braunstein Z, Chen J, Wei Y, Rao X, Dong L, Zhong J (2024) Precision medicine in rheumatic diseases: unlocking the potential of antibody-drug conjugates. Pharmacol Rev 76(4):579–598

Hui L, Ziyue Z, Chao L, Bin Y, Aoyu L, Haijing W (2023) Epigenetic regulations in autoimmunity and cancer: from basic science to translational medicine. Eur J Immunol 53(4):2048980

Ibáñez-Cabellos JS, Cosín-Roger J, Seco-Cervera M (2022) Epigenetic in precision medicine in autoimmune and inflammatory-mediated disorders. In: García-Giménez JL (ed) Epigenetics in precision medicine. Academic Press, pp 381–405. https://doi.org/10.1016/B978-0-12-823008-4.00006-8

Ismail FS, Meuth SG, Melzer N (2021) The role of dendritic cells and their interactions in the pathogenesis of antibody-associated autoimmune encephalitis. J Neuroinflammation 18(1):260. https://doi.org/10.1186/s12974-021-02310-z

Iyer VS (2022) Development of immunomodulatory RNA therapeutics for immune disorders. Karolinska Institutet (Sweden). Thesis. https://hdl.handle.net/10616/48123

Jiang Q, Duan J, Van Kaer L, Yang G (2024) The role of myeloid-derived suppressor cells in multiple sclerosis and its animal model. Aging Dis 15(3):1329–1343

Jin Q, Ren F, Song P (2024) Innovate therapeutic targets for autoimmune diseases: insights from proteome-wide mendelian randomization and Bayesian colocalization. Autoimmunity 57(1):2330392

Jung SM, Kim W-U (2022) Targeted immunotherapy for autoimmune disease. Immune Netw 22(1):e9. https://doi.org/10.4110/in.2022.22.e9

Kaliterna D, Barac B (2023) New approach for the treatment of autoimmune diseases. J Appl Cosmetol 41(1):4–11

Kiumarsi A (2024) Multi-omics insights into autoimmune thrombocytopenic purpura. Comprehensive Hematol Stem Cell Res:309–322

Klein K (2023) The role of epigenetics in autoimmune disorders. In: Tollefsbol TO (ed) Handbook of epigenetics, 3rd edn. Academic Press, pp 715–736. https://doi.org/10.1016/B978-0-323-91909-8.00004-9

Konforte D, Diamandis EP, van Venrooij WJ, Lories R, Ward MM (2012) Autoimmune diseases: early diagnosis and new treatment strategies. Clin Chem 58(11):1510–1514

Krienke C, Kolb L, Diken E, Streuber M, Kirchhoff S, Bukur T, Akilli-Öztürk Ö, Kranz LM, Berger H, Petschenka J (2021) A noninflammatory mRNA vaccine for treatment of experimental autoimmune encephalomyelitis. Science 371(6525):145–153

Kumar A, Das JK, Peng H-Y, Wang L, Ren Y, Xiong X, Song J (2021) Current development in iPSC-based therapy for autoimmune diseases. In: Birbrair (ed) Recent advances in iPSCs for therapy, Volume 3. Academic Press, pp 315–338. https://doi.org/10.1016/B978-0-12-822229-4.00001-2

LaBere B, Chu A, Platt CD, Chou J (2023) The integration of patient-reported quality of life and systemic biomarkers in patients with immune dysregulation. Research Square (Preprint) rs.3.rs-3270389. [Version 1] https://doi.org/10.21203/rs.3.rs-3270389/v1

Laigle L, Chadli L, Moingeon P (2023) Biomarker-driven development of new therapies for autoimmune diseases: current status and future promises. Expert Rev Clin Immunol 19(3):305–314

Lamendour L, Gilotin M, Deluce-Kakwata Nkor N, Lakhrif Z, Meley D, Poupon A, Laboute T, Di Tommaso A, Pin J-J, Mulleman D (2024) Bispecific antibodies tethering innate receptors induce human tolerant-dendritic cells and regulatory T cells. Front Immunol 15:1369117. https://doi.org/10.3389/fimmu.2024.1369117

Lee KH, Ahn BS, Cha D, Jang WW, Choi E, Park S, Park JH, Oh J, Park H, Park JH (2020) Understanding the immunopathogenesis of autoimmune diseases by animal studies using gene modulation: a comprehensive review. Autoimmun Rev 19(3):102469

Lee MH, Shin JI, Yang JW, Lee KH, Cha DH, Hong JB, Park Y, Choi E, Tizaoui K, Koyanagi A (2022) Genome editing using CRISPR-Cas9 and autoimmune diseases: a comprehensive review. Int J Mol Sci 23(3):1337. https://doi.org/10.3390/ijms23031337

Leighton P, Howells L, Bates J, Gibbons A, Cannings-John R, Collier F, Evans J, Harris C, Hood K, Howes R (2023) Research priorities in the management of hidradenitis suppurativa. Br J Dermatol 189(3):343–345

Liu X, Liu M, Zhao M, Li P, Gao C, Fan X, Lu Q, Chen X (2023) Fecal microbiota transplantation for the management of autoimmune diseases: potential mechanisms and challenges. J Autoimmun 103109. https://doi.org/10.1016/j.jaut.2023.103109

Liu Y, Dong M, Chu Y, Zhou L, You Y, Pang X, Yang S, Zhang L, Chen L, Zhu L (2024a) Dawn of CAR-T cell therapy in autoimmune diseases. Chin Med J 137(10):1140–1150

Liu Y, Tang S, Feng Y, Xue B, Cheng C, Su Y, Wei W, Zhang L, Huang Z, Shi X et al (2024b) Alteration in gut microbiota is associated with immune imbalance in graves' disease. Front Cell Infect Microbiol 14:1349397. https://doi.org/10.3389/fcimb.2024.1349397

Lordick F, Merz M, Büch E, Aigner A (2023) Antibody-drug conjugates as a targeted therapeutic approach across entities in oncology. Dtsch Arztebl Int 120(19):329–336

Macaron G, Willis MA (2021) Concepts of immune therapy and disease management. In: Piquet AL, Alvarez E (eds) Neuroimmunology: multiple sclerosis, autoimmune neurology and related diseases. Springer, pp 59–71. https://doi.org/10.1007/978-3-030-61883-4_5

Martin J, Hollowood Z, Chorlton J, Dyer C, Marelli-Berg F (2024) Modulating regulatory T cell migration in the treatment of autoimmunity and autoinflammation. Curr Opin Pharmacol 77:102466. https://doi.org/10.1016/j.coph.2024.102466

Mease PJ, Strand V, Furst D, Siegel E, Mcilraith M, Husni ME, Hay MC (2022) Are current patient reported outcomes tools optimized to capture the entire patient experience? Ann Rheum Dis 81(Suppl 1):1610. Scientific Abstracts (AB0966). https://doi.org/10.1136/annrheumdis-2022-eular.5325

Mehta JM, Hiremath SC, Chilimba C, Ghasemi A, Weaver JD (2024) Translation of cell therapies to treat autoimmune disorders. Adv Drug Deliv Rev 205:115161. https://doi.org/10.1016/j.addr.2023.115161

Mizui M, Kono M (2024) Novel therapeutic strategies targeting abnormal T-cell signaling in systemic lupus erythematosus. Clin Immunol 262:110182

Mo C, Bi J, Li S, Lin Y, Yuan P, Liu Z, Jia B, Xu S (2024) The influence and therapeutic effect of microbiota in systemic lupus erythematosus. Microbiol Res 281:127613

Nandal K, Kaur M, Rathee J, Sharma A (2023) Bioactive compounds of *Ganoderma lucidum* and their therapeutic applications. Int J Multidiscip Res 5(3): May-June 2023. https://doi.org/10.36948/ijfmr.2023.v05i03.3982

Nandal K, Rathee J, Dahiya P, Sharma A, Kaur M (2024) Navigating inflammatory bowel disease (IBD): insights into microbial dynamics, dietary interventions and probiotic strategies for gut health. J Exp Zool India 27(2):1493–1508. https://doi.org/10.51470/jez.2024.27.2.1493

Nasonov EL, Aleksandrova EN, Avdeeva AS, Rubtsov YP (2014) T-regulatory cells in rheumatoid arthritis. Rheumatol Sci Pract 52(4):430–437. (In Russ.). https://doi.org/10.14412/1995-4484-2014-430-437

Nayak RR, Orellana DA (2024) The impact of the human gut microbiome on the treatment of autoimmune disease. Immunol Rev 325(1):107–130. https://doi.org/10.1111/imr.13358

Nicolette C, DeBenedette M, Horvatinovich J (2018) Bispecific antibodies that modulate TLR-4 signaling and uses thereof. Canada Patents WO/2018/085617 A1, 11 May 2018

Nikolakis D, Garantziotis P, Sentis G, Fanouriakis A, Bertsias G, Frangou E, Nikolopoulos D, Banos A, Boumpas D (2023) Restoration of aberrant gene expression of monocytes in systemic lupus erythematosus via a combined transcriptome-reversal and network-based drug repurposing strategy. Ann Rheum Dis 82(Suppl 1):1246. Scientific Abstracts (AB0134). https://doi.org/10.1136/annrheumdis-2023-eular.2787

Nikolopoulos D, Lourenço MH, Depascale R, Triantafyllias K, Parodis I (2024) Evolving concepts in treat-to-target strategies for systemic lupus erythematosus. Mediterr J Rheumatol 35(Suppl 2):328–341

Ohno R, Nakamura A (2024) Advancing autoimmune rheumatic disease treatment: CAR-T cell therapies-evidence, safety, and future directions. Semin Arthritis Rheum 67:152479. https://doi.org/10.1016/j.semarthrit.2024.152479

Oktem EK, Yazar M (2022) Drug repositioning identifies six drug candidates for systemic autoimmune diseases by integrative analyses of transcriptomes from scleroderma, systemic lupus erythematosus, and Sjogren's syndrome. OMICS J Integr Biol 26(12):683–693

Ozhiganova NV, Belkovets AV, Kruchinina MV (2023) Early diagnosis of autoimmune gastritis. Exp Clin Gastroenterol 4:12–18

Pandey M, Gorain B, Kushwaha SP, Gulati N, Mayuren J, Butt AM (2023) Antibody–drug conjugate: emerging trend for targeted treatment. In: Madan J, Baldi A, Chaudhary M, Chopra N (eds) Polymer-drug conjugates: linker chemistry, protocols and applications. Academic Press, pp 347–387. https://doi.org/10.1016/B978-0-323-91663-9.00013-8

Pandey SP, Bhaskar R, Han SS, Narayanan KB (2024) Autoimmune responses and therapeutic interventions for systemic lupus erythematosus: a comprehensive review. Endocr Metab Immune Disord Drug Targets 24(5):499–518

Park Y, Kwok S-K (2022) Recent advances in cell therapeutics for systemic autoimmune diseases. Immune Netw 22(1):e10. https://doi.org/10.4110/in.2022.22.e10

Parodis I, Studenic P (2022) Patient-reported outcomes in systemic lupus erythematosus. can lupus patients take the driver's seat in their disease monitoring? J Clin Med 11(2):340

Patel KB, Luhar S, Srivastava DN (2024) Early diagnosis of autoimmune diseases through electrochemical biosensing using a modified plastic chip electrode. Anal Methods 16(29):4971–4980

Pathak L (2023) Personalized treatment for multiple sclerosis: the role of precision medicine. Neurology Letters 2(1):30–34

Peine KJ, Chen N, Bachelder EM, Ainslie KM (2019) Drug delivery strategies for tolerogenic therapy for autoimmune diseases in an antigen-specific manner. In: Chronic illness and long-term care: breakthroughs in research and practice, volume 1. IGI Global, pp 112–140. https://doi.org/10.4018/978-1-5225-7122-3.ch007

Perico L, Casiraghi F, Sônego F, Todeschini M, Corna D, Cerullo D, Pezzotta A, Isnard-Petit P, Faravelli S, Forneris F, et al (2023) A novel targeted immunotherapy for autoreactive B cell depletion in autoimmune diseases. Preprint from Research Square. https://doi.org/10.21203/rs.3.rs-3042909/v2

Peshkova MA, Korneev AA, Koteneva PI, Kosheleva NV, Timashev PS (2024) Prospects for microbiome modulation in autoimmune diseases: a literature review. Sechenov Med J 15(1):4–19. https://doi.org/10.47093/2218-7332.2024.15.1.4-19

Pickens CJ, Christopher MA, Leon MA, Pressnall MM, Johnson SN, Thati S, Sullivan BP, Berkland C (2019) Antigen-drug conjugates as a novel therapeutic class for the treatment of antigen-specific autoimmune disorders. Mol Pharm 16(6):2452–2461

Pignatti E, Maccaferri M, Pisciotta A, Carnevale G, Salvarani C (2024) A comprehensive review on the role of mesenchymal stromal/stem cells in the management of rheumatoid arthritis. Expert Rev Clin Immunol 20(5):463–484

Polis B, Cuda CM, Putterman C (2024) Animal models of neuropsychiatric systemic lupus erythematosus: deciphering the complexity and guiding therapeutic development. Autoimmunity 57(1):2330387

Pöysti S, Silojärvi S, Brodnicki TC, Catterall T, Liu X, Mackin L, Luster AD, Kay TWH, Christen U, Thomas HE, Hanninen A (2023) Gut dysbiosis promotes islet-autoimmunity by increasing T-cell attraction in islets via CXCL10 chemokine. J Autoimmun 140:103090

Pratsidou-Gertsi P (2018) Transition of pediatric patients with an auto-inflammatory disease: an alternative version of the Daedaulus and Icarus myth. Mediterr J Rheumatol 29(3):156–162
Puccetti M, Costantini C, Schoubben A, Giovagnoli S, Ricci M (2024) Strategies and delivery systems for cell-based therapy in autoimmunity. Front Drug Deliv 4:1436842
Qamar T, Mukherjee S (2021) Genetic approaches for the diagnosis and treatment of rheumatoid arthritis through personalized medicine. Gene Rep 23:101173
Quattrocchio G, Fervenza F, Roccatello D (2016) Conventional treatment of systemic lupus erythematosus. In: Roccatello D, Emmi L (eds) Connective tissue disease. Rare diseases of the immune system. Springer, pp 213–238. https://doi.org/10.1007/978-3-319-24535-5_16
Radygina TV, Kuptsova DG, Petrichuk SV, Potapov AS, Murashkin NN, Abdullaeva LM, Kurbatova OV, Tsvetkova VS (2023) Content of myeloid-derived suppressor cells in autoimmune diseases in children. Russ J Immunol 26(3):381–388
Rafeek RAM, Ketheesan N, Good MF, Pandey M, Lepletier A (2024) Immunotherapeutic strategy to prevent progression and complications of acute rheumatic fever. bioRxiv. https://doi.org/10.1101/2024.08.11.607519
Rajalingam A, Ganjiwale A (2024) Identification of common genetic factors and immune-related pathways associating more than two autoimmune disorders: implications on risk, diagnosis, and treatment. Genomics Informatics 22(1):10. https://doi.org/10.1186/s44342-024-00004-5
Ramo-Tello C (2022) Personalized diagnosis and therapy for multiple sclerosis. J Pers Mede 12(6):1017. https://doi.org/10.3390/jpm12061017
Rezaei N, Yazdanpanah N (2022) Introduction on therapeutic opportunities for autoimmunity. In: Rezaei N (ed) Translational autoimmunity: treatment of autoimmune diseases, Volume 2. Academic Press, pp 1–11
Rosato E, Pisarri S, Salsano F (2010) Current strategies for the treatment of autoimmune diseases. J Biol Regul Homeost Agents 24(3):251–259
Ruiz-Cerdá ML, Irurzun-Arana I, González-Garcia I, Hu C, Zhou H, Vermeulen A, Trocóniz IF, Gómez-Mantilla JD (2016) Towards patient stratification and treatment in the autoimmune disease lupus erythematosus using a systems pharmacology approach. Eur J Pharm Sci 94:46–58
Sadeghpour Heravi F (2024) Gut microbiota and autoimmune diseases: mechanisms, treatment, challenges, and future recommendations. Curr Clin Microbiol Rep 11(1):18–33
Sai A, Shetty GB, Shetty P, Nanjeshgowda HL (2024) Influence of gut microbiota on autoimmunity: a narrative review. Brain Behav Immunity Integr 5:100046
Saiki C, Kashiwado Y, Yokoyama T, Ayano M, Imabayashi K, Kawano S, Higashioka K, Kimoto Y, Fukata M, Mitoma H et al (2022) Successful transition from intravenous epoprostenol to oral selexipag and inhaled iloprost in a case of severe pulmonary arterial hypertension associated with systemic lupus erythematosus. Mod Rheumatol Case Rep 6(2):183–188
Sandhu G, Thelma BK (2022) New druggable targets for rheumatoid arthritis based on insights from synovial biology. Front Immunol 13:834247
Sarfraz A, Fatima SQ-U-A, Shehroz M, Ahmad I, Zaman A, Nishan U, Tayyab M, Sheheryar, Moura AA, Ullah R, Ali EA, Shah M (2024) Decrypting the multi-genome data for chimeric vaccine designing against the antibiotic resistant Yersinia pestis. Int Immunopharmacol 132:111952
Sármay G (2021) Biologia Futura: emerging antigen-specific therapies for autoimmune diseases. Biologia Futura 72(1):15–24
Saurin S, Meineck M, Erkel G, Opatz T, Weinmann-Menke J, Pautz A (2022) Drug candidates for autoimmune diseases. Pharmaceuticals (Basel) 15(5):503
Sebastiani GD, Prevete I, Iuliano A, Minisola G (2016) The importance of an early diagnosis in systemic lupus erythematosus. Israel Med Assoc J IMAJ 18(3–4):212–215
Segués A, Huang S, Sijts A, Berraondo P, Zaiss DM (2022) Opportunities and challenges of bi-specific antibodies. Int Rev Cell Mol Biol 369:45–70
Seida I, Al Shawaf M, Mahroum N (2024) Fecal microbiota transplantation in autoimmune diseases–an extensive paper on a pathogenetic therapy. Autoimmun Rev 23(7–8):103541

Shakya AK, Nandakumar KS (2018) Antigen-specific tolerization and targeted delivery as therapeutic strategies for autoimmune diseases. Trends Biotechnol 36(7):686–699

Shandil RK, Dhup S, Narayanan S (2022) Evaluation of the therapeutic potential of mesenchymal stem cells (MSCs) in preclinical models of autoimmune diseases. Stem Cells Int 2022(1):6379161

Shinde SS, Mate GD, Mali SS, Naikwadi ND, Abuzar M, Sarkate AP (2024) Systems biology approaches for autoimmune diseases. In: Joshi S, Ray RR, Nag M, Lahiri D (eds) Systems biology approaches: prevention, diagnosis, and understanding mechanisms of complex diseases. Springer, pp 299–319

Shukla AK, Yadav VK, Yadav VP, Mishra MK, Maurya R, Mishra RK, Dubey RS (2024) Autoimmune diseases in women and treatment options. In: Mishra S, Malviya R, Ojha S, Pandey M (eds) Women's health: a comprehensive guide to common health issues in women. Bentham, pp 247–260

Singh S (2024) Gene editing technologies: CRISPR/Cas9 and beyond for genetic disease therapy and research. Universal Research Reports 11(3):1–7. https://doi.org/10.36676/urr.v11.i3.1280

Stanway JA, Isaacs JD (2020) Tolerance-inducing medicines in autoimmunity: rheumatology and beyond. The Lancet Rheumatology 2(9):e565–e575

Su Q-Y, Li H-C, Jiang X-J, Jiang Z-Q, Zhang Y, Zhang H-Y, Zhang S-X (2024a) Exploring the therapeutic potential of regulatory T cell in rheumatoid arthritis: insights into subsets, markers, and signaling pathways. Biomed Pharmacother 174:116440

Su Q, Wang X, Li Y, Zhang J, Bai C, Wang X, Yang L, Zhang J, Zhang S-X (2024b) Efficacy, safety and the lymphocyte subset changes of low-dose IL-2 in patients with autoimmune rheumatic diseases: a systematic review and meta-analysis. Rheumatol Ther 11(1):79–96

Swain HN, Boyce PD, Bromet BA, Barozinksy K, Hance L, Shields D, Olbricht GR, Semon JA (2024) Mesenchymal stem cells in autoimmune disease: a systematic review and meta-analysis of pre-clinical studies. Biochimie 223:54–73

Taubmann J, Müller F, Fagni F, Bergmann C, Tur C, Bucci L, Völkl S, Aigner M, Hagen M, Grieshaber-Bouyer R et al (2024) Long-term safety and efficacy of CAR-T cell treatment in severe and treatment refractory autoimmune disease. Ann Rheum Dis 83(Suppl 1):137. Scientific Abstracts (OP0027). https://doi.org/10.1136/annrheumdis-2024-eular.1478

Teng X, Cornaby C, Li W, Morel L (2019) Metabolic regulation of pathogenic autoimmunity: therapeutic targeting. Curr Opin Immunol 61:10–16

Thatte AS, Billingsley MM, Weissman D, Melamed JR, Mitchell MJ (2024) Emerging strategies for nanomedicine in autoimmunity. Adv Drug Deliv Rev 207:115194

Tsigalou C, Konstantinidis T, Aloizou A-M, Bezirtzoglou E, Tsakris A (2023) Future therapeutic prospects in dealing with autoimmune diseases: treatment based on the microbiome model. In: Dwivedi MK, Sankaranarayanan A, Kemp EH, Shoenfeld Y (eds) Role of microorganisms in pathogenesis and management of autoimmune diseases: volume II: kidney, central nervous system, eye, blood, blood vessels & bowel. Springer, pp 489–520

Tsogka A, Kitsos DK, Stavrogianni K, Giannopapas V, Chasiotis A, Christouli N, Tsivgoulis G, Tzartos JS, Giannopoulos S (2023) Modulating the gut microbiome in multiple sclerosis management: a systematic review of current interventions. J Clin Med 12(24):7610

Tyndall A (2012) Modern therapeutic strategies for autoimmune diseases. Curr Pharm Des 18(29):4508–4512

Underwood DJ, Bettencourt J, Jawad Z (2022) The manufacturing considerations of bispecific antibodies. Expert Opin Biol Ther 22(8):1043–1065

Velikova T, Miteva D, Kokudeva M, Vasilev GH, Monov S, Shumnalieva R (2024) In the pursuit of precision: novel target therapies revolutionizing SLE care. Rheumato 4(3):120–136

Vitale AM, Paladino L, Caruso Bavisotto C, Barone R, Rappa F, Conway de Macario E, Cappello F, Macario AJL, Marino Gammazza A (2024) Interplay between the chaperone system and gut microbiota dysbiosis in systemic lupus erythematosus pathogenesis: is molecular mimicry the missing link between those two factors? Int J Mol Sci 25(11):5608

Wallace BI, Cooney L, Fox DA (2024) New molecular targets in the treatment of rheumatoid arthritis. Curr Opin Rheumatol 36(3):235–240

Xiang Y, Zhang M, Jiang D, Su Q, Shi J (2023) The role of inflammation in autoimmune disease: a therapeutic target. Front Immunol 14:1267091

Xiao F, Rui K, Shi X, Wu H, Cai X, Lui KO, Lu Q, Ballestar E, Tian J, Zou H, Lu L (2022) Epigenetic regulation of B cells and its role in autoimmune pathogenesis. Cell Mol Immunol 19(11):1215–1234

Xiong X, Zhang Y, Wen Y (2024) Diverse functions of myeloid-derived suppressor cells in autoimmune diseases. Immunol Res 72(1):34–49

Xu M, Du R, Xing W, Chen X, Wan J, Wang S, Xiong L, Nandakumar KS, Holmdahl R, Geng H (2023) Platelets derived citrullinated proteins and microparticles are potential autoantibodies ACPA targets in RA patients. Front Immunol 14:1084283

Yahsi B, Palaz F, Dincer P (2024) Applications of CRISPR epigenome editors in tumor immunology and autoimmunity. ACS Synth Biol 13(2):413–427

Yazdanpanah N, Rezaei N (2023) Introduction on challenges for autoimmune diseases. In: Rezaei N (ed) Translational autoimmunity: challenges for autoimmune diseases volume 5 in translational immunology. Academic Press, pp 1–8

Yu L, Fu Y, Miao R, Cao J, Zhang F, Liu L, Mei L, Ou M (2024a) Regulatory T cells and their derived cell pharmaceuticals as emerging therapeutics against autoimmune diseases. Adv Funct Mater 34(39):2405133

Yu X, Shang H, Shen X, Zhang J, Chang T, Ruan Z, Jia Y, Gao F (2024b) Application and research progress of CAR-T cell therapy in autoimmune diseases. Gene Protein Dis 3(2):2851. https://doi.org/10.36922/gpd.2851

Zanghì A, Di Filippo PS, Avolio C, D'Amico E (2024) Myeloid-derived suppressor cells and multiple sclerosis. Curr Neuropharmacol 23(1):36–57

Zeng L, Yang K, He Q, Zhu X, Long Z, Wu Y, Chen J, Li Y, Zeng J, Cui G et al (2024) Efficacy and safety of gut microbiota-based therapies in autoimmune and rheumatic diseases: a systematic review and meta-analysis of 80 randomized controlled trials. BMC Med 22(1):110

Zhanel GG, Keynan R, Keynan Y, Karlowsky JA (2023) The role of fecal microbiota transplantation (FMT) in treating patients with multiple sclerosis. Expert Rev Neurother 23(10):921–930

Zhang Z, Yao Z, Zhang R (2018) Noncoding RNA-targeted therapeutics in autoimmune diseases: from bench to bedside. In: Zhang R (ed) The epigenetics of autoimmunity, volume 5 in translational epigenetics, pp 359–386

Zhao Q (2020) Bispecific antibodies for autoimmune and inflammatory diseases: clinical progress to date. BioDrugs 34(2):111–119

Zhou F, Ben Y, Jiang H, Tan S, Mu G, Zha Z, Dong S, Huang S, Zhou Y, Jin Y, Chiu ML (2024) A novel dual-fc bispecific antibody with enhanced fc effector function. Biochemistry 63(8):958–968

Chapter 8
Molecular and Cellular Mechanisms Underlying Autoimmune Pathologies

Hahyeong Jang and Siyoung Yang

8.1 Introduction

Autoimmune diseases are conditions where the immune system mistakenly targets tissues of the body (Davidson and Diamond 2001), with an increasing prevalence observed globally (Conrad et al. 2023). These disorders range from systemic diseases such as systemic lupus erythematosus (SLE) to organ-specific conditions such as type 1 diabetes mellitus. Despite their diversity, they share common underlying mechanisms involving the loss of immune tolerance, which distinguishes the self from the nonself (Rosenblum et al. 2015).

The mechanisms underlying autoimmunity involve genetic predispositions, environmental triggers, and dysregulated immune responses (Rosenblum et al. 2015; Theofilopoulos et al. 2017). Genetic factors, particularly HLA alleles, influence immune regulation (Farh et al. 2015; Hedrich and Tsokos 2011), whereas infections, dietary antigens, and chemical exposures are environmental triggers (Pollard 2015). For example, molecular mimicry by pathogens such as the Epstein-Barr virus (EBV) can initiate autoimmune responses in diseases such as multiple sclerosis (Lanz et al. 2022; Welberg 2022).

Central and peripheral immune tolerance is essential for the prevention of autoimmunity. Central tolerance eliminates self-reactive lymphocytes in the thymus and bone marrow, while peripheral tolerance mechanisms, including regulatory T cells (Tregs) and immune checkpoints such as CTLA-4, suppress escaped autoreactive cells. Breakdown of these processes leads to autoimmunity (Theofilopoulos et al. 2017).

H. Jang · S. Yang (✉)
Department of Biological Sciences, Sungkyunkwan University, Suwon, Republic of Korea
e-mail: jhh7071@g.skku.edu; yangsy@skku.edu

Y.-O. Son, M. Ghosh (eds.), *Cutting-Edge Strategies in Drug Delivery and Immunotherapy for Autoimmune Disorders*,
https://doi.org/10.1007/978-981-95-0520-3_8

Recent advances in our understanding of molecular pathways such as NF-κB, JAK-STAT, and PI3K-Akt have led to the development of targeted therapies (Fruman et al. 2017; Liu et al. 2017; O'Shea et al. 2015). These include Janus kinase (JAK) inhibitors and biologics that target cytokine signaling. Precision medicine, which integrates genetic and proteomic data, offers personalized therapeutic strategies and highlights the potential of technologies such as CRISPR and single-cell sequencing to transform autoimmune disease management (Zhang et al. 2024a; Argyriou et al. 2022).

This chapter aims to elucidate the molecular and cellular mechanisms driving autoimmunity, while exploring advances in diagnostics and therapies that improve patient outcomes.

8.2 Part 1: Basic Concepts in Autoimmunity

8.2.1 Fundamentals of Autoimmune Pathogenesis

Autoimmunity occurs when the mechanisms of immune tolerance fail. Immune tolerance comprises two critical systems.

- *Central Tolerance*: This occurs in primary lymphoid organs, such as the thymus and bone marrow, where self-reactive lymphocytes are eliminated (Sakaguchi et al. 2008). In the thymus, the autoimmune regulator gene (AIRE) ensures the expression of tissue-specific antigens to facilitate negative selection of autoreactive T cells (Bluestone et al. 2015). Failure of this process can result in the survival of autoreactive clones that escape to the periphery. The role of medullary thymic epithelial cells (mTECs) in presenting self-antigens and facilitating clonal deletion is an area of active investigation with implications for understanding autoimmune susceptibility (Meyer et al. 2016).
- *Peripheral Tolerance*: Peripheral tolerance mechanisms, including Tregs and immune checkpoints (e.g., CTLA-4, PD-1), suppress autoreactive lymphocytes that escape central tolerance (Bluestone et al. 2015). Tregs play a vital role in maintaining immune homeostasis by suppressing the activity of effector T cells (Josefowicz et al. 2012). Dysfunctions in these checkpoints contribute to various autoimmune diseases such as multiple sclerosis (MS) and rheumatoid arthritis (RA). Additional mechanisms, including clonal anergy and deletion mediated by Fas/FasL interactions, are integral to the maintenance of tolerance (Sakaguchi et al. 2008). The key mechanisms, locations, and distinguishing features of central and peripheral tolerance are summarized in Table 8.1.
- *Autoantigens and Their Role*: Autoantigens, derived from the body's own tissues, are erroneously presented by antigen-presenting cells (APCs) during autoimmune responses (Sakaguchi et al. 2008). For instance, posttranslational modifications of self-proteins, such as citrullination, increase immunogenicity in RA. Additionally, the presentation of cryptic or neoantigens during cellular

Table 8.1 Overview of immune tolerance mechanisms: A comparative table highlighting the characteristics, locations, and key features of central and peripheral tolerance systems in the regulation of autoimmunity

Tolerance type	Location	Mechanisms	Key features
Central tolerance	Thymus, bone marrow	Negative selection of self-reactive lymphocytes; AIRE gene expression in thymus (Meyer et al. 2016)	Eliminates self-reactive lymphocytes during development (Sakaguchi et al. 2008)
Peripheral tolerance	Peripheral tissues (e.g., lymph nodes, spleen)	Regulatory T cells, anergy induction, clonal deletion (e.g., Fas/FasL pathway) (Sakaguchi et al. 2008; Josefowicz et al. 2012)	Suppresses escaped self-reactive lymphocytes, prevents autoimmunity (Bluestone et al. 2015)

damage or oxidative stress has been linked to the initiation of autoimmune diseases such as SLE (Aas-Hanssen et al. 2015).

- *Effector Roles of T and B Cells*: T helper cells, particularly the Th17 subset, drive inflammation through interleukin-17 (IL-17), whereas B cells produce autoantibodies (Sakaguchi et al. 2008). Formation of autoantibodies, such as antinuclear antibodies in SLE, amplifies tissue damage via immune complex deposition and complement activation (Aas-Hanssen et al. 2015). $CD8^+$ cytotoxic T cells directly attack self-antigen–expressing cells and exacerbate tissue injury. The interplay between follicular helper T cells (Tfh) and B cells in germinal centers perpetuates autoantibody production in multiple autoimmune diseases (Gensous et al. 2018).

8.2.2 Genetic and Environmental Factors

- *Genetic Factors*: Genome-wide association studies (GWAS) have revealed a strong correlation between specific HLA alleles and autoimmune susceptibility. For example, HLA-DR4 is associated with RA, whereas HLA-DQ2/DQ8 is associated with celiac disease (Schuppan et al. 2009). Non-HLA genes such as PTPN22, CTLA4, and STAT4 also influence immune regulation by modulating T and B cell signaling pathways (Cotsapas and Hafler 2013). Furthermore, epistatic interactions, such as gene–gene interactions, are emerging as critical contributors to autoimmune risk, with synergistic effects observed between HLA alleles and non-HLA genetic loci (Farh et al. 2015).
- *Environmental Triggers*: Environmental factors, including infection, diet, and chemical exposure, interact with genetic predispositions to trigger autoimmunity (Pollard 2015). Molecular mimicry, in which microbial antigens resemble self-antigens, is exemplified in conditions such as rheumatic fever. Chronic viral infections, particularly EBV, have been implicated in multiple sclerosis and SLE by altering host gene expression and promoting inflammatory cascades (Lanz

et al. 2022; Welberg 2022; Aas-Hanssen et al. 2015). Additionally, gut microbiota imbalances (dysbiosis) have been linked to autoimmune diseases (Pollard 2015), such as type 1 diabetes (Mariño et al. 2017) and inflammatory bowel disease. Dietary components, such as gluten, in celiac disease, provide direct antigenic triggers (Schuppan et al. 2009), whereas xenobiotics and heavy metals act indirectly by altering immune signaling pathways or epigenetic marks.

8.3 Part 2: Molecular Mechanisms in Autoimmune Diseases

8.3.1 Key Molecular Pathways

- *NF-κB*: Central to inflammatory responses, this pathway mediates cytokine production and immune cell survival (Li and Verma 2002). Dysregulation promotes chronic inflammation, as observed in RA. NF-κB signaling also enhances the survival of autoreactive B cells, perpetuating autoantibody production (Zhang et al. 2017). Targeting NF-κB's upstream activators, such as toll-like receptors (TLRs) and tumor necrosis factor receptors, has shown potential in preclinical models (Liu et al. 2017).
- *JAK-STAT*: Cytokine signaling through JAK-STAT regulates immune cell differentiation. Aberrant activation contributes to Th17 polarization and impairs regulatory T-cell development, thereby exacerbating autoimmunity. Therapeutic agents targeting JAK, such as tofacitinib and baricitinib, have shown significant efficacy against RA and ulcerative colitis (Philips et al. 2022). Research has focused on the development of selective inhibitors of STAT3, a key driver of Th17 differentiation, to minimize the side effects associated with broader JAK inhibitors (O'Shea et al. 2015).
- *PI3K-Akt*: This enhances lymphocyte survival and proliferation (Fruman et al. 2017). Dysregulated PI3K-Akt signaling contributes to the persistence of autoreactive immune cells and tissue damage. In autoimmune diseases, the hyperactivation of this pathway sustains the survival of autoreactive T and B cells, thereby amplifying the inflammatory response (Okkenhaug et al. 2014). Therapeutic efforts, including selective inhibitors, such as idelalisib, show promise for treating lupus and other conditions. Dual inhibitors targeting PI3K and mTOR offer the potential for a more comprehensive immune regulation. Figure 8.1 highlights the impact of PI3K-Akt dysregulation in autoimmune pathology, showing how the loss of PTEN-mediated inhibition can lead to uncontrolled immune activation and chronic inflammation.

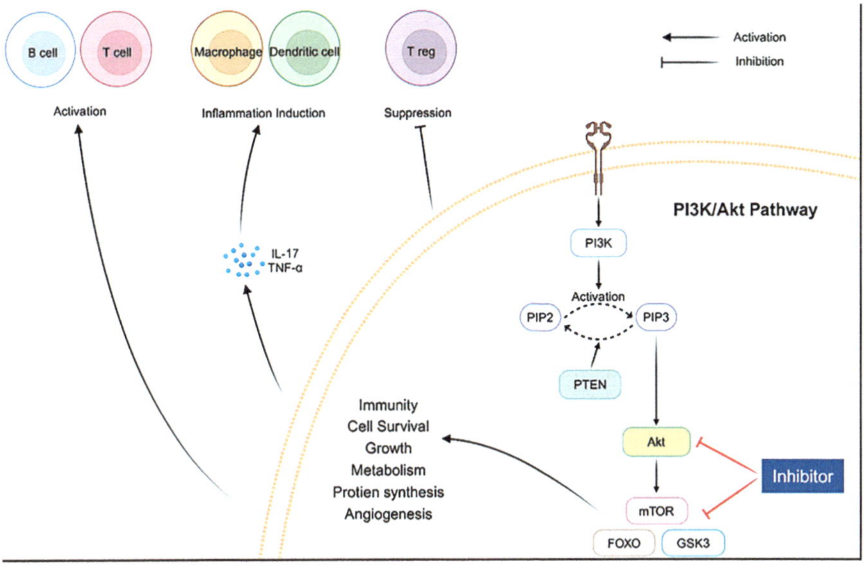

Fig. 8.1 Specific receptors, including RTKs (e.g., EGFR, FGFR, and VEGFR), BCRs, and GPCRs, initiate the activation of PI3K[1]. Once activated, PI3K phosphorylates PIP2 to produce PIP3, leading to the recruitment and activation of Akt. Activated Akt subsequently regulates key downstream targets such as mTOR and GSK3, which drive processes like cell survival, proliferation, and differentiation (Okkenhaug et al. 2014). PTEN functions as a crucial negative regulator by dephosphorylating PIP3 back to PIP2, thereby inhibiting Akt activation and maintaining signaling balance. Dysregulation of this pathway, including PTEN loss or PI3K overactivation, is implicated in autoimmune diseases, where it contributes to the survival of autoreactive immune cells and chronic inflammation. Therapeutic inhibitors targeting PI3K, Akt, or mTOR have demonstrated potential in mitigating these pathological effects, highlighting the pathway's significance in disease management. RTKs Receptor tyrosine kinases, EGFR epidermal growth factor receptor, FGFR fibroblast growth factor receptor, VEGFR vascular endothelial growth factor receptor, BCRs B-cell receptors, GPCRs G-protein–coupled receptors, PI3K phosphoinositide 3-kinase, PIP2 phosphatidylinositol 4,5-bisphosphate, PIP3 phosphatidylinositol 3,4,5-trisphosphate, PTEN phosphatase and tensin homolog, Akt protein kinase B, mTOR mammalian target of rapamycin, GSK3 glycogen synthase kinase 3

8.3.2 *Epigenetics and Autoimmunity*

Epigenetic modifications include DNA methylation, histone modifications, and noncoding RNAs:

- *DNA Methylation*: Aberrant methylation patterns are hallmarks of autoimmunity (Hedrich and Tsokos 2011). Hypomethylation of immune-related genes in T cells from patients with SLE leads to the hyperactivation and production of proinflammatory cytokines, driving disease progression (Surace and Hedrich 2019). This process disrupts immune homeostasis by altering the expression of key genes involved in immune regulation, including those encoding cytokines and signaling proteins. Epigenetic therapies that target DNA methyltransferases have been explored to reverse these changes. Such approaches aim to restore normal methylation patterns, reestablish immune tolerance, and reduce inflammation. For instance, decitabine, a hypomethylating agent, has shown preclinical efficacy in modulating T-cell activity under autoimmune conditions. Additionally, recent studies have highlighted the role of environmental factors such as diet and pollutants in influencing DNA methylation, suggesting that lifestyle interventions could complement therapeutic strategies (Brown and Wedderburn 2015; Rostami et al. 2024).

- *Histone Modifications*: Increased histone acetylation and methylation at proinflammatory gene loci promote their transcription and enhance the inflammatory response in autoimmune diseases (Hedrich and Tsokos 2011; Mu et al. 2024). Key enzymes, such as histone acetyltransferases (HATs) and histone deacetylases (HDACs), regulate these modifications. Dysregulation of these enzymes has been implicated in diseases such as psoriasis (Surace and Hedrich 2019) and lupus, in which excessive histone acetylation amplifies the expression of inflammatory cytokines. HDAC inhibitors, including vorinostat and trichostatin A, are being investigated as potential treatments for restoring epigenetic balance. These inhibitors can suppress an overactive immune response by downregulating proinflammatory gene expression and promoting regulatory T-cell development (Mu et al. 2024). Emerging research has also highlighted the interplay between histone modifications and noncoding RNAs, which may further modulate chromatin structure and gene expression, offering new therapeutic targets.
- *Noncoding RNAs*: MicroRNAs (miRNAs) and long noncoding RNAs (lncRNAs) play critical roles in regulating immune cell activation, cytokine production, and autoantibody formation (Hedrich and Tsokos 2011; Zhang et al. 2024b). The dysregulation of specific miRNAs, such as miR-155 and miR-146a, has been linked to increased inflammation and disease severity in RA and SLE (Chen et al. 2016). These miRNAs target mRNA transcripts of key inflammatory mediators, thereby altering their stability and translation. For example, overexpression of miR-155 enhances Th17 cell differentiation, promoting inflammation, while reduced levels of miR-146a impair the negative feedback loop controlling NF-κB signaling. LncRNAs such as NEAT1 and MALAT1 modulate chromatin accessibility and gene transcription, further influencing immune responses (Wu et al. 2015). Therapeutic strategies aimed at restoring normal levels of these noncoding RNAs, including miRNA mimics and inhibitors, are being developed. Circular RNAs (circRNAs), another class of noncoding RNAs, are emerging as sponges for miRNAs by sequestering them and modulating their activities. These findings suggest that targeting noncoding RNA networks may offer novel avenues for the management of autoimmune diseases.

8.4 Part 3: Cellular Players in Autoimmune Pathologies

8.4.1 Innate Immune Cells

- *Dendritic Cells*: These professional antigen-presenting cells are central to the initiation of autoimmune responses. Plasmacytoid dendritic cells (pDCs) are implicated in SLE through the production of type I interferons, which amplify inflammatory responses (Colonna et al. 2004). Recent studies have shown that dendritic cells can also influence the polarization of T helper cells, skewing them toward a proinflammatory Th1 or Th17 phenotype, depending on the cytokine milieu. Additionally, dendritic cell-derived exosomes have emerged as key players in autoimmune diseases and are capable of carrying and presenting autoantigens to T cells, thereby perpetuating chronic inflammation. Targeting dendritic cell activation pathways, such as TLR signaling, is being explored as a therapeutic strategy to dampen their proinflammatory roles while preserving their essential immune regulatory functions (Colonna et al. 2004).
- *Macrophages*: M1 macrophages secrete proinflammatory cytokines, including TNF-α and IL-6, driving chronic inflammation and tissue damage in autoimmune diseases such as RA (Murray and Wynn 2011). Conversely, M2 macrophages, which normally play an antiinflammatory role, are often dysregulated under these conditions and fail to effectively resolve inflammation. Emerging research has highlighted the plasticity of macrophages and their ability to switch between M1 and M2 phenotypes under specific signals, presenting opportunities for therapeutic interventions (Murray and Wynn 2011). Novel therapies targeting macrophage polarization to restore the balance include the use of small molecules or biologics to inhibit M1-associated pathways or enhance M2-associated antiinflammatory functions. Macrophage metabolism has been identified as a crucial factor influencing the inflammatory phenotype, and metabolic reprogramming has been explored as a potential therapeutic target. Targeting macrophage polarization is a promising strategy for mitigating tissue damage in RA.
- *Innate Lymphoid Cells (ILCs)*: These cells contribute to tissue homeostasis and become pathogenic under chronic inflammatory conditions. ILC3s secrete IL-17 and IL-22 and play pivotal roles in the pathogenesis of psoriasis and inflammatory bowel disease by promoting epithelial barrier dysfunction and sustaining inflammation (Artis and Spits 2015). Recent studies have identified crosstalk between ILCs and adaptive immune cells, such as T cells, in the amplification of autoimmune responses (Stockis et al. 2024). Furthermore, ILC1s, which produce IFN-γ, are implicated in the early stages of autoimmune diseases by activating macrophages and dendritic cells. Targeting ILCs with biologics that inhibit IL-17 and IL-22 or modulate their activation signals offers promising therapeutic avenues. Single-cell sequencing has provided insight into ILC heterogeneity, revealing novel subsets that may serve as potential therapeutic targets.

8.4.2 Adaptive Immune Cells

- *T Cells*: Th1 and Th17 cells drive inflammation in autoimmune diseases, with Th1 cells producing IFN-γ and Th17 cells secreting IL-17 to recruit neutrophils and amplify tissue damage. Dysfunctional regulatory T cells (Tregs), which fail to suppress autoreactive responses, are a hallmark of autoimmunity (Bluestone et al. 2015). Recent research has highlighted the role of metabolic pathways in T-cell differentiation and function, with glycolysis driving Th17 responses and fatty acid oxidation, supporting Treg stability (Ma et al. 2024). Therapeutic strategies targeting these metabolic pathways are being developed to restore the balance between proinflammatory and regulatory T cells. Additionally, tissue-resident memory T cells (TRM), which persist at sites of inflammation, have been implicated in chronic autoimmune conditions such as psoriasis (Li et al. 2025). Strategies for the depletion or reprogramming of TRM cells are currently being explored as potential treatments. TRM have recently been identified as persistent contributors to local inflammation in diseases such as psoriasis.
- *B Cells*: They play dual roles in autoimmunity by producing autoantibodies and acting as antigen-presenting cells for T cells (Shen and Fillatreau 2015). Autoantibodies such as antinuclear antibodies in SLE form immune complexes that are deposited in tissues and trigger complement activation and inflammation (Aas-Hanssen et al. 2015). Emerging therapies target B-cell subsets, including anti-CD20 monoclonal antibodies, such as rituximab, which deplete B cells to reduce autoantibody production (Xiao et al. 2022). Research has also focused on plasma cells, which are responsible for long-lived antibody production; novel agents, such as proteasome inhibitors, showing promise in clinical trials. Additionally, B–T-cell interactions within germinal centers are critical for sustaining autoimmunity, making these microenvironments potential therapeutic targets. Ongoing studies are evaluating the role of B-cell–activating factor (BAFF) inhibitors in modulating B-cell survival and function to improve disease control (Smulski and Eibel 2018; Giordano et al. 2023).

8.5 Part 4: Specific Autoimmune Diseases and Mechanistic Insights

8.5.1 Case Studies

- *Rheumatoid Arthritis*: RA is characterized by synovitis, joint destruction, and systemic inflammation (Smolen et al. 2016). Central to its pathogenesis is the overproduction of proinflammatory cytokines such as TNF-α, IL-6, and IL-1β, which drive immune cell activation and joint damage. Synovial fibroblasts, which are a hallmark of RA, exhibit a hyperactivated phenotype that contributes to cartilage degradation and bone erosion (Alivernini et al. 2020). The interaction between autoreactive B and T cells in synovium further amplifies inflammation (Shen and Fillatreau 2015). Advanced imaging technologies, such as ultrasound and MRI, enable the early detection of synovial inflammation and facilitate prompt intervention. Targeted therapies, including TNF inhibitors such as adalimumab and IL-6 receptor blockers such as tocilizumab, have transformed RA management by reducing disease activity and preventing joint damage. Emerging therapies such as Janus kinase (JAK) inhibitors offer new hope for patients with refractory RA and provide additional options for personalized treatment approaches (O'Shea et al. 2015).

- *Systemic Lupus Erythematosus*: SLE is a multisystem autoimmune disease characterized by the production of autoantibodies such as anti-dsDNA and anti-Smith antibodies, which form immune complexes that are deposited in tissues and cause inflammation and organ damage (Tsokos 2011). The role of type 1 interferons, particularly IFN-α, is central to SLE pathogenesis, promoting the activation of plasmacytoid dendritic cells and autoreactive lymphocytes (Chen et al. 2016; Colonna et al. 2004). Recent advances in biomarker discovery including the use of complementary and anti-C1q antibodies have improved disease monitoring. Treatment strategies for SLE range from corticosteroids and antimalarials such as hydroxychloroquine to biologics such as belimumab, which targets BAFF (Smulski and Eibel 2018). Experimental therapies, including CAR-T cells targeting $CD19^+$ B cells and type I interferon receptor inhibitors, are being investigated for refractory cases, offering the potential for more durable disease control (Alsuliman et al. 2024).
- *Multiple Sclerosis*: MS is a chronic autoimmune disease where the immune system attacks the central nervous system (CNS), leading to demyelination and axonal damage (Reich et al. 2018). Autoreactive T cells, particularly Th1 and Th17 subsets, infiltrate the CNS, producing cytokines like IL-17 and IFN-γ, which amplify inflammation (Bluestone et al. 2015). B-cells also produce autoantibodies and activate T cells (Shen and Fillatreau 2015). Genetic predispositions, such as HLA-DRB1*15:01, and environmental triggers, including the EBV (Lanz et al. 2022; Welberg 2022) and vitamin D deficiency, are critical in disease development. Current treatments focus on immune modulation with monoclonal antibodies, such as ocrelizumab, which target CD20 on B cells. Emerging therapies, including remyelination agents, aim to restore CNS function and slow disease progression.
- In addition to the previously discussed autoimmune diseases, type 1 diabetes and celiac disease also result from dysregulated immune responses, leading to chronic inflammation and tissue damage (Table 8.2).

Table 8.2 Comparison of selected autoimmune diseases: Summarizing key features, pathogenic mechanisms, and common therapeutic strategies for major autoimmune disorders

Disease	Key features	Pathogenic mechanisms	Common therapies
Rheumatoid Arthritis (RA)	Chronic inflammation of joints, synovitis, bone erosion (Alivernini et al. 2020)	TNF-α, IL-6, and autoantibodies (e.g., ACPAs)	TNF inhibitors, JAK inhibitors, IL-6 blockers
Systemic Lupus Erythematosus (SLE)	Multiorgan involvement, immune complex deposition (Aas-Hanssen et al. 2015)	Autoantibodies (e.g., anti-dsDNA), complement activation	Hydroxychloroquine, corticosteroids, B-cell depletion therapy
Multiple Sclerosis (MS)	Demyelination in CNS, neuroinflammation (Charabati et al. 2023)	Th1 and Th17 cells, autoreactive T cells attack myelin	Anti-CD20 monoclonal antibodies, interferon-β
Type 1 Diabetes (T1D) (Mariño et al. 2017; Kwon et al. 2022)	Destruction of insulin-producing β-cells in pancreas (Mariño et al. 2017; Kwon et al. 2022)	Autoreactive T cells, islet cell autoantibodies	Insulin therapy, immunomodulators (e.g., anti-CD3)
Celiac Disease	Gluten-triggered autoimmune response in small intestine (Schuppan et al. 2009)	T-cell–mediated response to gliadin (gluten component)	Gluten-free diet

8.6 Part 5: Therapeutic Approaches and Future Directions

8.6.1 Current Therapeutic Strategies

- *Biologics*: Biologics targeting proinflammatory cytokines, such as IL-6 and IL-17, have revolutionized the treatment of autoimmune diseases, including RA and psoriasis. Agents such as tocilizumab and secukinumab have demonstrated efficacy in reducing inflammation and preventing tissue damage. These therapies not only improve clinical outcomes but also reduce the progression of structural damage in diseases such as RA (Burmester and Pope 2017). Ongoing research into combination therapies, including dual cytokine blockade, aims to enhance the therapeutic efficacy for patients with refractory disease. In addition, studies on the long-term safety and cost-effectiveness of biologics are ongoing to address broader accessibility and sustainability issues.
- *Small Molecule Inhibitors*: Janus kinase (JAK) inhibitors, including tofacitinib and baricitinib, effectively reduce immune activation in conditions such as lupus and ulcerative colitis (O'Shea et al. 2015). A new class of TYK2 inhibitors, deucravacitinib, targets the suppression of inflammatory pathways with fewer side effects (Philips et al. 2022). These inhibitors selectively target signaling pathways critical for Th1 and Th17 cell activity, reducing inflammation while preserving broader immune functions. Ongoing trials are investigating their efficacy

in the treatment of additional autoimmune diseases, including psoriasis and multiple sclerosis, with promising results.

8.6.2 *Emerging Technologies*

- *CRISPR/Cas9*: This gene-editing technology holds promise for treating autoimmune diseases by correcting mutations associated with autoimmunity. Recent advances in delivery systems, including lipid nanoparticles (NPs), have enhanced the precision and safety of CRISPR-based therapies. For example, CRISPR has been used to disrupt proinflammatory cytokine genes, such as IL-23 or TNF-α, in experimental autoimmune models, showing reduced disease severity (Wu et al. 2019). Researchers are also exploring the potential of CRISPR to restore regulatory immune pathways by enhancing Treg activity through genetic modifications (Zhang et al. 2024a). These advances have paved the way for clinical trials targeting conditions such as lupus and RA.
- *Single-Cell Sequencing*: Single-cell RNA sequencing has provided unprecedented insights into the heterogeneity of immune cells in autoimmune diseases (Argyriou et al. 2022). This technology enables the identification of pathogenic cell subsets and guides the development of personalized therapies. Recent applications include the profiling of T- and B-cell populations in SLE to identify specific markers of disease activity and therapeutic responses (Nehar-Belaid et al. 2020). Integrating these data with multiomics approaches, including proteomics and epigenomics, holds promise for identifying novel therapeutic targets and stratifying patients for precision treatments.

8.7 Conclusion: Toward Precision Medicine in Autoimmunity

This chapter elucidates the molecular and cellular mechanisms that drive autoimmune diseases and presents a comprehensive overview of the interplay between genetic predispositions, environmental triggers, and dysregulated immune responses. Central to understanding autoimmunity is the breakdown of both central and peripheral immune tolerance mechanisms, which leads to the activation of autoreactive T and B cells, cytokine dysregulation, and tissue damage. Advances in molecular insights, such as the NF-κB, JAK-STAT, and PI3K-Akt pathways, have shed light on the key signaling cascades that perpetuate inflammatory responses and have opened new avenues for targeted therapeutic strategies.

The emergence of technologies, such as CRISPR/Cas9 and single-cell sequencing, has revolutionized autoimmune research, enabling the identification of disease-specific biomarkers and pathogenic cell subsets. These tools hold promise for uncovering novel therapeutic targets and tailoring precision medicine approaches for individual patients. Furthermore, the exploration of epigenetic modifications, including DNA methylation, histone acetylation, and noncoding RNAs, highlights the importance of gene-environment interactions in shaping immune responses, offering additional layers of therapeutic intervention.

Therapeutic advances, particularly in biologics and small-molecule inhibitors, have transformed disease management, improved clinical outcomes, and reduced disease progression in conditions, such as RA, systemic lupus erythematosus, and multiple sclerosis. However, achieving long-term remission and addressing treatment resistance remain challenging. Emerging approaches such as combination therapies, remyelination agents, and immune reconstitution strategies are paving the way for more durable and comprehensive disease control.

The integration of multiomics data, artificial intelligence, and systems biology is expected to further refine our understanding of autoimmune diseases. These advancements will enhance the development of personalized treatment regimens and ensure effective and sustainable patient care. With continued innovation and collaboration across disciplines, the ultimate goals of achieving long-term remission and improving the quality of life of individuals with autoimmune diseases are becoming an attainable reality.

References

Aas-Hanssen K, Thompson KM, Bogen B, Munthe LA (2015) Systemic lupus erythematosus: molecular mimicry between anti-dsDNA CDR3 Idiotype, microbial and self peptides—as antigens for Th cells. Front Immunol 6. https://doi.org/10.3389/fimmu.2015.00382. (Hypothesis and Theory) (In English)

Alivernini S, MacDonald L, Elmesmari A et al (2020) Distinct synovial tissue macrophage subsets regulate inflammation and remission in rheumatoid arthritis. Nat Med 26(8):1295–1306. https://doi.org/10.1038/s41591-020-0939-8

Alsuliman T, Marjanovic Z, Rimar D et al (2024) Harnessing the potential of CAR-T cell in lupus treatment: from theory to practice. Autoimmun Rev 23(12):103687. https://doi.org/10.1016/j.autrev.2024.103687

Argyriou A, Wadsworth MH, Lendvai A et al (2022) Single cell sequencing identifies clonally expanded synovial CD4+ TPH cells expressing GPR56 in rheumatoid arthritis. Nat Commun 13(1):4046. https://doi.org/10.1038/s41467-022-31519-6

Artis D, Spits H (2015) The biology of innate lymphoid cells. Nature 517(7534):293–301. https://doi.org/10.1038/nature14189

Bluestone JA, Bour-Jordan H, Cheng M, Anderson M (2015) T cells in the control of organ-specific autoimmunity. J Clin Invest 125(6):2250–2260. https://doi.org/10.1172/jci78089

Brown CC, Wedderburn LR (2015) Mapping autoimmune disease epigenetics: what's on the horizon? Nat Rev Rheumatol 11(3):131–132. https://doi.org/10.1038/nrrheum.2014.210

Burmester GR, Pope JE (2017) Novel treatment strategies in rheumatoid arthritis. Lancet 389(10086):2338–2348. https://doi.org/10.1016/s0140-6736(17)31491-5

Charabati M, Wheeler MA, Weiner HL, Quintana FJ (2023) Multiple sclerosis: Neuroimmune crosstalk and therapeutic targeting. Cell 186(7):1309–1327. https://doi.org/10.1016/j.cell.2023.03.008

Chen JQ, Papp G, Szodoray P, Zeher M (2016) The role of microRNAs in the pathogenesis of autoimmune diseases. Autoimmun Rev 15(12):1171–1180. https://doi.org/10.1016/j.autrev.2016.09.003

Colonna M, Trinchieri G, Liu Y-J (2004) Plasmacytoid dendritic cells in immunity. Nat Immunol 5(12):1219–1226. https://doi.org/10.1038/ni1141

Conrad N, Misra S, Verbakel J et al (2023) OP0007 incidence, prevalence and co-occurrence of autoimmune disorders, trends over time and by age, sex and socioeconomic status. A population-based study in 22 million individuals. Ann Rheum Dis 82(Suppl 1):5–6. https://doi.org/10.1136/annrheumdis-2023-eular.4269

Cotsapas C, Hafler DA (2013) Immune-mediated disease genetics: the shared basis of pathogenesis. Trends Immunol 34(1):22–26. https://doi.org/10.1016/j.it.2012.09.001

Davidson A, Diamond B (2001) Autoimmune Diseases. N Engl J Med 345(5):340–350. https://doi.org/10.1056/NEJM200108023450506

Farh KK-H, Marson A, Zhu J et al (2015) Genetic and epigenetic fine mapping of causal autoimmune disease variants. Nature 518(7539):337–343. https://doi.org/10.1038/nature13835

Fruman DA, Chiu H, Hopkins BD, Bagrodia S, Cantley LC, Abraham RT (2017) The PI3K pathway in human disease. Cell 170(4):605–635. https://doi.org/10.1016/j.cell.2017.07.029

Gensous N, Charrier M, Duluc D et al (2018) T follicular helper cells in autoimmune disorders. Front Immunol 9. https://doi.org/10.3389/fimmu.2018.01637

Giordano D, Kuley R, Draves KE, Elkon KB, Giltiay NV, Clark EA (2023) B cell-activating factor (BAFF) from dendritic cells, monocytes and neutrophils is required for B cell maturation and autoantibody production in SLE-like autoimmune disease. Front Immunol 14. https://doi.org/10.3389/fimmu.2023.1050528. (Original Research) (In English)

Hedrich CM, Tsokos GC (2011) Epigenetic mechanisms in systemic lupus erythematosus and other autoimmune diseases. Trends Mol Med 17(12):714–724. https://doi.org/10.1016/j.molmed.2011.07.005

Josefowicz SZ, Lu LF, Rudensky AY (2012) Regulatory T cells: mechanisms of differentiation and function. Annu Rev Immunol 30:531–564. https://doi.org/10.1146/annurev.immunol.25.022106.141623

Kwon BC, Anand V, Achenbach P et al (2022) Progression of type 1 diabetes from latency to symptomatic disease is predicted by distinct autoimmune trajectories. Nat Commun 13(1):1514. https://doi.org/10.1038/s41467-022-28909-1

Lanz TV, Brewer RC, Ho PP et al (2022) Clonally expanded B cells in multiple sclerosis bind EBV EBNA1 and GlialCAM. Nature 603(7900):321–327. https://doi.org/10.1038/s41586-022-04432-7

Li Q, Verma IM (2002) NF-κB regulation in the immune system. Nat Rev Immunol 2(10):725–734. https://doi.org/10.1038/nri910

Li J, Xiao C, Li C, He J (2025) Tissue-resident immune cells: from defining characteristics to roles in diseases. Signal Transduct Target Ther 10(1):12. https://doi.org/10.1038/s41392-024-02050-5

Liu T, Zhang L, Joo D, Sun S-C (2017) NF-κB signaling in inflammation. Signal Transduct Target Ther 2(1):17023. https://doi.org/10.1038/sigtrans.2017.23

Ma S, Ming Y, Wu J, Cui G (2024) Cellular metabolism regulates the differentiation and function of T-cell subsets. Cell Mol Immunol 21(5):419–435. https://doi.org/10.1038/s41423-024-01148-8

Mariño E, Richards JL, McLeod KH et al (2017) Gut microbial metabolites limit the frequency of autoimmune T cells and protect against type 1 diabetes. Nat Immunol 18(5):552–562. https://doi.org/10.1038/ni.3713

Meyer S, Woodward M, Hertel C et al (2016) AIRE-deficient patients harbor unique high-affinity disease-ameliorating autoantibodies. Cell 166(3):582–595. https://doi.org/10.1016/j.cell.2016.06.024

Mu S, Wang W, Liu Q et al (2024) Autoimmune disease: a view of epigenetics and therapeutic targeting. Front Immunol 15:15. https://doi.org/10.3389/fimmu.2024.1482728

Murray PJ, Wynn TA (2011) Protective and pathogenic functions of macrophage subsets. Nat Rev Immunol 11(11):723–737. https://doi.org/10.1038/nri3073

Nehar-Belaid D, Hong S, Marches R et al (2020) Mapping systemic lupus erythematosus heterogeneity at the single-cell level. Nat Immunol 21(9):1094–1106. https://doi.org/10.1038/s41590-020-0743-0

Okkenhaug K, Turner M, Gold MR (2014) PI3K signaling in B cell and T cell biology. Front Immunol 5. https://doi.org/10.3389/fimmu.2014.00557

O'Shea JJ, Schwartz DM, Villarino AV, Gadina M, McInnes IB, Laurence A (2015) The JAK-STAT pathway: impact on human disease and therapeutic intervention. Annu Rev Med 66:311–328. https://doi.org/10.1146/annurev-med-051113-024537

Philips RL, Wang Y, Cheon H et al (2022) The JAK-STAT pathway at 30: much learned, much more to do. Cell 185(21):3857–3876. https://doi.org/10.1016/j.cell.2022.09.023

Pollard KM (2015) Environment, autoantibodies, and autoimmunity. Front Immunol 6. https://doi.org/10.3389/fimmu.2015.00060

Reich DS, Lucchinetti CF, Calabresi PA (2018) Multiple sclerosis. N Engl J Med 378(2):169–180. https://doi.org/10.1056/NEJMra1401483

Rosenblum MD, Remedios KA, Abbas AK (2015) Mechanisms of human autoimmunity. J Clin Invest 125(6):2228–2233. https://doi.org/10.1172/jci78088

Rostami A, White K, Rostami K (2024) Pro and anti-inflammatory diets as strong epigenetic factors in inflammatory bowel disease. World J Gastroenterol 30(27):3284–3289. https://doi.org/10.3748/wjg.v30.i27.3284

Sakaguchi S, Yamaguchi T, Nomura T, Ono M (2008) Regulatory T cells and immune tolerance. Cell 133(5):775–787. https://doi.org/10.1016/j.cell.2008.05.009

Schuppan D, Junker Y, Barisani D (2009) Celiac disease: from pathogenesis to novel therapies. Gastroenterology 137(6):1912–1933. https://doi.org/10.1053/j.gastro.2009.09.008

Shen P, Fillatreau S (2015) Antibody-independent functions of B cells: a focus on cytokines. Nat Rev Immunol 15(7):441–451. https://doi.org/10.1038/nri3857

Smolen JS, Aletaha D, McInnes IB (2016) Rheumatoid arthritis. Lancet 388(10055):2023–2038. https://doi.org/10.1016/S0140-6736(16)30173-8

Smulski CR, Eibel H (2018) BAFF and BAFF-receptor in B cell selection and survival. Front Immunol 9. https://doi.org/10.3389/fimmu.2018.02285

Stockis J, Yip T, Moreno-Vicente J et al (2024) Cross-talk between ILC2 and Gata3^{high} T_{regs} locally constrains adaptive type 2 immunity. Sci Immunol 9(97):eadl1903. https://doi.org/10.1126/sciimmunol.adl1903

Surace AEA, Hedrich CM (2019) The role of epigenetics in autoimmune/inflammatory disease. Front Immunol 10. https://doi.org/10.3389/fimmu.2019.01525

Theofilopoulos AN, Kono DH, Baccala R (2017) The multiple pathways to autoimmunity. Nat Immunol 18(7):716–724. https://doi.org/10.1038/ni.3731

Tsokos GC (2011) Systemic lupus erythematosus. N Engl J Med 365(22):2110–2121. https://doi.org/10.1056/NEJMra1100359

Welberg L (2022) Strong evidence that Epstein–Barr virus infection triggers multiple sclerosis. Nat Neurosci 25(2):131–131. https://doi.org/10.1038/s41593-022-01017-5

Wu GC, Pan HF, Leng RX et al (2015) Emerging role of long noncoding RNAs in autoimmune diseases. Autoimmun Rev 14(9):798–805. https://doi.org/10.1016/j.autrev.2015.05.004

Wu J, Yang S, Yu D et al (2019) CRISPR/cas9 mediated knockout of an intergenic variant rs6927172 identified IL-20RA as a new risk gene for multiple autoimmune diseases. Genes Immun 20(2):103–111. https://doi.org/10.1038/s41435-018-0011-6

Xiao F, Rui K, Shi X et al (2022) Epigenetic regulation of B cells and its role in autoimmune pathogenesis. Cell Mol Immunol 19(11):1215–1234. https://doi.org/10.1038/s41423-022-00933-7

Zhang Q, Lenardo MJ, Baltimore D (2017) 30 years of NF-κB: a blossoming of relevance to human pathobiology. Cell 168(1):37–57. https://doi.org/10.1016/j.cell.2016.12.012

Zhang M-L, Li H-B, Jin Y (2024a) Application and perspective of CRISPR/Cas9 genome editing technology in human diseases modeling and gene therapy. Front Genet 15. https://doi.org/10.3389/fgene.2024.1364742

Zhang Y, Liu H, Niu M et al (2024b) Roles of long noncoding RNAs in human inflammatory diseases. Cell Death Discovery 10(1):235. https://doi.org/10.1038/s41420-024-02002-6

Chapter 9
Underlying the Cellular Regulations in Autoimmune Pathologies

Yoonkyung Won

9.1 Introduction

Autoimmune diseases are complex and heterogeneous disorders arising from a combination of genetic predisposition, hormonal influences, environmental exposures, and dysregulated immune responses. They are characterized by the loss of immunological tolerance to self-antigens, resulting in the generation of pathogenic autoantibodies and autoreactive immune cells that target normal tissues. Notable examples of disease-associated autoantibodies include anti-citrullinated protein antibodies (ACPA) in rheumatoid arthritis (RA), anti-tyrosine phosphatase (IA-2) antibodies in type 1 diabetes (T1D), and anti-myelin oligodendrocyte glycoprotein (MOG) antibodies in multiple sclerosis (MS) (Inshaw et al. 2018; Pisetsky 2023). These autoantibodies may directly contribute to tissue injury or initiate complement activation, immune complex formation, and aberrant intracellular signaling cascades, all of which perpetuate inflammation and organ damage (Pisetsky 2017). Target tissues vary across diseases, with the synovial joints, myelinated nerves, and pancreatic islets being particularly susceptible in RA, T1D, and MS, respectively (Fig. 9.1).

Autoimmune diseases affect approximately 3–5% of the global population, with over 75% of cases occurring in women, underscoring the potential role of sex hormones in disease susceptibility (Davidson and Diamond 2001). Genetic factors, especially those involving HLA alleles, are among the strongest contributors to disease risk, while infections, dysbiosis, and exposure to environmental chemicals (xenobiotics) can act as potential triggers in genetically susceptible individuals

Y. Won (✉)
Hard-Tissue Biointerface Research Center, School of Dentistry, Chonnam National University, Gwangju, Republic of Korea
e-mail: ykwon32@jnu.ac.kr

Y.-O. Son, M. Ghosh (eds.), *Cutting-Edge Strategies in Drug Delivery and Immunotherapy for Autoimmune Disorders*,
https://doi.org/10.1007/978-981-95-0520-3_9

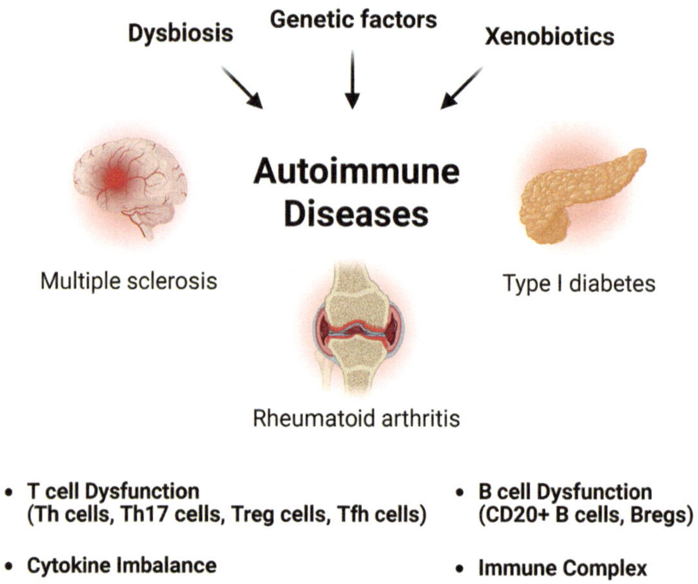

Fig. 9.1 Overview of autoimmune diseases

(Belbasis et al. 2018; Klareskog et al. 2006; Smolen et al. 2016). In some cases, patients present with multiple coexisting autoimmune diseases, highlighting common immunopathological mechanisms across different disorders (Selmi 2010).

A growing understanding of the cytokine networks involved in autoimmune diseases has revealed both overlapping and disease-specific cytokine signatures, which can vary significantly between patients even within the same diagnosis (Alivernini et al. 2022; Dendrou et al. 2015). This interpatient heterogeneity underscores the need for personalized therapeutic approaches targeting either proinflammatory cytokines, lymphocyte dysregulation, or pathways involved in the breakdown of self-tolerance (Tavakolpour 2017). Advances in molecular immunology are increasingly facilitating the development of targeted biologic therapies and small-molecule inhibitors, aimed at restoring immune balance and minimizing off-target immunosuppression (Smolen et al. 2018).

9.2 T Cells and Their Subsets in Autoimmunity

T cells play a central role in the initiation and propagation of autoimmune diseases through their ability to recognize self-antigens and orchestrate immune responses that result in tissue damage (Seiringer et al. 2022; Zhu et al. 2010). $CD4^{+}$ T helper (Th) cells, in particular, have been implicated in the pathogenesis of several autoimmune diseases, including RA, T1D, and MS (Bluestone et al. 2010; McInnes and Schett 2011). Aberrant differentiation and activation of proinflammatory Th subsets, especially Th1 and Th17 cells, contribute to excessive cytokine production and recruitment of inflammatory cells to affected tissues (Korn et al. 2009; Zhu et al. 2010). Th1 cells, which are characterized by their production of interferon-gamma (IFN-γ), are pivotal in promoting macrophage activation and cytotoxic responses in diseases such as T1D and MS (Bluestone et al. 2010; Kleinewietfeld and Hafler 2014). Th17 cells, which secrete interleukin-17 (IL-17), play a key role in neutrophil recruitment and chronic tissue inflammation, particularly in RA and MS (Cua and Tato 2010; Korn et al. 2009; McInnes and Schett 2011).

In contrast, regulatory T cells (Tregs), which normally act to suppress excessive immune responses and maintain self-tolerance, are often numerically or functionally impaired in autoimmunity (Goswami et al. 2022). Defective Treg-mediated suppression is observed in autoimmune diseases such as RA, where effector T cells escape regulatory control, leading to sustained inflammation and tissue destruction (McInnes and Schett 2011). $CD8^+$ T cells also contribute to autoimmunity, especially in T1D, where autoreactive $CD8^+$ T cells directly target and destroy insulin-producing beta cells in the pancreas (Bluestone et al. 2010).

Additionally, follicular helper T cells (Tfh) have emerged as key players in autoimmune pathogenesis by promoting aberrant B cell activation and autoantibody production within germinal centers (Crotty 2014). The balance between proinflammatory and regulatory T cell subsets is tightly controlled under normal conditions, but genetic susceptibility, environmental triggers, and chronic inflammation can disrupt this equilibrium, favoring pathogenic T cell responses (Schmidt et al. 2012; Zhu et al. 2010). Understanding the molecular mechanisms governing T cell differentiation, plasticity, and tolerance breakdown is essential for identifying novel therapeutic targets across autoimmune diseases (Goswami et al. 2022; Schmidt et al. 2012).

The loss of self-tolerance, a hallmark of autoimmune diseases, results from a combination of defects in central and peripheral immune tolerance mechanisms. Key processes implicated in the breakdown of self-tolerance include impaired clonal deletion of autoreactive $CD4^+$ Th lymphocytes in the thymus, increased peripheral activation of autoreactive T cells, and defective immunoregulatory function of $CD4^+$ Tregs and $CD8^+$ suppressor T cells (Baecher-Allan and Hafler 2006; Ohkura et al. 2013). Additionally, dysregulated intracellular signaling pathways—particularly those involved in cytokine production and inflammatory amplification—contribute to an imbalance favoring proinflammatory cytokines (e.g., IL-6, TNF-α, IFN-γ, and IL-17) over antiinflammatory mediators (Dendrou et al. 2015; McInnes and Schett 2011).

Several antigenic and molecular mechanisms can further contribute to the breach of self-tolerance. These include molecular mimicry, where the structural similarity between microbial antigens and self-antigens results in cross-reactive immune responses, and epitope spreading, in which tissue injury exposes previously hidden self-epitopes, propagating the autoimmune response (Cusick et al. 2012; Rose and Bona 1993). Furthermore, posttranslational modifications of self-proteins, such as citrullination or xenobiotic-induced neoantigen formation, can generate immunogenic epitopes that bypass normal tolerance checkpoints (Pollard et al. 2013; Steiner and Toes 2024).

The immune effector mechanisms involved in autoimmune diseases vary depending on the organ specificity and systemic involvement of the disease. Organ-specific autoimmune diseases (e.g., T1D, MS) are typically dominated by cell-mediated immunity, often driven by Th1 and Th17 cells, which produce IFN-γ and IL-17, respectively, promoting direct tissue destruction (Bluestone et al. 2010; Korn et al. 2009). In contrast, systemic or multiorgan autoimmune diseases (e.g., systemic lupus erythematosus) frequently involve a prominent autoantibody component, driven by Th2-polarized responses that support B cell activation and autoantibody production (Davidson and Diamond 2001; Dendrou et al. 2015). Understanding the cellular and molecular mechanisms underlying the loss of self-tolerance is critical for identifying therapeutic strategies aimed at restoring immune homeostasis in autoimmune diseases.

9.3 B Cells and Autoimmunity

B cells are integral to the onset and progression of autoimmune diseases, both through their antibody-dependent and antibody-independent functions. The production of autoantibodies, such as anti-cyclic citrullinated peptide (CCP) in RA and anti-glutamic acid decarboxylase (GAD) antibodies in T1D, contributes to immune complex formation, complement activation, and tissue injury (Herold et al. 2013). Beyond autoantibody production, B cells act as antigen-presenting cells (APCs), presenting autoantigens to T cells and promoting T cell activation and cytokine production, thereby amplifying inflammatory responses. Activated B cells also secrete proinflammatory cytokines, including IL-6 and TNF-α, which further enhance pathogenic immune responses (Mauri and Bosma 2012).

In many autoimmune diseases, B cell tolerance checkpoints—which normally eliminate autoreactive B cells during development—are compromised, allowing autoreactive clones to persist and expand (Küppers 2009). This is particularly evident in SLE, where loss of tolerance leads to widespread autoreactivity and immune complex deposition in multiple organs (Tsokos 2011). The germinal center reaction, where B cells undergo somatic hypermutation and affinity maturation, is also dysregulated in autoimmunity, often driven by excessive follicular helper T cell (Tfh) activity (Crotty 2011; Mintz and Cyster 2020). Dysregulated germinal center activity contributes to the generation of high-affinity autoantibodies with pathogenic potential.

Importantly, B cell depletion therapies, such as rituximab (anti-CD20 monoclonal antibody), have demonstrated efficacy in several autoimmune diseases, underscoring the central role of B cells in driving disease pathogenesis (Kessel et al. 2008). However, the heterogeneous roles of B cells across different autoimmune diseases, ranging from pathogenic antibody production to regulatory functions mediated by IL-10-producing regulatory B cells (Bregs), highlight the complex and context-dependent nature of B cell involvement in autoimmunity (Rosser and Mauri 2015). Understanding these diverse functions will be crucial for designing precision therapies targeting B cells in autoimmune diseases.

9.4 Molecular Mechanisms of Autoimmunity: Antibodies and Immune Complexes

Autoantibodies and immune complexes play a central role in the pathogenesis of many autoimmune diseases, particularly systemic and multiorgan disorders including RA. The production of autoantibodies against self-antigens reflects a breakdown in B cell tolerance mechanisms, allowing autoreactive B cells to escape immune checkpoints, undergo clonal expansion, and differentiate into autoantibody-secreting plasma cells (Mauri and Bosma 2012; Yanaba et al. 2008). Autoantibodies contribute to immune complex formation, complement activation, and tissue injury (Reindl and Waters 2019; Smolen et al. 2016). Once formed, immune complexes deposit in tissues such as joints, kidneys, blood vessels, and central nervous system (CNS), where they trigger complement cascades and activate Fc receptors (FcRs) on innate immune cells, including macrophages and neutrophils, promoting proinflammatory cytokine production and tissue damage (Lee et al. 2021; Pisetsky 2017). In multiple sclerosis, autoantibodies such as anti-MOG and antimyelin basic proteins can contribute to demyelination through complement-dependent cytotoxicity and activation of microglia and astrocytes (Reindl and Waters 2019). The pathogenic significance of autoantibodies varies between diseases, ranging from direct cytotoxicity and functional blockade to immune complex-driven inflammation. Understanding the specificity, pathogenic mechanisms, and diagnostic value of these autoantibodies has been crucial for developing biomarkers and targeted therapies for autoimmune diseases (Table 9.1).

Table 9.1 Autoantibodies and immune complexes in autoimmune diseases

Disease	Key autoantibodies	Pathogenic mechanisms
RA	Anti-citrullinated protein antibodies (ACPA), Rheumatoid Factor (RF)	Synovial immune complex formation Complement activation Proinflammatory cytokine production
T1D	Anti-glutamic acid decarboxylase (GAD), Tyrosine Phosphatase (IA-2), Anti-zinc transporter 8 (ZnT8)	Autoantibody binding to pancreatic beta cells Complement activation Beta-cell destruction
MS	Anti-myelin oligodendrocyte glycoprotein (MOG), myelin basic protein (MBP), Anti-neurofascin 155 (NF155)	Autoantibody-mediated demyelination Complement activation Microglial activation

9.5 Animal Models of Autoimmune Disease

Animal models have been indispensable in advancing our understanding of the pathogenesis of autoimmune diseases, including the identification of genetic risk factors, immune cell interactions, and the molecular pathways driving autoimmunity. These models provide controlled environments to study the complex interactions between genetics, environmental triggers, and the immune system, which is critical for uncovering disease mechanisms and evaluating novel therapeutic interventions (Taneja and David 2001; Yu and Petersen 2018). Autoimmune disease models can be broadly categorized into spontaneous models, which arise due to genetic susceptibility, induced models, triggered by immunization or environmental stimuli, and genetically engineered models, where specific genes are introduced, deleted, or modified (Robinson et al. 2018).

For example, the nonobese diabetic (NOD) mouse has been widely used to study T1D due to its genetic predisposition to spontaneous pancreatic beta-cell autoimmunity, closely mirroring human disease (Anderson and Bluestone 2005). Similarly, collagen-induced arthritis (CIA) and K/BxN arthritis models have been instrumental in elucidating the cellular and molecular pathways involved in RA (Brand et al. 2007). Experimental autoimmune encephalomyelitis (EAE), a model of MS, has provided critical insights into the roles of T cells, B cells, and the blood-brain barrier in autoimmune demyelination (Glatigny and Bettelli 2018) (Table 9.2).

Despite their value, animal models have limitations, including incomplete recapitulation of the genetic complexity, environmental factors, and heterogeneity observed in human autoimmune diseases. Nonetheless, these models remain indispensable tools for studying pathogenic mechanisms, biomarker discovery, and preclinical drug development, especially when combined with multiomics approaches and humanized mouse systems (Shoda et al. 2005; Taneja and David 2001).

Table 9.2 Animal models of autoimmune diseases

Disease	Animal model	Category	Key features
RA	Collagen-induced arthritis	Induced	Immunization with type II collagen T cell and autoantibody-mediated inflammation in joint tissues
	K/BxN serum transfer model	Induced	Autoantibody transfer drives arthritis FcγR and neutrophil-dependent inflammation
T1D	Nonobese Diabetic (NOD) Mouse	Spontaneous	Spontaneous pancreatic beta cell destruction Autoreactive T cell and B cell responses
MS	Experimental Autoimmune Encephalomyelitis (EAE)	Induced	Immunization with myelin antigens T-cell–mediated CNS demyelination and inflammation

9.6 Summary and Future Directions

Autoimmune diseases are complex and heterogeneous disorders driven by a combination of genetic susceptibility, environmental triggers, hormonal factors, and dysregulated immune responses (Davidson and Diamond 2001; Dendrou et al. 2015). A loss of self-tolerance at both central and peripheral levels leads to the activation of autoreactive T and B cells, the production of pathogenic autoantibodies, and the formation of immune complexes that contribute to tissue inflammation and organ damage (Lee et al. 2021; Smolen et al. 2016).

At the cellular level, innate immune cells, including dendritic cells, macrophages, and neutrophils, initiate and amplify inflammatory responses through cytokine secretion, antigen presentation, and tissue remodeling, bridging innate and adaptive immunity (McInnes and Schett 2011). The adaptive immune system, particularly $CD4^+$ T helper cells, cytotoxic $CD8^+$ T cells, B cells, and Tregs, orchestrates the sustained immune attack against self-tissues (Bluestone et al. 2010; Mauri and Bosma 2012). Each autoimmune disease features a unique cytokine signature and a distinctive balance between proinflammatory and regulatory pathways, but many share common immunopathogenic mechanisms, such as epitope spreading, molecular mimicry, and posttranslational modification of self-antigens (Cusick et al. 2012; Trouw et al. 2017) (Fig. 9.2).

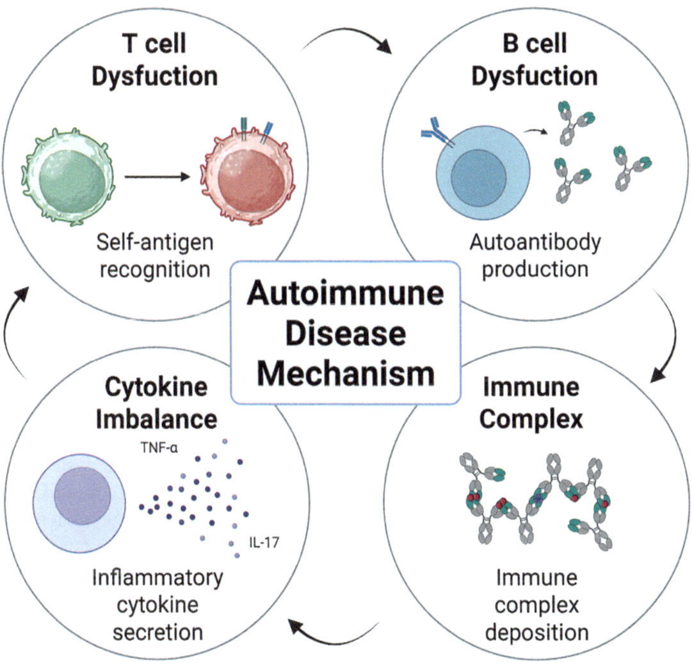

Fig. 9.2 Cellular mechanisms of autoimmune diseases

At the molecular level, key signaling pathways, such as NF-κB, JAK-STAT, and MAPK, are aberrantly activated, promoting the chronic inflammatory state characteristic of autoimmune diseases (McInnes and Schett 2011; O'Shea and Plenge 2012). Advances in genomic, transcriptomic, and epigenetic studies have identified critical susceptibility genes, such as HLA alleles, as well as non-HLA loci that influence immune regulation (Dendrou et al. 2015; Inshaw et al. 2018). Despite these advances, the heterogeneity of autoimmune diseases, both across individuals and within specific diseases, remains a major challenge in understanding disease mechanisms and developing effective therapies (Smolen et al. 2018).

Future research in autoimmune diseases will benefit from integrated multiomics approaches, combining genomics, transcriptomics, epigenomics, proteomics, and metabolomics to unravel the complex regulatory networks governing immune tolerance and autoimmunity (Barturen et al. 2021). Single-cell technologies and spatial transcriptomics will further clarify how immune cells interact with stromal and tissue-resident cells in affected organs, providing unprecedented insight into the local immune microenvironment in diseases such as RA, T1D, and MS (Binvignat et al. 2024; Collier et al. 2023; Schafflick et al. 2020; Smolen et al. 2018; Zhang et al. 2019).

A major frontier lies in the development of precision medicine strategies, tailoring therapies based on individual genetic risk, immune signatures, and environmental exposures (Toro-Domínguez and Alarcón-Riquelme 2021). Biologic therapies targeting specific cytokines, immune checkpoints, and autoreactive B cells have already transformed the treatment landscape, but there is a growing need to develop approaches that restore immune tolerance rather than broadly suppressing the immune system (Díaz-Peña 2021). Novel approaches, including tolerogenic vaccines, antigen-specific immunotherapies, engineered regulatory T cells (CAR-Tregs), and microbiome-based interventions, hold great promise for inducing durable remission with fewer side effects (Rampotas et al. 2025; Zeng et al. 2024).

Conflicts of Interest The author discloses no conflicts.

References

Alivernini S, Firestein GS, McInnes IB (2022) The pathogenesis of rheumatoid arthritis. Immunity 55(12):2255–2270. https://doi.org/10.1016/j.immuni.2022.11.009

Anderson MS, Bluestone JA (2005) The NOD mouse: a model of immune dysregulation. Annu Rev Immunol 23:447–485. https://doi.org/10.1146/annurev.immunol.23.021704.115643

Baecher-Allan C, Hafler DA (2006) Human regulatory T cells and their role in autoimmune disease. Immunol Rev 212:203–216. https://doi.org/10.1111/j.0105-2896.2006.00417.x

Barturen G, Babaei S, Català-Moll F, Martínez-Bueno M, Makowska Z, Martorell-Marugán J et al (2021) Integrative analysis reveals a molecular stratification of systemic autoimmune diseases. Arthritis Rheumatol 73(6):1073–1085. https://doi.org/10.1002/art.41610

Belbasis L, Dosis V, Evangelou E (2018) Elucidating the environmental risk factors for rheumatic diseases: an umbrella review of meta-analyses. Int J Rheum Dis 21(8):1514–1524. https://doi.org/10.1111/1756-185X.13356

Binvignat M, Miao BY, Wibrand C, Yang MM, Rychkov D, Flynn E et al (2024) Single-cell RNA-Seq analysis reveals cell subsets and gene signatures associated with rheumatoid arthritis disease activity. JCI Insight 9(16). https://doi.org/10.1172/jci.insight.178499

Bluestone JA, Herold K, Eisenbarth G (2010) Genetics, pathogenesis and clinical interventions in type 1 diabetes. Nature 464(7293):1293–1300. https://doi.org/10.1038/nature08933

Brand DD, Latham KA, Rosloniec EF (2007) Collagen-induced arthritis. Nat Protoc 2(5):1269–1275. https://doi.org/10.1038/nprot.2007.173

Collier JL, Pauken KE, Lee CAA, Patterson DG, Markson SC, Conway TS et al (2023) Single-cell profiling reveals unique features of diabetogenic T cells in anti-PD-1-induced type 1 diabetes mice. J Exp Med 220(10). https://doi.org/10.1084/jem.20221920

Crotty S (2011) Follicular helper CD4 T cells (TFH). Annu Rev Immunol 29:621–663. https://doi.org/10.1146/annurev-immunol-031210-101400

Crotty S (2014) T follicular helper cell differentiation, function, and roles in disease. Immunity 41(4):529–542. https://doi.org/10.1016/j.immuni.2014.10.004

Cua DJ, Tato CM (2010) Innate IL-17-producing cells: the sentinels of the immune system. Nat Rev Immunol 10(7):479–489. https://doi.org/10.1038/nri2800

Cusick MF, Libbey JE, Fujinami RS (2012) Molecular mimicry as a mechanism of autoimmune disease. Clin Rev Allergy Immunol 42(1):102–111. https://doi.org/10.1007/s12016-011-8294-7

Davidson A, Diamond B (2001) Autoimmune diseases. N Engl J Med 345(5):340–350. https://doi.org/10.1056/NEJM200108023450506

Dendrou CA, Fugger L, Friese MA (2015) Immunopathology of multiple sclerosis. Nat Rev Immunol 15(9):545–558. https://doi.org/10.1038/nri3871

Díaz-Peña R (2021) Personalized medicine in autoimmune diseases. J Pers Med 11(11). https://doi.org/10.3390/jpm11111181

Glatigny S, Bettelli E (2018) Experimental autoimmune encephalomyelitis (EAE) as animal models of multiple sclerosis (MS). Cold Spring Harb Perspect Med 8(11). https://doi.org/10.1101/cshperspect.a028977

Goswami TK, Singh M, Dhawan M, Mitra S, Emran TB, Rabaan AA et al (2022) Regulatory T cells (Tregs) and their therapeutic potential against autoimmune disorders – advances and challenges. Hum Vaccin Immunother 18(1):2035117. https://doi.org/10.1080/21645515.2022.2035117

Herold KC, Vignali DA, Cooke A, Bluestone JA (2013) Type 1 diabetes: translating mechanistic observations into effective clinical outcomes. Nat Rev Immunol 13(4):243–256. https://doi.org/10.1038/nri3422

Inshaw JRJ, Cutler AJ, Burren OS, Stefana MI, Todd JA (2018) Approaches and advances in the genetic causes of autoimmune disease and their implications. Nat Immunol 19(7):674–684. https://doi.org/10.1038/s41590-018-0129-8

Kessel A, Rosner I, Toubi E (2008) Rituximab: beyond simple B cell depletion. Clin Rev Allergy Immunol 34(1):74–79. https://doi.org/10.1007/s12016-008-8074-1

Klareskog L, Padyukov L, Ronnelid J, Alfredsson L (2006) Genes, environment and immunity in the development of rheumatoid arthritis. Curr Opin Immunol 18(6):650–655. https://doi.org/10.1016/j.coi.2006.06.004

Kleinewietfeld M, Hafler DA (2014) Regulatory T cells in autoimmune neuroinflammation. Immunol Rev 259(1):231–244. https://doi.org/10.1111/imr.12169

Korn T, Bettelli E, Oukka M, Kuchroo VK (2009) IL-17 and Th17 cells. Annu Rev Immunol 27:485–517. https://doi.org/10.1146/annurev.immunol.021908.132710

Küppers R (2009) The biology of Hodgkin's lymphoma. Nat Rev Cancer 9(1):15–27. https://doi.org/10.1038/nrc2542

Lee DSW, Rojas OL, Gommerman JL (2021) B cell depletion therapies in autoimmune disease: advances and mechanistic insights. Nat Rev Drug Discov 20(3):179–199. https://doi.org/10.1038/s41573-020-00092-2

Mauri C, Bosma A (2012) Immune regulatory function of B cells. Annu Rev Immunol 30:221–241. https://doi.org/10.1146/annurev-immunol-020711-074934

McInnes IB, Schett G (2011) The pathogenesis of rheumatoid arthritis. N Engl J Med 365(23):2205–2219. https://doi.org/10.1056/NEJMra1004965

Mintz MA, Cyster JG (2020) T follicular helper cells in germinal center B cell selection and lymphomagenesis. Immunol Rev 296(1):48–61. https://doi.org/10.1111/imr.12860

O'Shea JJ, Plenge R (2012) JAK and STAT signaling molecules in immunoregulation and immune-mediated disease. Immunity 36(4):542–550. https://doi.org/10.1016/j.immuni.2012.03.014

Ohkura N, Kitagawa Y, Sakaguchi S (2013) Development and maintenance of regulatory T cells. Immunity 38(3):414–423. https://doi.org/10.1016/j.immuni.2013.03.002

Pisetsky DS (2017) Antinuclear antibody testing – misunderstood or misbegotten? Nat Rev Rheumatol 13(8):495–502. https://doi.org/10.1038/nrrheum.2017.74

Pisetsky DS (2023) Pathogenesis of autoimmune disease. Nat Rev Nephrol 19(8):509–524. https://doi.org/10.1038/s41581-023-00720-1

Pollard KM, Cauvi DM, Toomey CB, Morris KV, Kono DH (2013) Interferon-γ and systemic autoimmunity. Discov Med 16(87):123–131

Rampotas A, Richter J, Isenberg D, Roddie C (2025) CAR-T cell therapy embarks on autoimmune disease. Bone Marrow Transplant 60(1):6–9. https://doi.org/10.1038/s41409-024-02429-6

Reindl M, Waters P (2019) Myelin oligodendrocyte glycoprotein antibodies in neurological disease. Nat Rev Neurol 15(2):89–102. https://doi.org/10.1038/s41582-018-0112-x

Robinson KA, Orent W, Madsen JC, Benichou G (2018) Maintaining T cell tolerance of alloantigens: lessons from animal studies. Am J Transplant 18(8):1843–1856. https://doi.org/10.1111/ajt.14984

Rose NR, Bona C (1993) Defining criteria for autoimmune diseases (Witebsky's postulates revisited). Immunol Today 14(9):426–430. https://doi.org/10.1016/0167-5699(93)90244-f

Rosser EC, Mauri C (2015) Regulatory B cells: origin, phenotype, and function. Immunity 42(4):607–612. https://doi.org/10.1016/j.immuni.2015.04.005

Schafflick D, Xu CA, Hartlehnert M, Cole M, Schulte-Mecklenbeck A, Lautwein T et al (2020) Integrated single cell analysis of blood and cerebrospinal fluid leukocytes in multiple sclerosis. Nat Commun 11(1):247. https://doi.org/10.1038/s41467-019-14118-w

Schmidt A, Oberle N, Krammer PH (2012) Molecular mechanisms of treg-mediated T cell suppression. Front Immunol 3:51. https://doi.org/10.3389/fimmu.2012.00051

Seiringer P, Garzorz-Stark N, Eyerich K (2022) T-cell–mediated autoimmunity: mechanisms and future directions. J Invest Dermatol 142(3 Pt B):804–810. https://doi.org/10.1016/j.jid.2021.04.032

Selmi C (2010) The worldwide gradient of autoimmune conditions. Autoimmun Rev 9(5):A247–A250. https://doi.org/10.1016/j.autrev.2010.02.004

Shoda LK, Young DL, Ramanujan S, Whiting CC, Atkinson MA, Bluestone JA et al (2005) A comprehensive review of interventions in the NOD mouse and implications for translation. Immunity 23(2):115–126. https://doi.org/10.1016/j.immuni.2005.08.002

Smolen JS, Aletaha D, Barton A, Burmester GR, Emery P, Firestein GS et al (2018) Rheumatoid arthritis. Nat Rev Dis Primers 4:18001. https://doi.org/10.1038/nrdp.2018.1

Smolen JS, Aletaha D, McInnes IB (2016) Rheumatoid arthritis. Lancet 388(10055):2023–2038. https://doi.org/10.1016/S0140-6736(16)30173-8

Steiner G, Toes REM (2024) Autoantibodies in rheumatoid arthritis – rheumatoid factor, anticitrullinated protein antibodies and beyond. Curr Opin Rheumatol 36(3):217–224. https://doi.org/10.1097/bor.0000000000001006

Taneja V, David CS (2001) Lessons from animal models for human autoimmune diseases. Nat Immunol 2(9):781–784. https://doi.org/10.1038/ni0901-781

Tavakolpour S (2017) Towards personalized medicine for patients with autoimmune diseases: opportunities and challenges. Immunol Lett 190:130–138. https://doi.org/10.1016/j.imlet.2017.08.002

Toro-Domínguez D, Alarcón-Riquelme ME (2021) Precision medicine in autoimmune diseases: fact or fiction. Rheumatology (Oxford) 60(9):3977–3985. https://doi.org/10.1093/rheumatology/keab448

Trouw LA, Rispens T, Toes REM (2017) Beyond citrullination: other post-translational protein modifications in rheumatoid arthritis. Nat Rev Rheumatol 13(6):331–339. https://doi.org/10.1038/nrrheum.2017.15

Tsokos GC (2011) Systemic lupus erythematosus. N Engl J Med 365(22):2110–2121. https://doi.org/10.1056/NEJMra1100359

Yanaba K, Bouaziz JD, Haas KM, Poe JC, Fujimoto M, Tedder TF (2008) A regulatory B cell subset with a unique CD1dhiCD5+ phenotype controls T cell-dependent inflammatory responses. Immunity 28(5):639–650. https://doi.org/10.1016/j.immuni.2008.03.017

Yu X, Petersen F (2018) A methodological review of induced animal models of autoimmune diseases. Autoimmun Rev 17(5):473–479. https://doi.org/10.1016/j.autrev.2018.03.001

Zeng L, Yang K, He Q, Zhu X, Long Z, Wu Y et al (2024) Efficacy and safety of gut microbiota-based therapies in autoimmune and rheumatic diseases: a systematic review and meta-analysis of 80 randomized controlled trials. BMC Med 22(1):110. https://doi.org/10.1186/s12916-024-03303-4

Zhang F, Wei K, Slowikowski K, Fonseka CY, Rao DA, Kelly S et al (2019) Defining inflammatory cell states in rheumatoid arthritis joint synovial tissues by integrating single-cell transcriptomics and mass cytometry. Nat Immunol 20(7):928–942. https://doi.org/10.1038/s41590-019-0378-1

Zhu J, Yamane H, Paul WE (2010) Differentiation of effector CD4 T cell populations (*). Annu Rev Immunol 28:445–489. https://doi.org/10.1146/annurev-immunol-030409-101212

Chapter 10
Computational Modeling in Understanding Autoimmune Disorders

Babita Pandey and Devendra Kumar Pandey

10.1 Introduction

Autoimmune diseases arise when the immune system mistakenly attacks the body's own tissues, leading to chronic inflammation and organ damage. Conditions such as rheumatoid arthritis (RA), systemic lupus erythematosus (SLE), type 1 diabetes, and multiple sclerosis (MS) exhibit complex pathophysiological mechanisms influenced by genetic, environmental, and immunological factors (Davidson and Diamond 2001). Given their heterogeneity and multifactorial nature, predicting disease onset, progression, and treatment response remains a significant challenge.

Computational modeling has emerged as a powerful tool for understanding autoimmune diseases by integrating diverse biological data, including genomics, proteomics, and clinical observations. Advanced approaches such as mechanistic modeling, machine learning, and network-based analyses enable researchers to identify biomarkers, predict disease trajectories, and explore novel therapeutic targets (Kollmann et al. 2017). These methodologies not only enhance our understanding of disease mechanisms but also contribute to precision medicine by tailoring treatments to individual patients.

Recent studies have highlighted the growing role of computational methods in autoimmune disease research. Vivas et al. (2024) provide a comprehensive review of the latest advancements in disease prediction, covering both traditional biomarkers and emerging AI-driven approaches. Their work underscores the importance of integrating AI techniques to improve diagnostic accuracy and early detection.

B. Pandey
Department of Computer Science, Babasaheb Bhimrao Ambedkar University, Lucknow, India

D. K. Pandey (✉)
School of BioSciences and Biotechnology, Lovely Professional University, Phagwara, Punjab, India

Y.-O. Son, M. Ghosh (eds.), *Cutting-Edge Strategies in Drug Delivery and Immunotherapy for Autoimmune Disorders*,
https://doi.org/10.1007/978-981-95-0520-3_10

Similarly, Danieli et al. (2024) examine the application of machine learning (ML) in systemic and organ-specific autoimmune diseases, such as SLE, RA, type 1 diabetes, and autoimmune thyroid disorders. Their findings emphasize the effectiveness of ML-based models in disease classification and risk assessment.

This chapter explores the role of computational approaches in autoimmune disease research. Section 10.2 provides an overview of computational modeling in biology, highlighting key methodologies and their applications. Section 10.3 delves into the use of computational models in autoimmune diseases, focusing on disease prediction, biomarker discovery, and treatment optimization. Finally, Sect. 10.4 discusses challenges, future directions, and the potential of computational strategies in advancing autoimmune disease research and clinical management.

10.2 Overview of Computational Modeling in Biology

This section presents the computational models used in biology. An overview of computational modeling in biology is shown in Fig. 10.1.

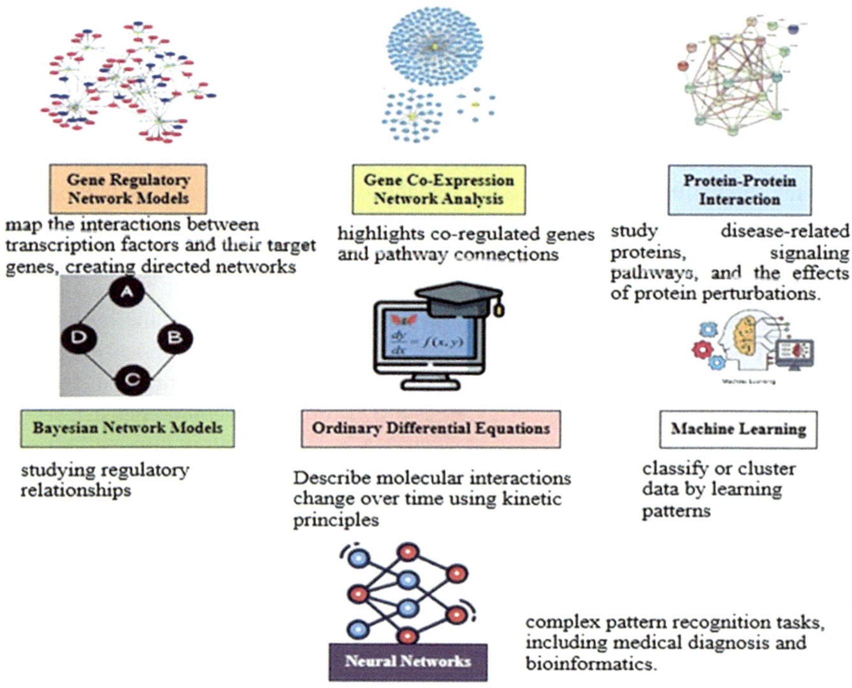

Fig. 10.1 Overview of computational modeling in biology

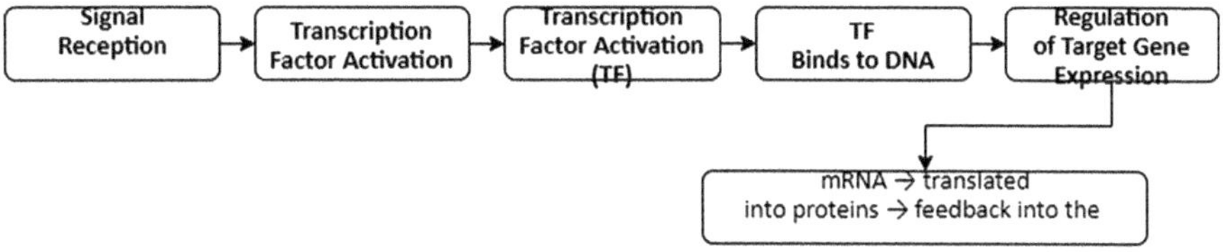

Fig. 10.2 GRN mechanism

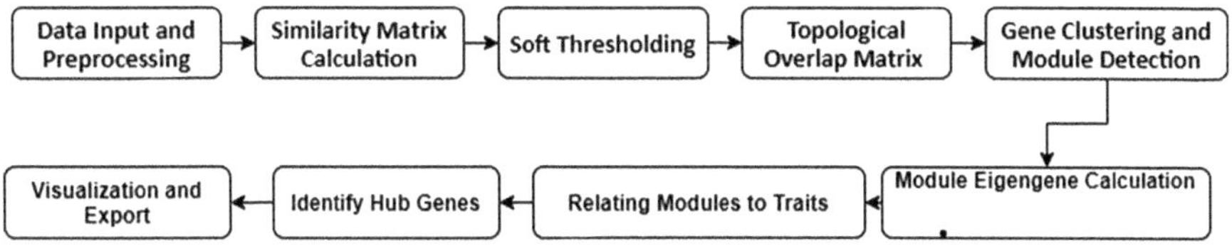

Fig. 10.3 Step-by-step workflow of WGCNA

Gene Regulatory Network Models These models map the interactions between transcription factors and their target genes, creating directed networks. They integrate data from transcription factor binding sites, ChIP-seq, and ATAC-seq to improve accuracy. GRN Mechanism is shown in Fig. 10.2.

Weighted Gene Coexpression Network Analysis (WGCNA) WGCNA identifies gene clusters based on expression correlations, linking them to phenotypes. It highlights coregulated genes and pathway connections by assigning higher weight to strong correlations while filtering out weak ones. Step-by-Step Workflow of WGCNA is shown in Fig. 10.3.

Protein–Protein Interaction (PPI) Networks PPIs are determined through experimental techniques (e.g., yeast two-hybrid, mass spectrometry) or computational methods (e.g., molecular docking). These networks help study disease-related proteins, signaling pathways, and the effects of protein perturbations. Step-by-step workflow of PPI network construction and analysis is shown in Fig. 10.4.

Bayesian Network Models Bayesian networks are directed acyclic graphs (DAGs) where nodes represent variables (e.g., genes) and edges define dependencies. Unlike

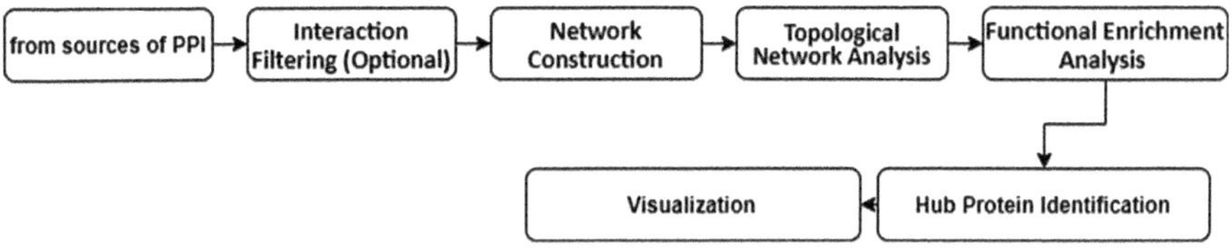

Fig. 10.4 Step-by-step workflow of PPI network construction and analysis

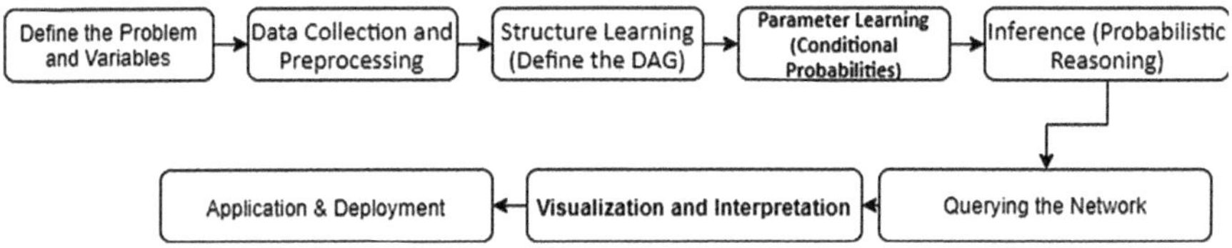

Fig. 10.5 Step-by-step workflow of Bayesian network modeling

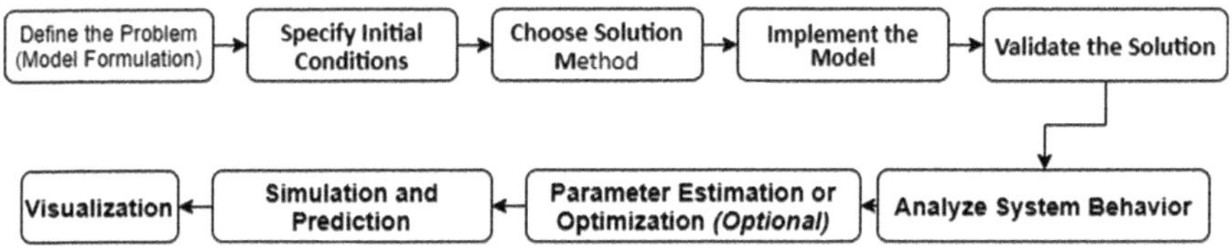

Fig. 10.6 Step-by-step workflow for solving ODEs

cyclic networks, these models allow causal inference, making them useful for studying regulatory relationships. Step-by-step workflow of Bayesian network modeling is shown in Fig. 10.5.

Ordinary Differential Equations (ODEs) ODEs describe how molecular interactions change over time using kinetic principles. They require accurate parameter values but are crucial for modeling drug development and metabolic pathways. Hybrid models combining ODEs and CBM can capture metabolic dynamics. Step-by-step workflow for solving ODEs shown in Fig. 10.6.

Machine Learning ML models classify or cluster data by learning patterns from training datasets. For example, patient records can be analyzed to develop diagnos-

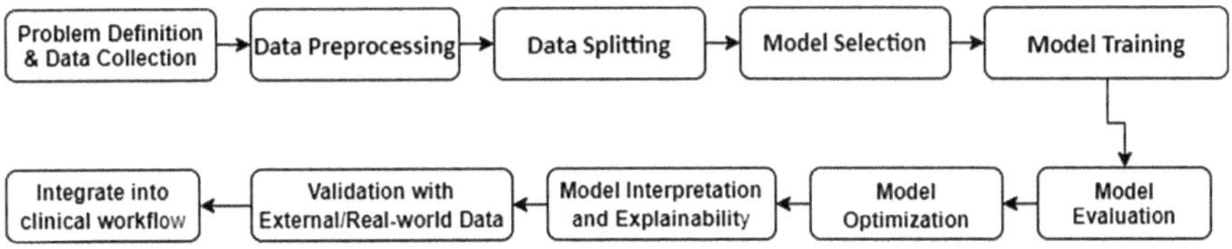

Fig. 10.7 Step-by-step workflow of ML in disease diagnosis tool development

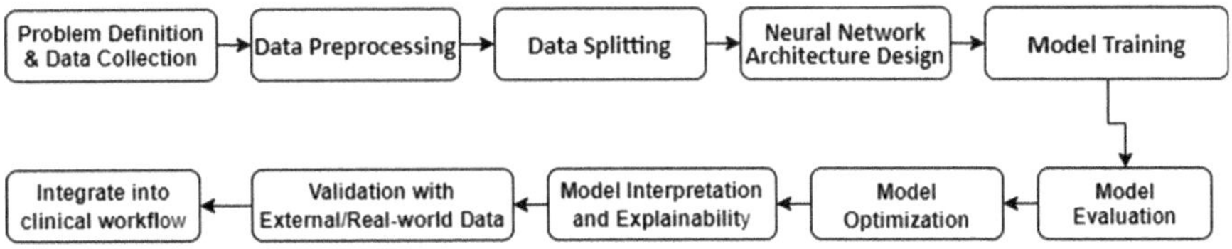

Fig. 10.8 Neural network workflow for disease diagnosis

tic tools for diseases like diabetes. Techniques include nearest neighbor, regression analysis, tree-based methods, and support vector machines. Step-by-step workflow of ML in disease diagnosis tool development is shown in Fig. 10.7.

Neural Networks Inspired by the brain, neural networks process input data through multiple layers to generate outputs. They are widely used for complex pattern recognition tasks, including medical diagnosis and bioinformatics. Neural network workflow for disease diagnosis is shown in Fig. 10.8.

10.3 Application of Computational Models in Autoimmune Diseases

Table 10.1 summarizes the key details from the mechanistic and mathematical models in autoimmune diseases.

Table 10.1 Summary of the key details from the computational models in autoimmune diseases

Authors	Model name	Key features
Sherman and Goessling (2021)	Mathematical modeling	Tracks enzyme levels. Helps in early detection of liver injury. Predicts therapy response.
Iwami et al. (2009)	Vector vaccine design	Examines viral infections triggering/preventing autoimmunity. Identifies immune response function, antigen affinity, and vaccine effectiveness.
Hara and Iwasa (2020)	Mathematical model	Models how viral infections trigger autoimmune diseases. Explores immune memory and cross-immunity effects. Identifies conditions under which infections cause autoimmunity.
Kanth and Devi (2022)	Fractal-fractional model	Uses fractal–fractional calculus to model disease progression. Employs the Caputo operator for immune system analysis. Captures recurrent autoimmune flare-ups.
Root-Bernstein (2015)	Integration of mathematical models	Advocates for integrating mathematical models, experimental data, and clinical observations. Highlights the need for interdisciplinary approaches to understand autoimmune mechanisms.
Menale and Travaglini (2024)	Nonconservative kinetic model	Introduces an external force field to simulate drug treatments in autoimmune responses. Provides theoretical results on model stability and uniqueness.
Kumar and Dhua (2025)	Artificial neural networks (ANNs)	Models interactions of TSH, FT4, antithyroid antibodies (Ab), and thyroid gland size. Validates ANN approach with mean squared error and loss functions
van der Vegt et al. (2022)	Mathematical models	Study immune checkpoint inhibitor-induced autoimmune myocarditis. Analyzes the role of immune cells and cytokines in disease progression. Explores conditions for resolution or chronic inflammation.
Kanth and Devi (2022)	Numerical simulations, mathematical modeling	Describe the dynamics of autoimmune diseases.
Huang et al. (2009)	Decision tree	Leveraged proteomics.

(continued)

Table 10.1 (continued)

Authors	Model name	Key features
Li et al. (2022)	ML algorithm	Combined scRNA-seq and proteomic data, identifying protein markers with strong predictive power for diagnosis and exacerbation.
Turner et al. (2017)	Neural network and random forest	Classification of systemic lupus erythematosus (SLE) using electronic health records (EHR)
Jorge et al. (2019)	Logistic regression and ensemble learning to	Identification of SLE from EHR data.
Barnado et al. (2022)	Random forest and XGBoost models	Electronic health record of three million subjects to identify SLE births.
Chung et al. (2021)	ML models from SNP data	First time used SHAP in ML on genome-wide association studies
Ma et al. (2022)	Random forest	Integrated large-scale scRNA-seq data for autoimmune disease classification.
Adamichou et al. (2021)	Random forest and least absolute shrinkage and selection operator-logistic regression	Created an SLE risk probability index using clinical data, achieving high diagnostic accuracy.
Moingeon et al. (2022)	AI and network-based approach	Identification, prioritization, and validation of therapeutic targets.
Chen et al. (2021)	XGBoost	Predict lupus nephritis (LN) flare-ups with significant risk stratification.
Tang et al. (2018)	Random forest	Classify renal damage histologically with moderate accuracy.
Yang et al. (2022)	Mask R-CNN	Kidney biopsy samples, achieving high precision.
Frangou et al. (2022)	Recursive feature elimination	Applied on transcriptomic data to distinguish LN patients with AUCs of 0.80–0.99.
Wang et al. (2022)	LASSO, SVM-RFE	Identified genetic biomarkers, achieving AUCs of 0.741–0.911.
Wang et al. (2023)	Multiple ML models, XGBoost	Best for noninvasive LN diagnosis.
Gu et al. (2021)	ML	Applied on immune cell data, identifying 13 key T-cell subsets in SLE patients with anxiety.
Qin et al. (2023)	ML	Ultrasonic radiomics, with MLP achieving 73.3% accuracy.
Zhuo et al. (2020).	SVM	To brain perfusion data, identifying cerebral blood flow changes in neuropsychiatric SLE (NPSLE) patients.
Yuan et al. (2022)	Noise-immune ML	Developed a noise-immune ML method for early detection.

(continued)

Table 10.1 (continued)

Authors	Model name	Key features
Li et al. (2020)	Combined H-MRS with BL-SVM	Diagnosis of early NPSLE with AUC 95%.
Tan et al. (2022)	SVM	Applied on H-MRS data, achieving 94.9% accuracy.
Simos et al. (2020)	RF	Applied to MRI data, distinguishing NPSLE with 73.5% accuracy.
Barraclough et al. (2022)	Similarity network fusion	Applied on cognitive function data to classify NPSLE subtypes.
Rumetshofer et al. (2022)	Cluster analysis	Identified white matter hyperintensity patterns
Fang and Sun (2024)	Relational neural networks and long short-term memory networks. Gene interaction network	Achieving higher AUC (0.7274) and AUPR (0.7599).
Hong et al. (2024)	PPI network: advance network toxicology strategies	Molecular docking/STRING Database.
Kim et al. (2025)	Random walk with restart, which is a network exploration-based analysis method	Intertissue communicators, disease-gene relation could be disease-related mutations, diagnostic biomarkers, or therapeutic targets.
Shanthamallu et al. (2024)	PPI network: Page rank algorithm	Predicting biomarkers area under the receiver operating characteristic.
Chen et al. (2015)	GRN: molecular deconvolution	Compartmentalizing and targeting molecular behaviors contributing to interactions between tissues.
Zong et al. (2025)	Modular-based network proximity approach	Explore drug repositioning and clinical advantages of formulae.
Li et al. (2024)	Weighted gene coexpression network analysis and machine learning	ROC curve analysis of the predictive model revealed moderate diagnostic value to discover important biomarkers associated with immune cell infiltration

10.4 Discussion, Challenges, and Future Directions

10.4.1 Discussion

Figure 10.9 illustrates an ML pipeline for biomarker identification and predictive modeling in healthcare using various data sources. The process begins with multiple data sources, including electronic health records (EHR), serum proteomics, lipid profiles, single-cell RNA sequencing (scRNA-seq), and genomics. These diverse datasets provide comprehensive biological and clinical information necessary for training ML models. The ML models applied include support vector machines

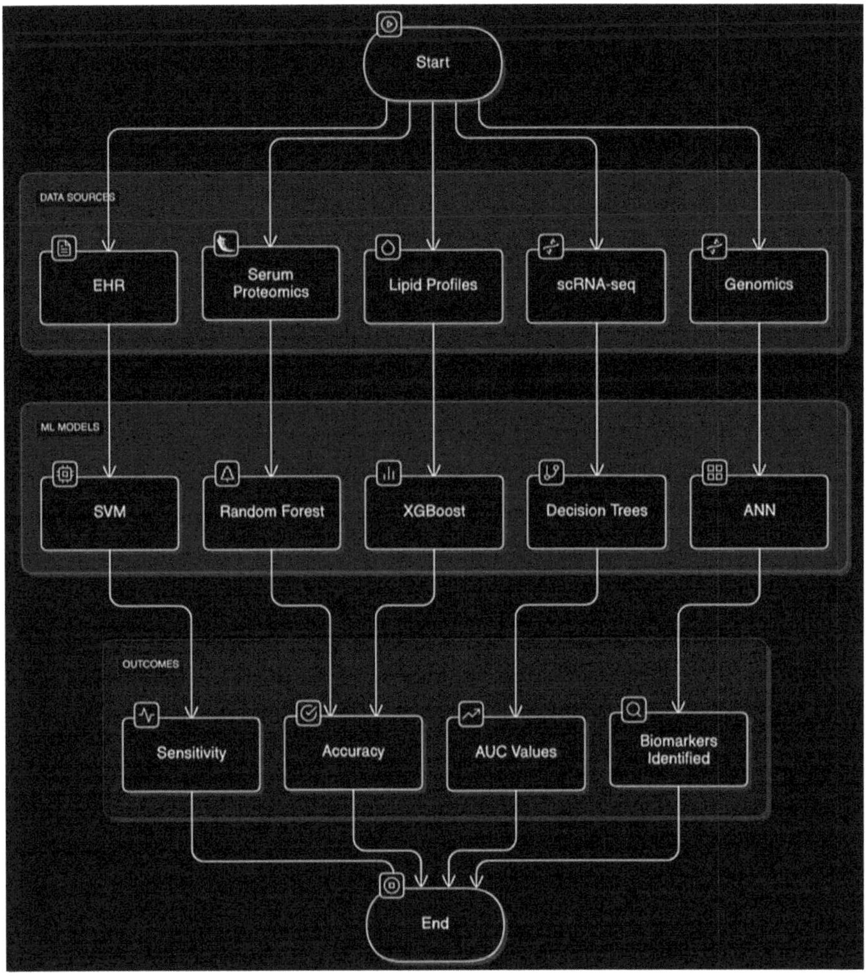

Fig. 10.9 Machine learning pipeline for biomarker identification and predictive modeling in healthcare

(SVM), random forest, XGBoost, decision trees, and artificial neural networks (ANN). Each model processes the input data to extract meaningful patterns and relationships that can aid in diagnosis, prognosis, or biomarker discovery. The final stage of the pipeline focuses on outcome evaluation, where key performance metrics such as sensitivity, accuracy, and area under the curve (AUC) values are assessed. Additionally, the identification of potential biomarkers is a critical outcome, providing insights for medical research and precision medicine. This structured pipeline ensures an efficient integration of multiomics data with ML techniques, enabling robust predictive modeling and biomarker discovery in healthcare applications.

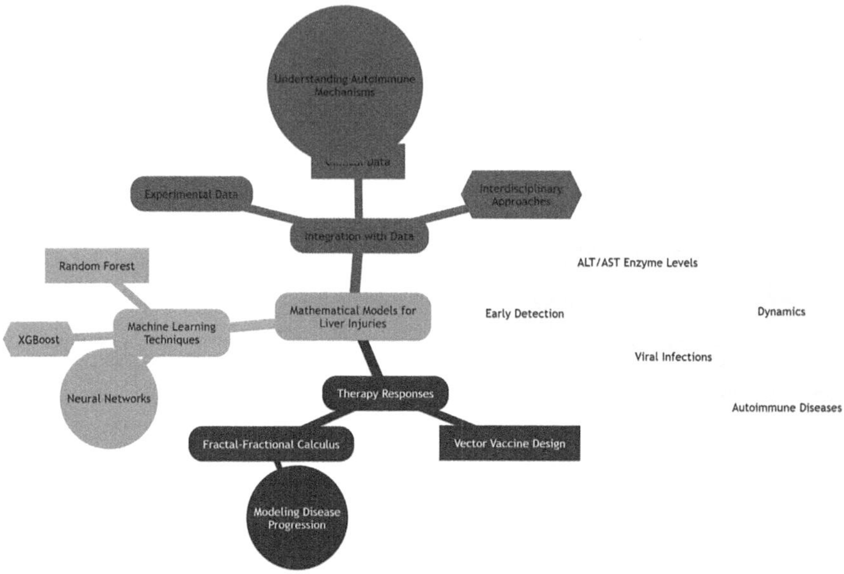

Fig. 10.10 Mathematical and AI-driven approaches in autoimmune disease modeling

Figure 10.10 illustrates an integrated approach to understanding autoimmune liver injuries using mathematical models, clinical and experimental data, and machine learning techniques. It highlights key components such as therapy responses, early detection, and disease progression modeling. Methods like fractal-fractional calculus and AI-based techniques (random forest, XGBoost, neural networks) are used to analyze liver injury dynamics, enzyme levels, and vaccine design.

10.4.2 Challenges

Despite significant advancements, computational modeling in autoimmune research faces several challenges. Figure 10.11 shows the challenges faced in deploying computational modeling in autoimmune research.

- *Data Heterogeneity:* Integrating diverse biological and clinical datasets requires standardized frameworks and interoperability (Wang and He 2021).
- *Solution:*
  - Adopt standardized formats such as FHIR (Fast Healthcare Interoperability Resources), HL7, OMOP (Observational Medical Outcomes Partnership), and CDISC (Clinical Data Interchange Standards Consortium) for clinical and biomedical data.
  - Use ontologies like Gene Ontology (GO), SNOMED CT, UMLS (Unified Medical Language System), and MeSH to standardize terminologies across datasets.

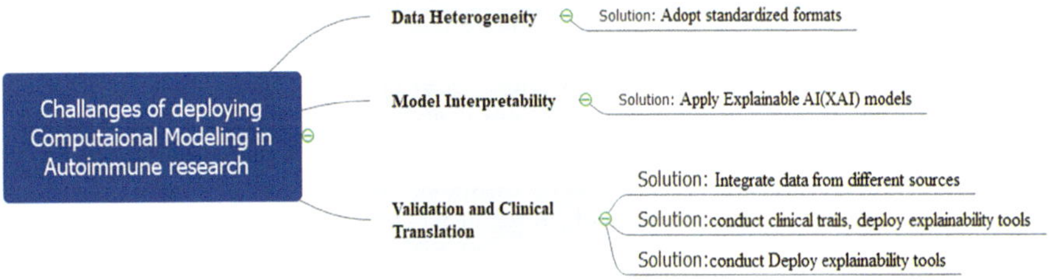

Fig. 10.11 Challenges in deploying Computational modeling in autoimmune research

- *Model Interpretability:* Complex ML models often lack explainability, necessitating transparent and interpretable approaches (Lipton 2018).
- *Solution:* Apply various explainable AI (XAI) models to make models transparent and interpretable.
- *Validation and Clinical Translation:* Computational predictions must be rigorously validated through experimental and clinical studies before being adopted in clinical practice.
- *Solution:* Integrate data from different sources, conduct clinical trials, and deploy explainability tools.

10.4.3 Future Directions

- Expansion of AI-driven biomarker discovery and precision medicine applications (Wang et al. 2021).
- Collaborative efforts between computational scientists and clinicians to bridge the gap between theoretical modeling and real-world clinical applications (Li et al. 2022).

10.5 Conclusion

Computational modeling has revolutionized autoimmune disease research by enhancing diagnosis, biomarker discovery, and treatment strategies. Mathematical models aid in early detection, treatment optimization, and drug efficacy evaluation. Machine learning improves SLE and LN classification, risk assessment, and imaging-based diagnostics. Network toxicology strategies, including PPI networks and GRNs, refine biomarker identification and drug repositioning. Integrating these computational approaches with clinical research advances precision medicine, optimizing autoimmune disease management.

References

Adamichou C, Genitsaridi I, Nikolopoulos D, Nikoloudaki M, Repa A, Bortoluzzi A et al (2021) Lupus or not? SLE risk probability index (SLERPI): a simple, clinician-friendly machine learning-based model to assist the diagnosis of systemic lupus erythematosus. Ann Rheum Dis 80:758–766. https://doi.org/10.1136/ANNRHEUMDIS-2020-219069

Barnado A, Eudy AM, Blaske A, Wheless L, Kirchoff K, Oates JC et al (2022) Developing and validating methods to assemble systemic lupus erythematosus births in the electronic health record. Arthritis Care Res 74:849–857. https://doi.org/10.1002/ACR.24522

Barraclough M, Erdman L, Diaz-Martinez JP, Knight A, Bingham K, Su J et al (2022) Systemic lupus erythematosus phenotypes formed from machine learning with a specific focus on cognitive impairment. Rheumatology (Oxford). https://doi.org/10.1093/RHEUMATOLOGY/KEAC653

Chen JC, Cerise JE, Jabbari A, Clynes R, Christian AM (2015) Master regulators of infiltrate recruitment in autoimmune disease identified through network based molecular deconvolution. Cell Syst 1:326–337

Chen Y, Huang S, Chen T, Liang D, Yang J, Zeng C et al (2021) Machine learning for prediction and risk stratification of lupus nephritis renal flare. Am J Nephrol 52:152–160. https://doi.org/10.1159/000513566

Chung C-W et al (2021) Machine learning approaches for the genomic prediction of rheumatoid arthritis and systemic lupus erythematosus. BioData Min 14:52. https://doi.org/10.1186/S13040-021-00284-5

Danieli G, Brunetto S, Gammeri L, Palmeri D, Claudi I, Shoenfeld Y, Gangemi S (2024) Machine learning application in autoimmune diseases: state of art and future prospectives. Autoimmun Rev 23(2):103496

Davidson A, Diamond B (2001) Autoimmune diseases. N Engl J Med 345(5):340–350

Fang F, Sun Y (2024) Prediction of systemic lupus erythematosus-related genes based on graph attention network and deep neural network. Comput Biol Med 175:108371

Frangou E, Garantziotis P, Grigoriou M, et al. (2022). Ann Rheum Dis 81:1409–1419.

Gu XX, Jin Y, Fu T, Zhang XM, Li T, Yang Y et al (2021) Relevant characteristics analysis using natural language processing and machine learning based on phenotypes and T cell subsets in systemic lupus erythematosus patients with anxiety. Front Psychol 12:793505. https://doi.org/10.3389/FPSYT.2021.793505

Hara and Iwasa (2020) Autoimmune diseases initiated by pathogen infection: mathematical modeling. J Theor Biol 498:110296. ISSN 0022-5193, https://doi.org/10.1016/j.jtbi.2020.110296

Hong Y, Wang D, Lin Y, Yang Q, Wang Y, Xie Y, Wanyi S, Gao S, Hua C (2024) Environmental triggers and future risk of developing autoimmune diseases: molecular mechanism and network toxicology analysis of bisphenol A. Ecotoxicol Environ Saf 288:117352

Huang Z, Shi Y, Cai B, Wang L, Wu Y, Ying B et al (2009) MALDI-TOF MS combined with magnetic beads for detecting serum protein biomarkers and establishment of boosting decision tree model for diagnosis of systemic lupus erythematosus. Rheumatology (Oxford) 48:626–631. https://doi.org/10.1093/RHEUMATOLOGY/KEP058

Iwami S, Takeuchi Y, Iwamoto K, Naruo Y, Yasukawa M (2009) A mathematical design of vector vaccine against autoimmune disease. J Theor Biol 256(3):382–392. ISSN 0022-5193, https://doi.org/10.1016/j.jtbi.2008.09.038

Jorge A, Castro VM, Barnado A, Gainer V, Hong C, Cai T et al (2019) Identifying lupus patients in electronic health records: development and validation of machine learning algorithms and application of rule-based algorithms, Semin. Arthritis Rheum 49:84–90. https://doi.org/10.1016/J.SEMARTHRIT.2019.01.002

Kanth ASVR, Devi S (2022) A computational approach for numerical simulations of the fractal–fractional autoimmune disease model. Chaos, Solitons Fractals 165(Part 2):112829. ISSN 0960-0779, https://doi.org/10.1016/j.chaos.2022.112829

Kim K, Han M, Lee D (2025) InTiCAR: network-based identification of significant inter-tissue communicators for autoimmune diseases. Comput Struct Biotechnol J 27:333–345

Kollmann TR et al (2017) Systems biology in vaccine research. Immunol Rev 285(1):247–262

Kumar R, Dhua S (2025) Dynamic analysis of Hashimoto's thyroiditis bio-mathematical model using artificial neural network. Math Comput Simul 229:235–245. ISSN 0378-4754, https://doi.org/10.1016/j.matcom.2024.10.001

Li B et al (2020) Predicting autoimmune disease-associated genes using PPI networks. Front Immunol 11:1234

Li Y, Ma C, Liao S, Qi S, Meng S, Cai W et al (2022) Combined proteomics and single cell RNA-sequencing analysis to identify biomarkers of disease diagnosis and disease exacerbation for systemic lupus erythematosus. Front Immunol 13:969509. https://doi.org/10.3389/FIMMU.2022.969509

Li Z, Miao L, Ren G, Wang H, Shangguan L, Zhao H, Li X (2024) Weighted gene coexpression network analysis and machine learning for the determination of tfh cell and B cell infiltrating biomarkers in thymoma-associated myasthenia gravis. Heliyon 10(14):e34364

Lipton ZC (2018) The mythos of model interpretability. Queue 16(3):31–57

Ma Y, Chen J, Wang T, Zhang L, Xu X, Qiu Y et al (2022) Accurate machine learning model to diagnose chronic autoimmune diseases utilizing information from B cells and monocytes. Front Immunol 13:870531. https://doi.org/10.3389/FIMMU.2022.870531

Menale M, Romina Travaglini A (2024) Nonconservative kinetic model under the action of an external force field for modeling the medical treatment of autoimmune response. Commun Nonlinear Sci Numer Simul 137:108126. ISSN 1007-5704, https://doi.org/10.1016/j.cnsns.2024.108126

Moingeon P, Kuenemann M, Guedj M (2022) Artificial intelligence-enhanced drug design and development: toward a computational precision medicine. Drug Discov Today 27(1):215–222

Qin X, Xia L, Zhu C, Hu X, Xiao W, Xie X et al (2023) Noninvasive evaluation of lupus nephritis activity using a Radiomics machine learning model based on ultrasound. J Inflamm Res 16:433–441. https://doi.org/10.2147/JIR.S398399

Root-Bernstein R (2015) Towards an integration of mathematical models, theories and observations concerning autoimmune diseases. J Theor Biol 375:1–3. ISSN 0022-5193, https://doi.org/10.1016/j.jtbi.2015.04.003

Rumetshofer T, Inglese F, de Bresser J, Mannfolk P, Strandberg O, Jönsen A et al (2022) Tract-based white matter hyperintensity patterns in patients with systemic lupus erythematosus using an unsupervised machine learning approach. Sci Rep:12. https://doi.org/10.1038/S41598-022-25990-W

Shanthamallu US, Kilpatrick C, Jones A, Rubin J, Saleh A, Barabási A-L, Akmaev VR, Ghiassian SD (2024) A network-based framework to discover treatment-responsee predicting biomarkers for complex diseases. J Mol Diagn 26(10)

Sherman and Goessling (2021) Fr 397 mathematical modeling of ALT and AST trajectories enables early recognition of liver injury resolution and predicts response to therapy in autoimmune hepatitis. Gastroenterology 160(6):S-811. ISSN 0016-5085, https://doi.org/10.1016/S0016-5085(21)02656-1

Simos NJ, Dimitriadis SI, Kavroulakis E, Manikis GC, Bertsias G, Simos P et al (2020) Quantitative identification of functional connectivity disturbances in neuropsychiatric lupus based on resting-state fMRI: a robust machine learning approach. Brain Sci 10:1–18. https://doi.org/10.3390/BRAINSCI10110777

Tan G, Huang B, Cui Z, Dou H, Zheng S, Zhou T (2022) A noise-immune reinforcement learning method for early diagnosis of neuropsychiatric systemic lupus erythematosus. Math Biosci Eng 19:2219–2239. https://doi.org/10.3934/MBE.2022104

Tang Y, Zhang W, Zhu M, Zheng L, Xie L, Yao Z et al (2018) Lupus nephritis pathology prediction with clinical indices. Sci Rep 8:10231. https://doi.org/10.1038/S41598-018-28611-7

Turner CA, Jacobs AD, Marques CK, Oates JC, Kamen DL, Anderson PE et al (2017) Word2Vec inversion and traditional text classifiers for phenotyping lupus. BMC Med Inform Decis Mak 17:126. https://doi.org/10.1186/S12911-017-0518-1

van der Vegt SA, Polonchuk L, Wang K, Waters SL, Baker RE (2022) Mathematical modelling of autoimmune myocarditis and the effects of immune checkpoint inhibitors. J Theor Biol 537:111002. ISSN 0022-5193, https://doi.org/10.1016/j.jtbi.2021.111002

Vivas AJ, Boumediene S, Tobón GJ (2024) Predicting autoimmune diseases: a comprehensive review of classic biomarkers and advances in artificial intelligence. Autoimmun Rev 23(9):103611

Wang Z, He Y (2021) Precision omics data integration and analysis with interoperable ontologies and their application for COVID-19 research, Brief Funct Genomics 20(4):235–248. https://doi.org/10.1093/bfgp/elab029

Wang L, Yang Z, Yu H, Lin W, Wu R, Yang H et al (2022) Predicting diagnostic gene expression profiles associated with immune infiltration in patients with lupus nephritis. Front Immunol 13:839197. https://doi.org/10.3389/FIMMU.2022.839197

Wang DC, Xu WD, Wang SN, Wang X, Leng W, Fu L et al (2023) Lupus nephritis or not? A simple and clinically friendly machine learning pipeline to help diagnosis of lupus nephritis. Inflamm Res 72:1315–1324. https://doi.org/10.1007/S00011-023-01755-7

Yang CK, Lee CY, Wang HS, Huang SC, Liang PI, Chen JS et al (2022) Glomerular disease classification and lesion identification by machine learning. Biom J 45:675–685. https://doi.org/10.1016/J.BJ.2021.08.011

Yuan Y, Quan T, Song Y, Guan J, Zhou T, Wu R (2022) Noise-immune extreme ensemble learning for early diagnosis of neuropsychiatric systemic lupus erythematosus. IEEE J Biomed Health Inform 26:3495–3506. https://doi.org/10.1109/JBHI.2022.3164937

Zhuo Z, Su L, Duan Y, Huang J, Qiu X, Haller S et al (2020) Different patterns of cerebral perfusion in SLE patients with and without neuropsychiatric manifestations. Hum Brain Mapp 41:755–766. https://doi.org/10.1002/HBM.24837

Zong W, Tiana S, Niua Q, Lia X, Wanga P, Tongb L, Zhanga S, Zhengb D, Zhanga Y, Xiongc W, Caid Q, Zenga Z, Wanga J'a, Xua H, Zhange H, Lia B (2025) Comparable clinical advantages identification of three formulae on rheumatic disease using a modular-based network proximity approach. J Ethnopharmacol 337:118764

Chapter 11
Novel Compounds and Drug Delivery System in Autoimmune Therapy

Surender Jangra and Anu Bansal

11.1 Introduction

Autoimmune diseases cover a variety of body conditions characterized by an aberrant immune response that targets healthy cells and tissues of the body. Consequently, this inappropriate immune response causes chronic inflammation, tissue damage, and malfunction of the affected organ/s. Research suggests that systemic lupus erythematosus (SLE), inflammatory bowel diseases (IBD), rheumatoid arthritis (RA), type 1 diabetes (T1D), and multiple sclerosis (MS) are among the common autoimmune diseases, and collectively, they are estimated to affect 5–10% of the global population (Hayter and Cook 2012; Li et al. 2023). These conditions are major contributor to morbidity and mortality worldwide, with their prevalence rising in recent decades, possibly due to changing environmental factors such as dietary shift, pollution and toxins, and infection patterns (Rosenblum et al. 2015). The etiology of autoimmune diseases is indeed complex, and characterized by complex interactions among various factors, including genetic, environmental, and immunological factors. The interplay of genetic factors, such as variations in the human leukocyte antigen (HLA) genes, and epigenetic modifications, including DNA methylation, histone modification, and noncoding RNAs, play a significant role in determining susceptibility to autoimmune diseases, as well as their onset and progression (Farh et al. 2015). Environmental factors, including viral infections,

S. Jangra
Department of Chemistry and Biochemistry, Sharda School of Engineering and Sciences, Sharda University, Greater Noida, Uttar Pradesh, India

A. Bansal (✉)
School of Bioengineering & Biosciences, Lovely Professional University, Phagwara, Punjab, India
e-mail: anu.21971@lpu.co.in

Y.-O. Son, M. Ghosh (eds.), *Cutting-Edge Strategies in Drug Delivery and Immunotherapy for Autoimmune Disorders*,
https://doi.org/10.1007/978-981-95-0520-3_11

smoking, and gut microbiota dysbiosis, are increasingly recognized as key contributors to the pathogenesis of autoimmune disorders (Belkaid and Hand 2014). Dysregulation in immune pathways, particularly involving B cells, T cells, and proinflammatory cytokines like tumor necrosis factor-alpha (TNF-α) and various interleukins, plays a crucial role in exacerbating autoimmune pathology (Firestein and McInnes 2017).

The chronic and often progressive nature of autoimmune diseases poses significant challenges for both individuals and healthcare systems. Patients with autoimmune diseases face a myriad of challenges that significantly diminish their quality of life, such as chronic pain, fatigue, organ dysfunction, and psychological effects. Whereas, healthcare systems bear the burden of managing these lifelong conditions, including their frequent complications (Cao et al. 2023). The economic impact is substantial, encompassing direct medical costs as well as indirect costs, including lost productivity due to absenteeism and presenteeism (Alamanos and Drosos 2005). Current therapeutic strategies for managing autoimmune diseases encompass a range of pharmaceuticals such as corticosteroids, nonsteroidal antiinflammatory drugs (NSAIDs), and disease-modifying antirheumatic drugs (DMARDs), which include both conventional synthetic and biologic agents (Smolen et al. 2016). Although these therapies can mitigate symptoms and delay the disease progression, their systemic nature often leads to serious side effects, including immunosuppression, increased infection risk, and potential malignancies (Curtis et al. 2017). Additionally, a substantial proportion of patients fail to achieve long-term remission or experience disease recurrence, highlighting the unmet need for more effective and targeted therapies.

Recent advances in immunology and molecular biology have provided critical insights into autoimmunity mechanisms, paving the way for developing targeted therapeutic agents. These include biologics like monoclonal antibodies and small molecules that specifically modulate key immune pathways, such as Janus kinase (JAK) inhibitors and TNF-α inhibitors (O'Shea et al. 2004). Concurrently, advancements in drug delivery systems, including nanotechnology-based carriers and controlled-release formulations, have enabled the precise delivery of therapeutic agents while minimizing systemic exposure and side effects (Serra and Santamaria 2015). Integrating interdisciplinary approaches, encompassing molecular biology, bioengineering, and materials science, is indeed revolutionizing the treatment landscape for autoimmune diseases. This chapter delves into the scientific foundations of these breakthroughs, focusing on novel therapeutic agents and advanced drug delivery platforms designed to enhance the efficacy and safety of autoimmune therapies.

11.2 Novel Therapeutic Compounds

11.2.1 Biologics

Biologics are complex medications that are derived from living cells or organisms, including bacteria, yeast, and mammalian cells. These treatments are designed to interact and modulate specific components of the immune system, making them effective for conditions characterized by immune dysregulation. This group of treatment includes monoclonal antibodies, fusion proteins, and recombinant cytokines, which target proinflammatory cytokines, immune cell receptors, or costimulatory molecules. Recent progress in molecular understanding and production has significantly expanded the arsenal for treating autoimmune diseases, featuring novel anti-TNF blockers, interleukin (IL) inhibitors, and B-cell therapies. The development of tumor necrosis factor-alpha (TNF-α) inhibitors has been a significant advancement in the treatment of autoimmune disorders. The researchers studied and found that TNF-α is a key proinflammatory cytokine that plays a key role in the inflammatory processes linked with autoimmune disorders, such as RA, IBD, and psoriasis. Anti-TNF biologics, including soluble TNF receptor fusion proteins and anti-TNF antibodies, have transformed the treatment of autoimmune disorders but their effectiveness is limited by nonresponse rate of up to 40% in patients and adverse effects, including increased susceptibility to infections or potential development of new autoimmune conditions. The selective inhibition of TNF receptors (TNFR1 or TNFR2) presents a promising avenue for developing safer and potentially more effective treatments for autoimmune disorders. Biologics are indeed at the forefront of treatment strategies for autoimmune diseases, largely due to their specificity and relatively low toxicity compared to traditional systemic therapies. The advancing research in understanding the pathophysiological processes of autoimmune diseases plays a crucial role in the development of novel biologics (Zarros et al. 2017).

Unlike traditional small-molecule drugs, biologics are complex macromolecules, typically proteins, engineered to interact with distinct immune system components, such as cytokines, cell surface markers, or intracellular signaling pathways. Biologic therapies for rheumatologic diseases indeed represent a significant advancement in medical treatment, specifically for conditions such as RA, lupus, psoriatic arthritis, and ankylosing spondylitis. By targeting specific immune system molecules, biologics reduce inflammation and halt disease progression more effectively than traditional immunosuppressive treatments. Epratuzumab, sifalimumab, belimumab, rituximab, golimumab, abatacept, ofatumumab, tocilizumab, and certolizumab are common biologics that are used in conditions like vasculitis, SLE, and RA. Belimumab is notable as the first FDA-approved targeted therapy for SLE. It also shows promise for off-label uses in various resistant autoimmune conditions (Rosman et al. 2013). Some of the biologic agents for autoimmune diseases are shown in Table 11.1. Despite their efficacy, several challenges, such as high costs, intravenous administration, and adverse events, restrict their widespread use as first-line options.

Table 11.1 Biologic agents for the treatment of autoimmune diseases

Biologic	Target	Indications	Reference
Infliximab	TNF-α	Ankylosing Spondylitis (AS) Psoriatic Arthritis (PsA), IBD, RA	Liu et al. (2022), Feldmann and Maini (2003)
Adalimumab	TNF-α	RA, IBD, PsA, AS, Hidradenitis Suppurativa (HS), Ulcerative Colitis (UC), Uveitis, Crohn's disease (CD)	Burmester et al. (2019)
Etanercept	TNF-α	RA, PsA, AS, Juvenile Idiopathic Arthritis (JIA), Plaque Psoriasis	Burness and Duggan (2016), Greenberg and Kishimoto (2006)
Rituximab	CD20	RA, SLE, Chronic Lymphocytic Leukemia (CLL), Non-Hodgkin's Lymphoma (NHL)	Machado et al. (2014), Plosker and Figgitt (2003), Iwata et al. (2018)
Tocilizumab	IL-6	RA, Cytokine Release Syndrome (CRS), Systemic JIA, Giant Cell Arteritis (GCA)	Nishimoto et al. (2004), Nishimoto and Kishimoto (2008)
Sarilumab	IL-6	RA	Burmester et al. (2017)
Secukinumab	IL-17A	Psoriasis, AS, PsA	Baeten et al. (2015), Langley et al. (2014)
Ixekizumab	IL-17A	Psoriasis, PsA, AS	Gordon et al. (2016), Deodhar et al. (2020)
Ustekinumab	IL-12/IL-23	Crohn's Disease, Psoriasis, PsA	Kimball et al. (2013), Feagan et al. (2016)
Guselkumab	IL-23	Psoriasis, PsA	McGeachy et al. (2019)
Belimumab	BAFF	SLE	Navarra et al. (2011)
Abatacept	CTLA-4	RA, PsA, JIA	Pombo-Suarez and Gomez-Reino (2019), Brunner et al. (2020)
Dupilumab	IL-4Rα	Chronic Rhinosinusitis with Nasal Polyps, Atopic Dermatitis, Asthma,	McGeachy et al. (2019), Wenzel et al. (2016)
Anifrolumab	Type I IFN receptor	SLE	Navarra et al. (2011)
Eculizumab	Complement C5	Atypical Hemolytic Uremic Syndrome (aHUS), Myasthenia Gravis (MG), Paroxysmal Nocturnal Hemoglobinuria (PNH)	Hillmen et al. (2004), Howard et al. (2017)
Ocrelizumab	CD20	Multiple Sclerosis (MS)	Montalban et al. (2017)
Natalizumab	α4-integrin	MS, Crohn's Disease	Armuzzi and Felice (2013), Polman et al. (2006)
Vedolizumab	α4β7-integrin	CD, UC	Feagan et al. (2013)
Tildrakizumab	IL-23p19	Psoriasis	Reich et al. (2017)
Brodalumab	IL-17RA	Psoriasis	Lebwohl et al. (2015)

Medications such as infliximab, adalimumab, and etanercept have shown effectiveness in various autoimmune conditions, including RA, IBD, and ankylosing spondylitis (AS) (McInnes and Schett 2017). Monoclonal antibodies like infliximab and adalimumab are designed to neutralize proinflammatory cytokines, particularly TNF-α. This cytokine has a critical role in the inflammatory processes linked with several autoimmune conditions, including RA, IBD, and psoriasis (McInnes and Schett 2017). These agents have effectively reduced disease activity, improved physical function, and prevented structural joint damage. New biologics targeting specific interleukins are being developed for the treatment of autoimmune disorders. For example, secukinumab and ixekizumab specifically inhibit IL-17A, a cytokine associated with psoriasis and AS, whereas ustekinumab targets IL-12 and IL-23 and is effective for psoriasis and Crohn's disease (Smolen et al. 2016).

Rituximab, a chimeric monoclonal antibody, selectively eliminates the B-cells from the immune system by targeting their CD20 antigen. B-cells have a crucial role in the pathogenesis of autoimmune conditions through two main mechanisms, namely, production of autoantibodies and facilitation of antigen presentation to T cells. Rituximab is effective in conditions such as SLE and multiple sclerosis (MS), where B-cell–mediated mechanisms are implicated (Randall 2016). Similarly, inhibitors of the B-cell activating factor (BAFF), such as belimumab, have shown promise in reducing disease flares in SLE by modulating B-cell survival, differentiation, and function (Navarra et al. 2011). Targeting T-cell activation and function has also been a successful strategy in the creation of biological treatments for various autoimmune disorders. Abatacept is a fusion protein comprising the extracellular domain of cytotoxic T-lymphocyte–associated antigen-4 (CTLA-4) and the Fc portion of human IgG1. Its unique structure allows it to specifically prevent T-cell activation by interfering with costimulatory signals via CD80 and CD86 receptors on antigen-presenting cells. Abatacept is approved for the treatment of RA and juvenile idiopathic arthritis (JIA). By interfering with the activation of T-lymphocytes, it has shown benefits in reducing disease progression while maintaining a favorable safety profile (Pombo-Suarez and Gomez-Reino 2019).

Interleukin (IL) inhibitors represent a vital class of biologic therapies used to manage various autoimmune and inflammatory processes. IL-6, a proinflammatory cytokine, is implicated in the pathogenesis of several conditions, including RA and giant cell arteritis. Tocilizumab is a monoclonal antibody that targets IL-6 receptors, effectively blocking the action of IL-6. By blocking IL-6 signaling, tocilizumab reduces the inflammation, alleviating symptoms and disease progression in conditions where IL-6 is instrumental (Tanaka et al. 2014). Similarly, IL-17 and IL-23 inhibitors, including secukinumab and ustekinumab, have transformed the landscape of treatment for psoriasis and ankylosing spondylitis. IL-17 and IL-23 play crucial roles in the Th17-mediated inflammatory pathway, which is central to the pathogenesis of psoriasis and ankylosing spondylitis (McGeachy et al. 2019).

While biologics have revolutionized the management of many autoimmune and inflammatory conditions, they are also associated with several limitations. Immunogenicity, where the patient's immune system develops antibodies against the biologic (antidrug antibodies), can reduce therapeutic effectiveness or cause adverse reactions (Strand et al. 2021). Additionally, the high cost of biologic therapies presents a significant barrier to access, especially in low- and middle-income countries. Furthermore, the risk of infections, particularly opportunistic infections, remains a concern due to the immunosuppressive nature of these agents (Curtis et al. 2017; de La Forest Divonne et al. 2017). Biologic therapies have revolutionized the treatment of autoimmune rheumatic diseases (ARD) by targeting cytokines and immune cells involved in inflammation. However, their use can induce autoimmune phenomena, ranging from isolated antibody production (e.g., ANA, anti-dsDNA) to full-blown autoimmune diseases, including lupus, vasculitis, and paradoxical inflammation (e.g., psoriasis). Neurological manifestations, such as demyelinating disorders and progressive multifocal leukoencephalopathy (PML), and organ-specific conditions, like interstitial lung disease, are important considerations when using biologic therapies. The underlying mechanisms associated with these risks include cytokine shifts, immune dysregulation, and reactivation of latent infections. Early recognition and management of these adverse events are crucial to prevent misdiagnoses and to optimize patient outcomes, highlighting the need for close monitoring of patients and interdisciplinary care (Drosos et al. 2021).

Ongoing research focuses on optimizing biologic therapy by identifying novel therapeutic targets, improving drug delivery mechanisms, and developing biosimilars—cost-effective alternatives with comparable efficacy and safety. The emergence of personalized medicine, which utilizes biomarkers to predict patient response, is anticipated to further refine the applications of biologics, improving treatment outcomes while minimizing associated risks. Biologics provide significant advantages, including high specificity and effectiveness in targeting diseases. However, obstacles such as high production costs, the risk of immunogenic responses, and the requirement for injectable administration hinder their widespread application. Advancements in protein engineering and formulation technology seek to address these challenges, improving the therapeutic capabilities of biologics. Continuous research into new biologic agents reveals additional targets, including IL-6 inhibitors and therapies that block costimulatory signals, thereby broadening the treatment options for autoimmune conditions.

11.2.2 Small Molecules

Low-molecular-weight compounds, also known as small molecules, have a major role in the regulation of intracellular signaling pathways, which are vital in the onset and advancement of autoimmune conditions. A study explores the significant advancements made between 2018 and 2020 in the creation of small molecule (SM) therapeutics, particularly kinase inhibitors, for the management of autoimmune and inflammatory conditions. Highlighting their enhanced target and tissue selectivity, the study aims to improve the therapeutic index of these treatments, thereby reducing side effects while increasing efficacy. As a first part of a series, SMs have been categorized into three classes, including protein kinases, nucleic acid-sensing pathways, and soluble ligands/receptors. These advancements signify a pivotal development in creating safer and more effective complementary treatment options, addressing the increasing healthcare challenges posed by such diseases (Zheng et al. 2021).

The recent study on small molecule drugs targeting specific intracellular signaling pathways highlights a significant advancement in the development of therapeutics for various challenging conditions, including inflammatory diseases, autoimmune disorders, genetic diseases, and neoplastic diseases. Based on evidence from 114 studies, this study presents a thorough analysis of key protein kinase families, including JAK, Src, Syk, MAPK, and BTK. The study highlights their physiological roles and details the small molecule drugs designed to modulate their activity. The study also addresses these drugs' cytokines and their clinical implications in dermatology, focusing on conditions like psoriasis, psoriatic arthritis, atopic dermatitis, alopecia areata, and vitiligo. Even though small molecule drugs may exhibit less specificity compared to traditional immunobiological therapies, they present viable treatment alternatives for diseases that currently lack effective therapies. This advancement not only broadens the therapeutic landscape but also signifies a substantial progression in dermatological care, giving hope to patients facing these challenging conditions (Criado et al. 2023).

Janus kinase (JAK) inhibitors represent a major advancement in the management of various autoimmune and inflammatory diseases, particularly through their ability to interrupt the JAK-STAT signaling pathway, which is crucial for the function of cytokines involved in immune responses. Approved therapies such as tofacitinib and baricitinib have received endorsements for use in conditions including RA, PsA, and UC due to their effectiveness in managing symptoms and improving patient outcomes (Taylor et al. 2017). Cytokines are essential for hematopoietic cell functions, and their dysregulation is often implicated in diseases such as psoriasis and RA. Janus kinases (JAKs) are essential for transmitting signals from cytokine receptors, making them prime targets for therapeutic intervention. The early JAK inhibitors, tofacitinib and ruxolitinib, have demonstrated significant efficacy in RA and myelofibrosis, respectively, establishing a foundation for further development in this class of medications. Several additional JAK inhibitors, such as baricitinib, VX-509, and GLPG0634, are presently being investigated for their possible use in RA, while SAR302503 and others are under investigation for various myeloproliferative disorders. These drugs have the advantage of modulating multiple cytokine signals, offer hope for refractory cases. With advancements in specificity and safety, JAK inhibitors may revolutionize treatment, echoing the success of biologics (Ghoreschi and Gadina 2014).

Sphingosine-1-phosphate (S1P) receptor modulators, including fingolimod, prevent lymphocytes from leaving lymph nodes, decreasing their presence in the bloodstream and mitigating autoimmunity. This approach is especially effective in multiple sclerosis (MS), where fingolimod has demonstrated significant success in reducing relapses and delaying the progression of disease (Chen et al. 2022). New small molecules are emerging that target different pathways, such as Bruton's tyrosine kinase (BTK) inhibitors and compounds that inhibit protein-protein interactions, providing innovative therapeutic options. Recent advancements in structure-based drug design and high-throughput screening methods have expedited the identification of these compounds, significantly expanding the range of small-molecule therapies.

11.2.3 Cell-Based Therapies

Autoimmune diseases, characterized by immune system dysregulation that leads to damage to self-tissues, are complex conditions influenced by a myriad of factors, including genetic predisposition, environmental exposures, and hormonal fluctuations. These diseases impact various organ systems, manifesting in illnesses such as RA, SLE, MS, psoriasis, and vitiligo. Recent therapeutic advancements in autoimmune disease treatment focus on targeting autoreactive antibodies secreted by B and plasma cells, which are pivotal in the development of autoimmune disorders. Traditional therapies like rituximab aim to deplete B-cells but often struggle to target long-lived plasma cells (LLPCs) that persist and continue to secrete autoantibodies. One of the promising strategies involves the use of chimeric antigen receptor (CAR)-T cells that are designed to specifically target these autoreactive LLPCs. CD19-targeting CAR-T cells have exhibited promising short-term effectiveness in SLE, demonstrating an ability to effectively deplete autoantibody-secreting cells and successfully infiltrate tissue niches where cells reside. However, further controlled clinical trials and long-term effectiveness and safety data are necessary to ascertain their potential for widespread use. In addition to CD19-targeting CAR T cells, there are emerging therapies that focus on specific plasma cell antigens, such as B-cell maturation antigen (BCMA), offering refined targeting strategies. Furthermore, chimeric autoantibody receptors (CAARs) represent another innovative approach, allowing for direct targeting of pathogenic B cells responsible for disease symptoms. These cellular therapies hold transformative potential but require innovations to enhance affordability and accessibility (Wagner and Ostendorf 2023).

While existing treatments focus on symptom management, cellular treatments have become evident as a hopeful avenue. By targeting autoreactive B and T cells via markers like CD19, BCMA, CD20, and CTLA-4, cellular treatments have shown potential for autoimmune disease management. Cell therapies seek to restore immune tolerance by influencing the behavior of immune cells. A key area of therapeutic research is focused on regulatory T cells (Tregs), which inhibit inappropriate immune responses. Approaches include the ex vivo expansion of autologous Tregs, followed by their infusion into individuals suffering from autoimmune diseases such as T1D and SLE (Bluestone et al. 2008).

Emerging cellular therapies, such as chimeric antigen receptor T cells (CART), mesenchymal stromal cells (MSCs), and regulatory T cells (Tregs), have shown significant promise in treating autoimmune diseases. A global expert panel comprising members from the European Society for Blood and Marrow Transplantation (EBMT) and the International Society for Cell and Gene Therapy (ISCG) reviewed existing evidence and expert practices on cellular therapies for gastroenterological, neurological, and rheumatological autoimmune diseases. Their findings culminated in a set of consensus-based recommendations aimed at guiding clinicians, scientists, and multidisciplinary teams, ensuring high-quality care for patients (Greco et al. 2024).

A recent study highlights advancements in cellular therapies for autoimmune disorders, emphasizing the necessity for further authorization in terms of safety, effectiveness, and cost-effectiveness. Future research objectives include refining personalized immunocyte therapies, integrating cell therapies with traditional therapies, and developing holistic approaches to improve patient outcomes. With ongoing innovation, cellular therapies have the capacity to greatly enhance both the long-term health and quality of life for patients suffering from autoimmune conditions. (Fu et al. 2024).

11.3 Advanced Drug Delivery Systems

In recent decades, advancements in therapies and treatment strategies have significantly benefited many patients with autoimmune diseases. However, the primary goals of modern treatment remain focused on symptom relief and slowing disease progression. Commonly used medications include biological agents, immunosuppressants, and glucocorticoids. However, glucocorticoids provide quick symptom relief; they are unable to stop the progression of the disease, and prolonged drug use often results in severe side effects and financial burdens on patients. This emphasizes the importance of developing novel or advanced drug delivery methods to accomplish successful targeted therapy. In this section, we will discuss various advanced drug delivery systems for autoimmune diseases as shown in Fig. 11.1..

11.3.1 Nanoparticles

Nanoparticles (NPs) offer a versatile platform for drug administration, improving the stability, solubility, and targeted delivery of therapeutic drugs. Different kinds of NPs, including polymeric, lipid-based, and inorganic forms like gold nanoparticles, have been investigated for treating autoimmune diseases. For example, polymeric NPs infused with methotrexate have shown enhanced antiinflammatory properties in models of rheumatoid arthritis (Zhang et al. 2008). By functionalizing NPs with specific ligands or antibodies, researchers can achieve targeted delivery, thereby reducing off-target interactions. Furthermore, studies have highlighted the capacity of nanoparticle-based drug delivery systems to promote the generation of tolerogenic dendritic cells (Tol DCs), which play a pivotal role in inducing immune tolerance and inhibiting autoreactivity. (Li et al. 2022).

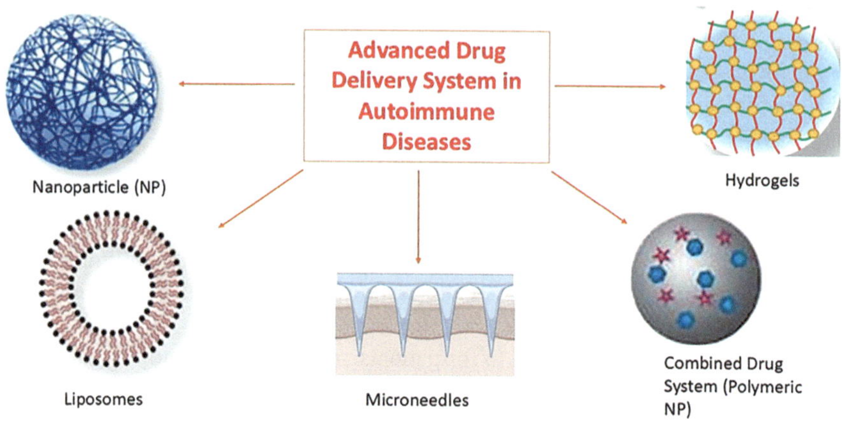

Fig. 11.1 Demonstrating different advanced drug delivery systems in autoimmune diseases

11.3.2 Liposomes

Liposomes are round vesicles that can encapsulate both hydrophilic and hydrophobic medications, allowing for regulated and targeted drug release. Formulations of corticosteroids and biologics within liposomes have shown extended circulation times and improved delivery to inflamed tissues. Innovations in surface modifications, such as PEGylation, have further enhanced the pharmacokinetics of these liposomal formulations. Notably, phosphatidylserine-rich liposomes loaded with disease-specific autoantigens have been shown to induce tolerogenic dendritic cells and mitigate autoimmunity in a multiple sclerosis mouse model, leading to reduced disease severity, delayed onset, and increased regulatory T-cell subsets. This study validates phosphatidylserine (PS)-liposomes as a promising therapeutic tool for reestablishing tolerance in multiple autoimmune diseases, including MS and T1D (Pujol-Autonell et al. 2017). PS-liposomes encapsulating autoantigens restore self-tolerance in autoimmune diseases by inducing tolerogenic and immunoregulatory effects. PS-liposomes can be adapted for various conditions by altering the autoantigen, effectively treating RA and myasthenia gravis. This offers a safe, antigen-specific platform with long-term efficacy for the management of autoimmune diseases (Almenara-Fuentes et al. 2023).

In another study, liposomes coated with antibodies were assessed as a therapy for immune thrombocytopenic purpura (ITP) in a mouse model. Liposomes coated with anti-methotrexate antibodies and liposomes coated with intravenous immunoglobulin (IVIG) have demonstrated a dose-dependent attenuation of thrombocytopenia. Notably, these liposomes provided efficacy similar to traditional IVIG treatment while allowing for the use of lower doses of immunoglobulin, potentially reducing the danger of adverse effects linked with higher doses. Unlike TER119, an anti-RBC monoclonal antibody known for causing severe hemolysis, liposomes have been beneficial as they selectively increase platelet counts without adversely affecting red blood cells. This characteristic positions AMI-coated and IVIG-coated liposomes as a safer alternative for managing thrombocytopenia, providing a promising therapeutic strategy with less risk of hemolytic events. (Deng and Balthasar 2006).

11.3.3 Hydrogels

Hydrogels are three-dimensional networks of hydrophilic polymers capable of holding significant amounts of water. They are highly versatile materials that have gained considerable attention in the field of drug delivery, particularly for the therapy of autoimmune diseases. They encapsulate drugs within their structure, allowing for sustained, controlled, and localized drug release, making them best for administering immunomodulatory drugs in autoimmune conditions. Recent advancements include stimuli-responsive hydrogels that can release medications in response to changes in pH, temperature, or enzymatic activity, thereby improving their therapeutic accuracy (Wei et al. 2017).

11.3.4 Microneedles

Microneedle patches are innovative and minimally invasive devices, designed to facilitate efficient and painless transdermal administration of drugs through the skin (Table 11.2). These patches utilize arrays of tiny needles (25–1000 mm in length) that create temporary openings in the outer layer of the skin (the stratum corneum), allowing therapeutic agents to be delivered directly into the dermis. This method offers significant advantages in the therapy of autoimmune disorders through their capacity to deliver medications locally, enhancing therapeutic effects while minimizing systemic exposure and the risk of adverse effects. Microneedles can administer antiinflammatory medications, such as corticosteroids or biologics like TNF-α inhibitors, directly to the site of inflammation, particularly in conditions like RA. This targeted strategy minimizes systemic immunosuppression while increasing the effectiveness of the treatment (Avcil and Çelik 2021). In contrast to conventional injections, microneedle patches provide a nearly painless experience. They are simple to use, leading to better patient compliance, particularly for chronic autoimmune diseases where consistent medication administration is crucial. The use of microneedles for the simultaneous delivery of multiple therapeutic agents, such as a combination of small molecules like methotrexate and biologics like tofacitinib, has the potential to significantly enhance disease management in RA (Garhwal et al. 2024). Additionally, microneedles that incorporate nanocarriers, such as lipid nanoparticles or hydrogels, constitute a noteworthy development in the field of drug delivery, particularly for biological agents. This approach not only enhances the stability and bioavailability of drugs but also allows for more controlled and

Table 11.2 Summarizing microneedles in autoimmune disease therapy

Microneedle type	Role in autoimmune disease	Mechanism of action	Key advantages	Challenges	References
Solid microneedles	Management of pain in autoimmune diseases (e.g., rheumatoid arthritis)	Facilitates antigen penetration into skin, triggering local immune response	Noninvasive, easy to administer, enhances immune activation	Limited penetration depth, may not induce long-term tolerance	Priya and Singhvi (2022)
Hollow microneedles + nanotechnology (TER-SLNs)	Used to lessen the adverse effects of systemic drug administration in the therapy of rheumatoid arthritis (RA)	Encapsulates teriflunomide (TER) in solid lipid nanoparticles (TER-SLNs) and delivers it transdermally via hollow microneedles directly to the affected joint	Noninvasive, effective drug delivery to the joint, sustained release for nearly 96 h, reduced side effects	Challenges include optimizing transdermal drug delivery in different tissue types and ensuring long-term efficacy	Abd-El-Azim et al. (2023)
Coated microneedles	Used for treating dermatological diseases like vitiligo, particularly effective for localized and stable lesions	Microneedles penetrate the skin layers, enhancing therapeutic delivery and modulating immune disorders	Targeted delivery for localized vitiligo lesions, increased repigmentation effectiveness, noninvasive	Treating diseases with indistinct lesion boundaries, like atopic dermatitis, and maintaining microneedle consistency for lesion types, like psoriasis	Yu et al. (2024)
Microneedles (MNs) + nanoformulations	Used in rheumatoid arthritis (RA) treatment for transdermal drug delivery, enhancing targeting, solubility, stability, and biocompatibility of drugs	It bypasses the barrier and delivers medications into the lymphatic system by temporarily opening micropores in the stratum corneum (SC); nanoformulations improve drug targeting and stability	Painless drug delivery, improved drug targeting, enhanced biocompatibility, stability, and solubility	Incompatible for drugs having less solubility and stability in water; safety concerns with cytotoxic drugs, and suboptimal targeting efficiency	Wendong et al. (2024)

Drug-loaded microneedles	Used for the treatment of RA by delivering triptolide (TP) locally to the affected joints	TP is loaded into microneedles with polyvinylpyrrolidone and hyaluronic acid as matrix materials, enhancing penetration and reducing systemic toxicity	Noninvasive, enhanced drug penetration, reduced organ toxicity, improved therapeutic efficacy	No significant difference in acupoint vs. nonacupoint delivery, and challenges with optimizing local drug distribution	S. Li et al. (2024)
Microneedle arrays with immunomodulators	Used for the localized delivery of anti–TNF-α antibodies to treat inflammatory diseases such as psoriasis	Anti–TNF-α antibodies are integrated into dissolvable microneedles (MNAs), which deliver the antibodies to the dermis of the skin, avoiding systemic dosing	Localized and efficient delivery, reduced systemic immunosuppression, preserved antibody activity	Ensure consistent delivery and targeting in the dermis, as there are potential issues with large-scale production	Korkmaz et al. (2015)
Dissolving microneedles	Autoimmune therapy through the controlled release of immunomodulatory agents (e.g., rheumatoid arthritis)	Microneedles dissolve upon insertion, releasing drugs or antigens into the skin	No needle removal required reduces discomfort	Dissolution rate variability, limited for large molecules	Guo et al. (2023)
Dissolvable microneedles (dMNAs) + curcumin-albumin-EVs (CA-EVs)	Used for the treatment of inflammatory skin diseases by delivering curcumin-loaded extracellular vesicles (CA-EVs) locally	CA-EVs are encapsulated into dissolvable microneedles (dMNAs) for local delivery, enhancing curcumin stability and antiinflammatory activity	Improved stability and bioactivity of curcumin, extended storage for up to 12 months, effective local delivery, and antiinflammatory effects	Potential issues with large-scale production, ensuring consistent EV encapsulation, and the need for further optimization of the system for various inflammatory conditions	Yerneni et al. (2022)

sustained drug release. Research investigating antigen-loaded microneedles to promote immune tolerance in autoimmune disorders, such as type 1 diabetes (T1D), shows great potential, as it seeks to restore immune balance (Sang et al. 2019).

11.3.5 Combination Delivery Systems

Combination delivery systems merge various drug delivery methods, including liposomes, nanoparticles, hydrogels, microneedles, or more, to address the limitations found in single-platform strategies. For instance, microneedles coated with liposomes represent an innovative method in drug delivery, merging the advantages of transdermal administration with the beneficial properties of liposomal encapsulation. This innovative approach allows for the direct and efficient delivery of therapeutic agents through the skin barrier while ensuring the stability and bioavailability of the encapsulated drugs. Likewise, nanoparticles incorporated into hydrogels represent an advanced strategy in drug delivery systems, providing several advantages for localized and controlled therapy. This dual system allows for the simultaneous encapsulation of therapeutic agents within the hydrogel matrix while utilizing nanoparticles to modulate the release kinetics (Cao et al. 2023).

Latest developments in nanotechnology have enabled researchers to develop a diverse array of nanomaterials with exceptional optical, magnetic, electrical, catalytic, and mechanical properties compared with their bulk counterparts. The integration of these nanomaterials with liposomes has unlocked new possibilities in biological and biomedical applications, leveraging the unique properties of both components to enhance biocompatibility and stability. Liposomes are versatile, self-assembling nanoscale structures formed by phospholipid bilayers enclosing an aqueous core. They are capable of encapsulating drug molecules and biomolecules, accommodating both hydrophobic and hydrophilic substances. The ability of liposomes to mimic biological membranes makes them an ideal vehicle for drug delivery. A recent study explores the preparation methods for nanoparticle (NP)-loaded liposomes and highlights the significance of their innovative hybrid nanoscale architectures, where NPs are incorporated in the aqueous core, lipid bilayer, or vesicle surface (De Leo et al. 2022).

A recent investigation highlights the novel combination of drug-loaded liposomes within nanofibrous scaffolds, creating a multiunit nanoscale system that leverages the individual advantages of both liposomes and nanofibers (NFs). This integration significantly enhances drug stability and facilitates controlled release mechanisms, ultimately leading to improved therapeutic efficacy. A systematic review identified three key fabrication methods for producing liposomal nanofibrous scaffolds and highlighted the superior performance of these scaffolds spanning a range of uses, such as wound healing, tissue engineering, cancer therapy, and oromucosal drug delivery. These systems ensure sustained drug release while minimizing the risk of significant burst release, which is often a challenge in conventional drug delivery systems. Overall, this versatile platform presents a revolutionary method for drug delivery and has the potential to greatly advance the field of nanomedicine. Its ability to ensure effective and controlled release of medications might open the door to novel therapies (Abboud et al. 2024).

Nanoparticle (NP)-based therapeutics are emerging as a significant advancement in the therapy of SLE, especially in addressing the impaired clearance of apoptotic cells, a critical factor in SLE pathogenesis. A notable study focused on creating a gold nanocage (AuNC)-based drug delivery system that imitates apoptotic cells. This was achieved by conjugating phosphatidylserine (PS) to the surface of liposome-coated AuNCs, facilitating targeted delivery of the liver X receptor (LXR) agonist T0901317. The design of this NP system enhances the clearance of apoptotic cells by increasing expression of Mer receptor on macrophages. This increased receptor expression contributes to the reduction of anti-dsDNA autoantibodies, mitigates inflammation, and improves kidney function in lupus model mice, highlighting its potential therapeutic effects in SLE. Furthermore, the system incorporates photoacoustic imaging capabilities, providing a means to track biodistribution of the NPs in real-time. Overall, this innovative NP-based delivery strategy addresses a vital pathological mechanism in SLE, offering a promising therapeutic approach to manage this complex autoimmune disorder and potentially improve patient outcomes through enhanced apoptotic cell clearance and modulation of immune responses. Further studies will be essential to evaluate the long-term effectiveness and safety of this treatment modality (Xu et al. 2019).

11.4 Conclusion

Autoimmune diseases pose significant treatment challenges due to their complex etiology and systemic effects. Conventional immunosuppressive therapies often cause substantial side effects, leading to a growing interest in exploration of novel compounds and drug delivery systems. Recent advancements have shifted focus toward targeted small molecules, biologics, and nanotechnology-based drug delivery systems, which aim to improve therapeutic effectiveness while reducing adverse effects. These novel approaches include controlled-release formulations that allow for steady drug delivery over time and personalized medicine approaches that leverage genomic and immunological profiling to tailor treatments to individual patients' needs. Continued research is crucial to validate these innovations and improve patient outcomes. Biotherapies, such as engineered antigen-presenting cell (APC) therapies, provide precise and targeted interventions for autoimmune diseases by inducing immune tolerance. They offer long-term efficacy and reduced side effects through personalized approaches. However, challenges such as scalability, safety, and regulatory issues necessitate ongoing research and optimization to ensure their effective implementation in clinical practice. Additionally, nanomedicine has significantly advanced in the context of autoimmune disease therapy, transitioning from a basic drug delivery system to the engineering of nanoparticles (NPs) with diverse functional capabilities. Initial studies highlighted the potential of NPs in targeting drug delivery to sites of inflammation, improving therapeutic outcomes. Although autoantigen-loaded NPs and microparticles (MPs) have shown promising results in experimental models, they encountered difficulties, particularly when addressing complex and spontaneous autoimmune disorders, primarily due to their limited targeting of APCs. One innovative development in the area is the use of peptide-MHC (pMHC)-NPs, which can induce disease-specific regulatory T cells. This approach allows for targeted immunoregulation while also preserving normal immune functions, making it a compelling option for managing autoimmune conditions. Despite the promise shown in experimental contexts, clinical trials involving these nanomedicine approaches remain limited and have yielded mixed outcomes, leaving the long-term success of these therapies uncertain. Moreover, immunotherapies for autoimmune diseases have gained significant attention, with trials exploring both monotherapies and combination treatments. A thorough understanding of their molecular mechanisms, efficacy, safety, and tolerance is crucial for optimizing these therapies and ensuring their success in clinical settings.

References

Abboud HA, Zelkó R, Kazsoki A (2024) A systematic review of liposomal nanofibrous scaffolds as a drug delivery system: a decade of progress in controlled release and therapeutic efficacy. Drug Deliv 32(1):2445259. https://doi.org/10.1080/10717544.2024.2445259

Abd-El-Azim H, Abbas H, El Sayed NS, Fayez AM, Zewail M (2023) Non-invasive management of rheumatoid arthritis using hollow microneedles as a tool for transdermal delivery of teriflunomide loaded solid lipid nanoparticles. Int J Pharm 644:123334. https://doi.org/10.1016/j.ijpharm.2023.123334

Alamanos Y, Drosos AA (2005) Epidemiology of adult rheumatoid arthritis. Autoimmun Rev 4(3):130–136. https://doi.org/10.1016/j.autrev.2004.09.002

Almenara-Fuentes L, Rodriguez-Fernandez S, Rosell-Mases E, Kachler K, You A, Salvado M, Andreev D, Steffen U, Bang H, Bozec A, Schett G, Le Panse R, Verdaguer J, Dalmases M, Rodriguez-Vidal S, Barneda-Zahonero B, Vives-Pi M (2023) A new platform for autoimmune diseases. Inducing tolerance with liposomes encapsulating autoantigens. Nanomedicine 48:102635. https://doi.org/10.1016/j.nano.2022.102635

Armuzzi A, Felice C (2013) Natalizumab in Crohn's disease: past and future areas of applicability. Ann Gastroenterol 26(3):189–190. https://pmc.ncbi.nlm.nih.gov/articles/PMC3959451

Avcil M, Çelik A (2021) Microneedles in drug delivery: progress and challenges. Micromachines 12(11):1321. https://doi.org/10.3390/mi12111321

Baeten D, Sieper J, Braun J, Baraliakos X, Dougados M, Emery P, Deodhar A, Porter B, Martin R, Andersson M, Mpofu S, Richards HB (2015) Secukinumab, an Interleukin-17A Inhibitor, in Ankylosing Spondylitis. N Engl J Med 373(26):2534–2548. https://doi.org/10.1056/nejmoa1505066

Belkaid Y, Hand T (2014) Role of the microbiota in immunity and inflammation. Cell 157(1):121–141. https://doi.org/10.1016/j.cell.2014.03.011

Bluestone JA, Tang Q, Sedwick CE (2008) T regulatory cells in autoimmune diabetes: past challenges, future prospects. J Clin Immunol 28(6):677–684. https://doi.org/10.1007/s10875-008-9242-z

Brunner HI, Wong R, Nys M, Kou TD, Dominique A, Martini A, Lovell DJ, Ruperto N (2020) Abatacept: a review of the treatment of Polyarticular-Course Juvenile Idiopathic Arthritis. Pediatr Drugs 22(6):653–672. https://doi.org/10.1007/s40272-020-00422-2

Burmester GR, Lin Y, Patel R, Van Adelsberg J, Mangan EK, Graham NMH, Van Hoogstraten H, Bauer D, Ignacio Vargas J, Lee EB (2017) Efficacy and safety of sarilumab monotherapy versus adalimumab monotherapy for the treatment of patients with active rheumatoid arthritis (MONARCH): a randomised, double-blind, parallel-group phase III trial. Ann Rheum Dis 76(5):840–847. https://doi.org/10.1136/annrheumdis-2016-210310

Burmester GR, Gordon KB, Rosenbaum JT, Arikan D, Lau WL, Li P, Faccin F, Panaccione R (2019) Long-term safety of Adalimumab in 29,967 adult patients from global clinical trials across multiple indications: an updated analysis. Adv Ther 37(1):364–380. https://doi.org/10.1007/s12325-019-01145-8

Burness CB, Duggan ST (2016) Etanercept (SB4): a review in autoimmune inflammatory diseases. BioDrugs 30(4):371–378. https://doi.org/10.1007/s40259-016-0188-z

Cao F, He YS, Wang Y, Zha CK, Lu JM, Tao LM, Jiang ZX, Pan HF (2023) Global burden and cross-country inequalities in autoimmune diseases from 1990 to 2019. Autoimmun Rev 22(6):103326. https://doi.org/10.1016/j.autrev.2023.103326

Chen H, Wang J, Zhang C, Ding P, Tian S, Chen J, Ji G, Wu T (2022) Sphingosine 1-phosphate receptor, a new therapeutic direction in different diseases. Biomed Pharmacother 153:113341. https://doi.org/10.1016/j.biopha.2022.113341

Criado PR, Lorenzini D, Miot HA, Bueno-Filho R, Carneiro FRO, Ianhez M (2023) New small molecules in dermatology: for the autoimmunity, inflammation and beyond. Inflamm Res 72(6):1257–1274. https://doi.org/10.1007/s00011-023-01744-w

Curtis JR, Chen L, Greenberg JD, Harrold L, Kilgore ML, Kremer JM, Solomon DH, Yun H (2017) The clinical status and economic savings associated with remission among patients with rheumatoid arthritis: leveraging linked registry and claims data for synergistic insights. Pharmacoepidemiol Drug Saf 26(3):310–319. https://doi.org/10.1002/pds.4126

de La Forest Divonne M, Gottenberg JE, Salliot C (2017) Safety of biologic DMARDs in RA patients in real life: A systematic literature review and meta-analyses of biologic registers. Joint Bone Spine 84(2):133–140. https://doi.org/10.1016/j.jbspin.2016.02.028

De Leo V, Maurelli AM, Giotta L, Catucci L (2022) Liposomes containing nanoparticles: preparation and applications. Colloids Surf B: Biointerfaces 218:112737. https://doi.org/10.1016/j.colsurfb.2022.112737

Deng R, Balthasar JP (2006) Comparison of the effects of antibody-coated liposomes, IVIG, and anti-RBC immunotherapy in a murine model of passive chronic immune thrombocytopenia. Blood 109(6):2470–2476. https://doi.org/10.1182/blood-2006-04-018093

Deodhar A, van der Heijde D, Gensler LS, Kim TH, Maksymowych WP, Østergaard M, Poddubnyy D, Marzo-Ortega H et al (2020) Ixekizumab for patients with non-radiographic axial spondyloarthritis (COAST-X): a randomised, placebo-controlled trial. Lancet (London, England) 395(10217):53–64. https://doi.org/10.1016/s0140-6736(19)32971-x

Drosos AA, Pelechas E, Kaltsonoudis E, Markatseli TE, Voulgari PV (2021) Biologic therapies and autoimmune phenomena. Mediterr J Rheumatol 32(2):96–103. https://doi.org/10.31138/mjr.32.2.96

Farh KKH, Marson A, Zhu J, Kleinewietfeld M, Housley WJ, Beik S, Shoresh N, Whitton H et al (2015) Genetic and epigenetic fine mapping of causal autoimmune disease variants. Nature 518(7539):337–343. https://doi.org/10.1038/nature13835

Feagan BG, Rutgeerts P, Sands BE, Hanauer S, Colombel J-F, Sandborn WJ, Van Assche G, Axler J, Kim H-J, Danese S, Fox I, Milch C, Sankoh S, Wyant T, Xu J, Parikh A (2013) Vedolizumab as induction and maintenance therapy for ulcerative colitis. N Engl J Med 369(8):699–710. https://doi.org/10.1056/nejmoa1215734

Feagan BG, Sandborn WJ, Gasink C, Jacobstein D, Lang Y, Friedman JR, Blank MA, Johanns J, Gao L-L, Miao Y, Adedokun OJ, Sands BE, Hanauer SB, Vermeire S et al (2016) Ustekinumab as induction and maintenance therapy for Crohn's Disease. N Engl J Med 375(20):1946–1960. https://doi.org/10.1056/nejmoa1602773

Feldmann M, Maini RN (2003) TNF defined as a therapeutic target for rheumatoid arthritis and other autoimmune diseases. Nat Med 9(10):1245–1250. https://doi.org/10.1038/nm939

Firestein GS, McInnes IB (2017) Immunopathogenesis of Rheumatoid Arthritis. Immunity 46(2):183–196. https://doi.org/10.1016/j.immuni.2017.02.006

Fu Y, Feng C, Qin S, Xing Z, Liu C, Liu Z, Yu H (2024) Breaking barriers: advancing cellular therapies in autoimmune disease management. Front Immunol 15:1503099. https://doi.org/10.3389/fimmu.2024.1503099

Garhwal A, Kendya P, Soni S, Kori S, Soni V, Kashaw SK (2024) Drug delivery system approaches for rheumatoid arthritis treatment: a review. Mini Rev Med Chem 24(7):704–720. https://doi.org/10.2174/1389557523666230913105803

Ghoreschi K, Gadina M (2014) JAKpot! New small molecules in autoimmune and inflammatory diseases. Exp Dermatol 23(1):7–11. https://doi.org/10.1111/exd.12265

Gordon KB, Blauvelt A, Papp KA, Langley RG, Luger T, Ohtsuki M, Reich K, Amato D, Ball SG, Braun DK et al (2016) Phase 3 Trials of Ixekizumab in Moderate-to-Severe Plaque Psoriasis. N Engl J Med 375(4):345–356. https://doi.org/10.1056/nejmoa1512711

Greco R, Alexander T, Del Papa N, Müller F, Saccardi R, Sanchez-Guijo F, Schett G et al (2024) Innovative cellular therapies for autoimmune diseases: expert-based position statement and clinical practice recommendations from the EBMT practice harmonization and guidelines committee. eClinicalMedicine 69:102476. https://doi.org/10.1016/j.eclinm.2024.102476

Greenberg JD, Kishimoto M (2006) Etanercept in rheumatology. In: Weinberg JM, Buchholz R (eds) TNF-alpha inhibitors. Milestones in drug therapy. Birkhäuser, Basel, pp 45–54. https://doi.org/10.1007/3-7643-7438-1_4

Guo P, Huang C, Yang Q, Zhong G, Zhang J, Qiu M, Zeng R, Gou K, Zhang C, Qu Y (2023) Advances in formulations of microneedle system for Rheumatoid Arthritis treatment. Int J Nanomedicine 18:7759–7784. https://doi.org/10.2147/ijn.s435251

Hayter SM, Cook MC (2012) Updated assessment of the prevalence, spectrum and case definition of autoimmune disease. Autoimmun Rev 11(10):754–765. https://doi.org/10.1016/j.autrev.2012.02.001

Hillmen P, Hall C, Marsh JCW, Elebute M, Bombara MP, Petro BE, Cullen MJ, Richards SJ, Rollins SA, Mojcik CF, Rother RP (2004) Effect of eculizumab on hemolysis and transfusion requirements in patients with paroxysmal nocturnal hemoglobinuria. N Engl J Med 350(6):552–559. https://doi.org/10.1056/nejmoa031688

Howard JF, Utsugisawa K, Benatar M, Murai H, Barohn RJ, Illa I, Jacob S, Vissing J, Burns TM, Kissel JT et al (2017) Safety and efficacy of eculizumab in anti-acetylcholine receptor antibody-positive refractory generalised myasthenia gravis (REGAIN): a phase 3, randomised, double-blind, placebo-controlled, multicentre study. Lancet Neurol 16(12):976–986. https://doi.org/10.1016/s1474-4422(17)30369-1

Iwata S, Saito K, Hirata S, Ohkubo N, Nakayamada S, Nakano K, Hanami K, Kubo S, Miyagawa I, Yoshikawa M, Miyazaki Y, Yoshinari H, Tanaka Y (2018) Efficacy and safety of anti-CD20 antibody rituximab for patients with refractory systemic lupus erythematosus. Lupus 27(5):802–811. https://doi.org/10.1177/0961203317749047

Kimball AB, Papp KA, Wasfi Y, Chan D, Bissonnette R, Sofen H, Yeilding N, Li S, Szapary P, Gordon KB (2013) Long-term efficacy of ustekinumab in patients with moderate-to-severe psoriasis treated for up to 5 years in the PHOENIX 1 study. J Eur Acad Dermatol Venereol 27(12):1535–1545. https://doi.org/10.1111/jdv.12046

Korkmaz E, Friedrich EE, Ramadan MH, Erdos G, Mathers AR, Burak Ozdoganlar O, Washburn NR, Falo LD (2015) Therapeutic intradermal delivery of tumor necrosis factor-alpha antibodies using tip-loaded dissolvable microneedle arrays. Acta Biomater 24:96–105. https://doi.org/10.1016/j.actbio.2015.05.036

Langley RG, Elewski BE, Lebwohl M, Reich K, Griffiths CEM, Papp K, Puig L, Nakagawa H, Spelman L, Sigurgeirsson B, Rivas E, Tsai T-F, Wasel N, Tyring S, Salko T, Hampele I, Notter M, Karpov A, Helou S, Papavassilis C (2014) Secukinumab in plaque psoriasis–results of two phase 3 trials. N Engl J Med 371(4):326–338. https://doi.org/10.1056/nejmoa1314258

Lebwohl M, Strober B, Menter A, Gordon K, Weglowska J, Puig L, Papp K, Spelman L, Toth D, Kerdel F, Armstrong AW et al (2015) Phase 3 studies comparing Brodalumab with Ustekinumab in Psoriasis. N Engl J Med 373(14):1318–1328. https://doi.org/10.1056/nejmoa1503824

Li H, Yang YG, Sun T (2022) Nanoparticle-based drug delivery systems for induction of tolerance and treatment of autoimmune diseases. Front Bioeng Biotechnol 10:889291. https://doi.org/10.3389/fbioe.2022.889291

Li DP, Han YX, He YS, Wen Y, Liu YC, Fu ZY, Pan HF, Cao F (2023) A global assessment of incidence trends of autoimmune diseases from 1990 to 2019 and predicted changes to 2040. Autoimmun Rev 22(10):103407. https://doi.org/10.1016/j.autrev.2023.103407

Li S, Chen Q, Zhang Y, Wang D, Hu H, Li J, Zhang C, Zhang J (2024) Hyaluronic acid dissolving microneedle patch-assisted acupoint transdermal delivery of triptolide for effective rheumatoid arthritis treatment. Sci Rep 14(1):1–16. https://doi.org/10.1038/s41598-024-76341-w

Liu X, Wu W, Fang L, Liu Y, Chen W (2022) TNF-α Inhibitors and other biologic agents for the treatment of immune checkpoint inhibitor-induced myocarditis. Front Immunol 13:922782. https://doi.org/10.3389/fimmu.2022.922782

Machado RIL, Scheinberg MA, de Queiroz MYCF, de Brito DCSE et al (2014) Use of rituximab as a treatment for systemic lupus erythematosus: retrospective review. Einstein 12(1):36. https://doi.org/10.1590/s1679-45082014ao2706

McGeachy MJ, Cua DJ, Gaffen SL (2019) The IL-17 Family of cytokines in health and disease. Immunity 50(4):892–906. https://doi.org/10.1016/j.immuni.2019.03.021

McInnes IB, Schett G (2017) Pathogenetic insights from the treatment of rheumatoid arthritis. Lancet (London, England) 389(10086):2328–2337. https://doi.org/10.1016/s0140-6736(17)31472-1

Montalban X, Hauser SL, Kappos L, Arnold DL, Bar-Or A, Comi G, de Seze J et al (2017) Ocrelizumab versus placebo in primary progressive multiple sclerosis. N Engl J Med 376(3):209–220. https://doi.org/10.1056/nejmoa1606468

Navarra SV, Guzmán RM, Gallacher AE, Hall S, Levy RA, Jimenez RE, Li EKM, Thomas M, Kim HY et al (2011) Efficacy and safety of belimumab in patients with active systemic lupus erythematosus: a randomised, placebo-controlled, phase 3 trial. Lancet (London, England) 377(9767):721–731. https://doi.org/10.1016/s0140-6736(10)61354-2

Nishimoto N, Kishimoto T (2008) Humanized antihuman IL-6 receptor antibody, Tocilizumab. In: Chernajovsky Y, Nissim A (eds) Therapeutic antibodies. Handbook of experimental pharmacology, vol 181, pp 151–160. https://doi.org/10.1007/978-3-540-73259-4_7

Nishimoto N, Yoshizaki K, Miyasaka N, Yamamoto K, Kawai S, Takeuchi T, Hashimoto J, Azuma J, Kishimoto T (2004) Treatment of rheumatoid arthritis with humanized anti-interleukin-6 receptor antibody: a multicenter, double-blind, placebo-controlled trial. Arthritis Rheum 50(6):1761–1769. https://doi.org/10.1002/art.20303

O'Shea JJ, Pesu M, Borie DC, Changelian PS (2004) A new modality for immunosuppression: targeting the JAK/STAT pathway. Nat Rev Drug Discov 3(7):555–564. https://doi.org/10.1038/nrd1441

Plosker GL, Figgitt DP (2003) Rituximab: a review of its use in non-Hodgkin's lymphoma and chronic lymphocytic leukaemia. Drugs 63(8):803–843. https://doi.org/10.2165/00003495-200363080-00005

Polman CH, O'Connor PW, Havrdova E, Hutchinson M, Kappos L, Miller DH, Phillips JT, Lublin FD, Giovannoni G, Wajgt A, Toal M, Lynn F, Panzara MA, Sandrock AW (2006) A randomized, placebo-controlled trial of natalizumab for relapsing multiple sclerosis. N Engl J Med 354(9):899–910. https://doi.org/10.1056/nejmoa044397

Pombo-Suarez M, Gomez-Reino JJ (2019) Abatacept for the treatment of rheumatoid arthritis. Expert Rev Clin Immunol 15(4):319–326. https://doi.org/10.1080/1744666x.2019.1579642

Priya S, Singhvi G (2022) Microneedles-based drug delivery strategies: a breakthrough approach for the management of pain. Biomed Pharmacother 155:113717. https://doi.org/10.1016/j.biopha.2022.113717

Pujol-Autonell I, Mansilla MJ, Rodriguez-Fernandez S, Cano-Sarabia M, Navarro-Barriuso J, Ampudia RM, Rius A, Garcia-Jimeno S, Perna-Barrull D, Caceres EM, Maspoch D, Vives-Pi

M (2017) Liposome-based immunotherapy against autoimmune diseases: therapeutic effect on multiple sclerosis. Nanomedicine (Lond) 12(11):1231–1242. https://doi.org/10.2217/nnm-2016-0410

Randall KL (2016) Rituximab in autoimmune diseases. Aust Prescr 39(4):131–134. https://doi.org/10.18773/austprescr.2016.053

Reich K, Papp KA, Blauvelt A, Tyring SK, Sinclair R, Thaçi D, Nograles K, Mehta A, Cichanowitz N, Li Q, Liu K, La Rosa C, Green S, Kimball AB (2017) Tildrakizumab versus placebo or etanercept for chronic plaque psoriasis (reSURFACE 1 and reSURFACE 2): results from two randomised controlled, phase 3 trials. Lancet (London, England) 390(10091):276–288. https://doi.org/10.1016/s0140-6736(17)31279-5

Rosenblum MD, Remedios KA, Abbas AK (2015) Mechanisms of human autoimmunity. J Clin Invest 125(6):2228–2233. https://doi.org/10.1172/jci78088

Rosman Z, Shoenfeld Y, Zandman-Goddard G (2013) Biologic therapy for autoimmune diseases: an update. BMC Med 11(1):88. https://doi.org/10.1186/1741-7015-11-88

Sang W, Zhang Z, Dai Y, Chen X (2019) Recent advances in nanomaterial-based synergistic combination cancer immunotherapy. Chem Soc Rev 48(14):3771–3810. https://doi.org/10.1039/c8cs00896e

Serra P, Santamaria P (2015) Nanoparticle-based autoimmune disease therapy. Clin Immunol 160(1):3–13. https://doi.org/10.1016/j.clim.2015.02.003

Smolen JS, Breedveld FC, Burmester GR, Bykerk V, Dougados M, Emery P, Kvien TK et al (2016) Treating rheumatoid arthritis to target: 2014 update of the recommendations of an international task force. Ann Rheum Dis 75(1):3–15. https://doi.org/10.1136/annrheumdis-2015-207524

Strand V, Goncalves J, Isaacs JD (2021) Immunogenicity of biologic agents in rheumatology. Nat Rev Rheumatol 17(2):81–97. https://doi.org/10.1038/s41584-020-00540-8

Tanaka T, Narazaki M, Kishimoto T (2014) IL-6 in inflammation, immunity, and disease. Cold Spring Harb Perspect Biol 6(10):a016295. https://doi.org/10.1101/cshperspect.a016295

Taylor PC, Keystone EC, van der Heijde D, Weinblatt ME, del Carmen Morales L, Reyes Gonzaga J, Yakushin S et al (2017) Baricitinib versus placebo or Adalimumab in Rheumatoid Arthritis. N Engl J Med 376(7):652–662. https://doi.org/10.1056/nejmoa1608345

Wagner DL, Ostendorf L (2023) New dawn of cellular therapies in autoimmune diseases. Mol Ther Methods Clin Dev 31:101141. https://doi.org/10.1016/j.omtm.2023.101141

Wei L, Chen J, Zhao S, Ding J, Chen X (2017) Thermo-sensitive polypeptide hydrogel for locally sequential delivery of two-pronged antitumor drugs. Acta Biomater 58:44–53. https://doi.org/10.1016/j.actbio.2017.05.053

Wendong Y, Xingxing Y, Xianze X, Qiaomei F, Yujun S, Shanshan Z, Zheng S, Hairu X (2024) Nanoformulation-assisted microneedle transdermal drug delivery system: An innovative platform enhancing rheumatoid arthritis treatment. Biomed Pharmacother 178:117219. https://doi.org/10.1016/j.biopha.2024.117219

Wenzel S, Castro M, Corren J, Maspero J, Wang L, Zhang B, Pirozzi G, Sutherland ER, Evans RR, Joish VN et al (2016) Dupilumab efficacy and safety in adults with uncontrolled persistent asthma despite use of medium-to-high-dose inhaled corticosteroids plus a long-acting β2 agonist: a randomised double-blind placebo-controlled pivotal phase 2b dose-ranging trial. Lancet (London, England) 388(10039):31–44. https://doi.org/10.1016/s0140-6736(16)30307-5

Xu N, Li J, Gao Y, Zhou N, Ma Q, Wu M, Zhang Y, Sun X, Xie J, Shen G, Yang M, Tu Q, Xu X, Zhu J, Tao J (2019) Apoptotic cell-mimicking gold nanocages loaded with LXR agonist for attenuating the progression of murine systemic lupus erythematosus. Biomaterials 197:380–392. https://doi.org/10.1016/j.biomaterials.2019.01.034

Yerneni SS, Yalcintas EP, Smith JD, Averick S, Campbell PG, Ozdoganlar OB (2022) Skin-targeted delivery of extracellular vesicle-encapsulated curcumin using dissolvable microneedle arrays. Acta Biomater 149:198–212. https://doi.org/10.1016/j.actbio.2022.06.046

Yu Y, Li W, Li Q, Liu W, Zhang S, Zhang X, Liang J (2024) Review of microneedle technology for targeted therapeutics in vitiligo: design principles, application prospects. Drug Des Devel Ther 18:4901–4914. https://doi.org/10.2147/dddt.s449381

Zarros A, Kallikourdis M, Università H, Fuchs D, Chen X, Mo X, Li P, Zheng Y (2017) Drugs for autoimmune inflammatory diseases: from small molecule compounds to anti-TNF biologics. Front Pharmacol 8:460. https://doi.org/10.3389/fphar.2017.00460

Zhang L, Gu FX, Chan JM, Wang AZ, Langer RS, Farokhzad OC (2008) Nanoparticles in medicine: therapeutic applications and developments. Clin Pharmacol Ther 83(5):761–769. https://doi.org/10.1038/sj.clpt.6100400

Zheng J, Wu J, Ding X, Shen HC, Zou G (2021) Small molecule approaches to treat autoimmune and inflammatory diseases (Part I): Kinase inhibitors. Bioorg Med Chem Lett 38:127862. https://doi.org/10.1016/j.bmcl.2021.127862

Chapter 12
Targeted Next-Generation Nanoparticles in Autoimmune Disease Treatment

Md. Meraj Ansari, Godagama Gamaarachchige Dinesh Suminda, and Young-Ok Son

12.1 Introduction

Rheumatoid arthritis (RA) is a chronic inflammatory autoimmune disease that primarily affects the joints (Guo et al. 2018). In 2019, approximately 18 million people worldwide were living with RA. Its prevalence is higher in industrialized countries, possibly because of demographics, environmental toxins, and lifestyle risk factors. About 70% of people with RA are women and 55% are older than 55 years. The number of RA cases has increased over the past three decades and this trend is expected to continue. In 2020, an estimated 17.6 million people had RA globally, with an age-standardized prevalence rate of 208.8 cases per 100,000 population. Its prevalence is projected to reach 31.7 million people by 2050. The prevalence of RA varies by region, with higher rates in industrialized countries than in low- and middle-income countries. Underdiagnosis in low- and middle-income countries may contribute to the lower reported prevalence rates. As RA is an autoimmune disease, the immune system plays a significant role in disease onset and pathogenesis. The human immune system is a complex network designed to protect the body from harmful invaders such as bacteria,

Md. M. Ansari · Y.-O. Son (✉)
Department of Animal Biotechnology, Faculty of Biotechnology, College of Applied Life Sciences, Jeju National University, Jeju, Korea (Republic of)

Interdisciplinary Graduate Program in Advanced Convergence Technology and Science, Jeju National University, Jeju-si, Republic of Korea
e-mail: ansarimm@jejunu.ac.kr; sounagi@jejunu.ac.kr

G. G. D. Suminda
Interdisciplinary Graduate Program in Advanced Convergence Technology and Science, Jeju National University, Jeju-si, Republic of Korea
e-mail: godagama@jejunu.ac.kr

Y.-O. Son, M. Ghosh (eds.), *Cutting-Edge Strategies in Drug Delivery and Immunotherapy for Autoimmune Disorders*,
https://doi.org/10.1007/978-981-95-0520-3_12

viruses, and other pathogens. It comprises various components, each of which plays a crucial role. This intricate system works tirelessly to keep us healthy and protect us from a myriad of potential threats. The human body is protected from external invasion by the immune system (Vojdani et al. 2020). The immune system can be classified into two subsystems: innate and adaptive (Vivier and Malissen 2005). When an immune response is triggered by a pathogen or a hazardous chemical, both subsystems collaborate closely. The innate immune system is the first line of defense against pathogens. It is a crucial component of the overall immune system and is designed to respond quickly and effectively to a wide range of invaders. The adaptive immune system is a sophisticated part of the body's defense mechanism that provides a targeted and long-lasting response to specific pathogens. The adaptive immune system is essential for long-term immunity and for protection against recurrent infections. The body can combat evolving germs and viruses because the adaptive immune system is constantly learning and changing. Currently, more than 80 distinct autoimmune disorders are plaguing people worldwide. The core concept of autoimmune diseases is the failure of the immune system to differentiate between self and nonself and to treat self-tissues as antigens. RA is an example of a chronic autoimmune disease. Approximately 1% of the global population is affected by RA (Aletaha and Smolen 2018; Cross et al. 2014).

Women are more prone to RA than men. It mostly affects symmetrical small joints, leading to synovial hyperplasia, joint synovitis, bone and cartilage degeneration, and eventually labor deficiency (Sparks 2019). Medication and surgery are the mainstay of treatment for RA to effectively reduce or control its progression (Burmester and Pope 2017). However, it is critical to look for novel therapeutic protocols because of the severe side effects of conventional RA treatment strategies, which can be useless for certain people. Numerous cutting-edge treatments for RA have been introduced, including gene therapy, radiotherapy, immunotherapy, and energy conversion therapy (Chen et al. 2019; Kumar et al. 2016; Liu and Maeyama 2016; Zhao et al. 2021).

With the aid of nanobiotechnology and nanomedicine, these therapies have helped patients with RA have more remissions. In recent years, NP-based nanotherapy has emerged as a viable treatment option for RA, with the potential to enhance patient quality of life and reduce the social burden. In contrast to alternative therapeutic approaches, it can deliver therapeutic drugs at the site of inflammation, but also perform a subclinical diagnosis to manage inflammation and minimize adverse effects with lower dosages of medication. Additionally, next-generation NPs nanotherapy can help patients with RA avoid developing treatment resistance. In addition to clarifying the present difficulties and upcoming advancements concerning the development of various functional NPs with capabilities for effective RA treatment, this chapter highlights and analyzes the most recent advancements in next-generation NP-based nanotherapy.

12.2 Pathophysiology of RA

The hallmark of RA, a chronic inflammatory illness, is inflammation of the synovial joints, which leads to pain, swelling, and eventually, joint degeneration (Gibofsky 2014). RA is caused by inappropriate immunological reactions in which the body's immune system mistakenly attacks its own tissues. The thickening and inflammation of the synovium and the lining of the joints result in the formation of a pannus or a layer of inflamed tissue. The release of inflammatory cytokines such as tumor necrosis factor (TNF), interleukin-1 (IL-1), and interleukin-6 (IL-6) facilitates inflammatory processes. Enzymes released by the inflammatory synovium break down bone and cartilage, causing joint injury and deformity. Systemic inflammation caused by RA can also affect other organs and result in comorbidities such as cardiovascular disease. A higher risk of RA is associated with specific genes, including HLA-DR4. In genetically predisposed individuals, RA can develop as a result of factors such as smoking, infections, and hormonal changes. Imaging investigations, blood tests for rheumatoid factor (RF) and anti-cyclic citrullinated peptide (anti-CCP) antibodies, and clinical symptoms were performed to diagnose RA (Fig. 12.1). Biologic agents, disease-modifying antirheumatic medications (DMARDs), and nonpharmacological therapies, including physical therapy and lifestyle changes, are all part of the treatment. Here, we present an overview of RA pathophysiology.

12.2.1 Immune System Activation

Genetic Factors An important part of the development of autoimmune disorders is genetic. The chance of getting an autoimmune disease is increased if there is a family history of it. For instance, your chances of developing RA may increase if a close relative already has it (Yamamoto and Okada 2019). Autoimmune illnesses are closely linked to specific genetic markers, especially those found in the major histocompatibility complex (MHC) area (Yamamoto and Okada 2019). Variations in these genes can result in immune system malfunction because these markers aid the immune system in differentiating between self and nonself. Autoimmune illnesses are highly heritable, according to twin studies. For example, it is estimated that about 60% of RA is heritable (Yamamoto and Okada 2019). RA is significantly influenced by genetic predisposition, with some genes (such as HLA-DR4) making people more susceptible. In addition to providing valuable insights into the mechanisms underlying the disease, genetic research holds promise for improving RA diagnosis, treatment, and prevention (Fig. 12.1).

Environmental Factors The onset and worsening of autoimmune illnesses are significantly influenced by environmental variables. Long-term stress can affect the immune system and play a role in the emergence of autoimmune disorders. Cortisol and other stress chemicals can change how the immune system works and make

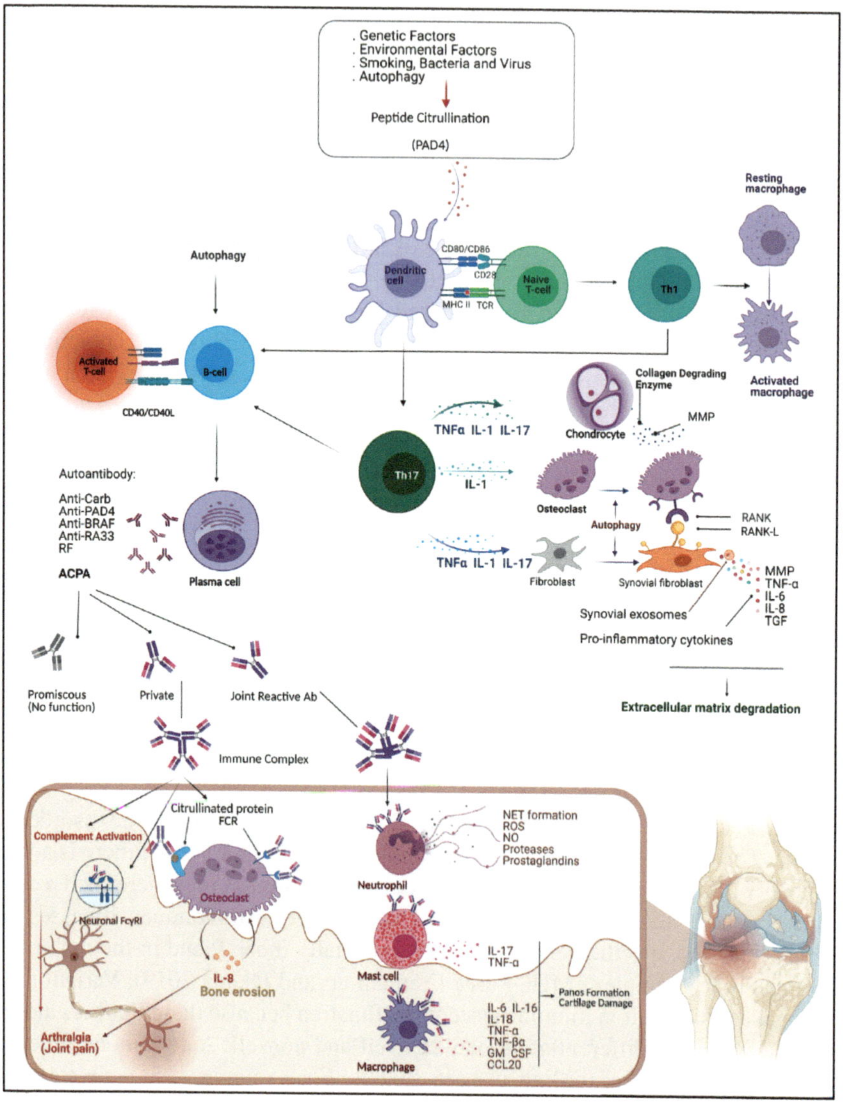

Fig. 12.1 Rheumatoid arthritis etiology. Several variables, including genetic background, smoking, bacterial and viral infections, and autophagy, contribute to the PADI4 enzyme's catalytic conversion of arginine to citrulline. Naive T cells and Th1, Th17, and Th2 cells are activated by antigen presentation by antigen-presenting cells (APC). This figure was adapted from a study by Mueller et al. (2021). This article is open access to MDPI publishers and articles distributed under the terms and conditions of the Creative Commons Attribution (CC BY) license (https://creativecommons.org/licenses/by/4.0/)

people more prone to autoimmune reactions. Autoimmune illnesses can be brought on by hormonal changes, such as those that occur during pregnancy, menopause, or hormone replacement therapy. Pesticides, heavy metals, and industrial chemicals are examples of environmental pollutants that might impair immune function and raise the risk of autoimmune illnesses. Immune responses may be impacted by the gut microbiome's makeup. Autoimmune conditions including RA and inflammatory bowel disease have been connected to dysbiosis, or an imbalance in the gut microbiota. One well-known risk factor for RA and other autoimmune illnesses is tobacco smoke (Harel-Meir et al. 2007). It can raise inflammation and change immunological responses. Smoking, infections, and hormonal changes can all contribute to the development of RA in people who are genetically prone to it. By comprehending these environmental triggers, solutions to lower the risk of RA and improve disease management can be developed.

12.2.2 *Inflammatory Response*

Synovial Inflammation A common feature of many autoimmune illnesses, especially RA, is synovial inflammation, sometimes referred to as synovitis. Joint cartilage and bones are lubricated and nourished by synovial fluid, which is produced by the synovial membrane lining the inside of joints. When the immune system unintentionally targets the synovial membrane in autoimmune disorders like RA, inflammation results (Takeuchi et al. 2019). Chronic inflammation results from immune cells such as T and B cells attacking the synovial membrane in RA. TNF-α, IL-1, and IL-6 are examples of inflammatory cytokines produced by inflammatory mediators in the inflamed synovium that lead to joint injury (Takeuchi et al. 2019). The synovium's fibroblast-like synoviocytes (FLSs) cells become activated and are essential for promoting inflammation and joint degeneration (Takeuchi et al. 2019). Comprehending the function of synovial inflammation in autoimmune disorders is essential for creating efficient therapies and enhancing patient outcomes.

Cytokines and Chemokines In the inflammatory response, cytokines and chemokines are essential mediators (Lindell and Lukacs 2010). Cells, especially immune cells, emit tiny proteins called cytokines, which serve as signaling molecules to control inflammation, immunity, and hematopoiesis. Interleukins (ILs), growth factors, interferons (IFNs), and tumor necrosis factors (TNFs) are important cytokines. They support immunological responses, cell proliferation, differentiation, and cell signaling. For instance, proinflammatory cytokines such as TNF-α, IL-1, and IL-6 are important in both causing and maintaining inflammation (Turner et al. 2014). A subclass of cytokines known as chemokines has chemotactic qualities, which means that they draw cells to areas where they are more abundant. Based on how the cysteine residues are arranged, chemokines are divided into four categories: CC, CXC, CX3C, and XC. The main function of chemokines is to attract leukocytes, or white blood cells, to areas of infection or damage (Turner et al. 2014). For instance, neutrophils are drawn to the site of inflammation by the CXC chemokine IL-8. Developing targeted treatments for inflammatory illnesses requires an understanding of the functions of cytokines and chemokines.

12.2.3 Pannus Formation

Pannus One of the main characteristics of RA is pannus development, which is the aberrant proliferation of synovial tissue (Sharif et al. 2024). It is very cellular and includes immunological cells such as synoviocytes and macrophages. In RA, the synovium is wrongly attacked by the immune system, resulting in persistent inflammation. The synovial tissue grows and infiltrates the surrounding joint structures as a result of this inflammation. The production of inflammatory cytokines and chemokines, which stimulate the formation of synovial tissue, is triggered by persistent synovial inflammation. The production of pannus is facilitated by the infiltration of T cells, B cells, and macrophages into the synovium. Joint degradation and loss of function may result from the invasion and injury of cartilage, bone, tendons, and ligaments by the growing pannus tissue. In the afflicted joints, pannus development results in discomfort, edema, and stiffness. Bone and cartilage may be eroded over time by the invasive pannus, resulting in joint abnormalities and functional disability. Fatigue and fever are examples of systemic symptoms that RA can cause in other sections of the body. In order to manage RA and avoid pannus development and joint degeneration, early diagnosis and vigorous treatment are essential.

Angiogenesis A key factor in the development of RA is angiogenesis, which is the creation of new blood vessels from preexisting ones (Elshabrawy et al. 2015). More oxygen and nutrients are needed by the RA-inflamed synovium, which causes new blood vessels to develop (Paleolog 2002). The pannus, an aberrant tissue that invades and kills joint structures, grows thanks to the new blood vessels (Paleolog 2002). The synovium produces proangiogenic factors that promote angiogenesis, such as fibroblast growth factor (FGF) and vascular endothelial growth factor (VEGF) (Elshabrawy et al. 2015). The basement membrane is broken down by matrix metalloproteinases (MMPs) enzymes, which permits endothelial cells to move and create new blood vessels (Elshabrawy et al. 2015). Comprehending the function of angiogenesis in RA can lead to new therapy options.

12.2.4 Joint Destruction

Proteolytic Enzymes In RA, proteolytic enzymes in particular, matrix metalloproteinases, or MMPs are crucial for joint degradation (Rengel et al. 2007). Collagen and proteoglycans, two elements of the extracellular matrix (ECM) that give cartilage its structural support, are broken down by MMPs (Hauw et al. 2022). MMPs are extracellularly activated in the joint environment after being generated as dormant proenzymes (Hauw et al. 2022). Cartilage is destroyed when MMPs are active because they degrade collagen and other ECM proteins. By breaking down bone matrix proteins, MMPs also aid in bone degradation. Inflammatory cytokines like TNF-α and IL-1β stimulate the production of MMPs, exacerbating joint damage

(Rengel et al. 2007). Developing targeted medicines to treat RA and other rheumatic disorders requires an understanding of how proteolytic enzymes contribute to joint deterioration.

Osteoclast Activation In RA, osteoclast activation is essential for joint degradation. During the process of bone remodeling, specialized cells called osteoclasts break down bone tissue, a process known as bone resorption. By binding to its receptor RANK on osteoclast precursors, the receptor activator of nuclear factor kappa-B ligand (RANKL)/RANK Pathway promotes the development of osteoclast precursors into mature osteoclasts (Zheng et al. 2024). Inflammatory Immune cells in the inflammatory synovium release cytokines including TNF-α and IL-1, which promote osteoclastogenesis and raise osteoclast activity. Activated osteoclasts contribute to the distinctive bone erosion observed in RA by resorbing bone at the joint edges. Rheumatoid factor (RF) and anti-citrullinated protein antibodies (ACPAs) are examples of autoantibodies that can increase osteoclast activity and production, which further accelerates bone loss (Steffen et al. 2019). Cytokines that stimulate osteoclastogenesis and raise osteoclastogenesis activity include TNF-α, IL-1, and IL-17 (Fig. 12.1). One factor contributing to the distinctive bone erosions seen in RA is the degradation of bone at the joint edges caused by overactive osteoclasts. In order to effectively prevent bone degradation and manage RA, it is imperative to comprehend and target osteoclast activation.

12.2.5 Systemic Inflammation

Extra-Articular Manifestations Extra-articular manifestations (ExRA) refer to symptoms and complications that occur outside of the joints. A patient's quality of life may be greatly affected by these manifestations, which can damage different body systems and organs. In addition to affecting other organs, RA can cause problems, such as rheumatoid nodules, lung involvement, and cardiovascular disease. Comprehensive therapy for patients with RA requires understanding of and attention to extra-articular symptoms.

Autoantibodies The diagnosis and pathophysiology of RA are significantly influenced by autoantibodies, which are the hallmarks of the disease. An important marker of RA and a contributing factor to the chronic inflammatory state is the presence of autoantibodies, such as rheumatoid factor and anti-citrullinated protein antibodies (Steffen et al. 2019). Anti-cyclic citrullinated peptide (anti-CCP) antibodies are highly specific for RA and are associated with a more severe course of the disease (Steffen et al. 2019). Although it is not unique to RA and can occur in other autoimmune illnesses, RF is one of the most widely used indicators of RA. When paired with other laboratory tests and clinical symptoms, the presence of these autoantibodies can help in the diagnosis of RA. Some autoantibodies, such as ACPAs, can be used to predict outcomes and are associated with a more severe disease course. Accurate diagnosis, prognosis, and therapeutic planning for RA depend on an understanding of the involvement of these autoantibodies.

12.2.6 Clinical Features

Symmetric Joint Involvement Symmetric joint involvement is a characteristic feature of RA (Sanchez-Castillo et al. 2021). This indicates that joints on both sides of the body are usually affected by RA. The hands, wrists, and feet are frequently affected, and RA usually affects joints symmetrically. Although RA may start asymmetrically in some cases, it typically becomes symmetrical as the disease progresses. Symmetry is one of the key criteria used in diagnosing RA, along with other factors, such as morning stiffness, joint swelling, and the presence of specific autoantibodies. Understanding symmetric joint involvement is important for effective management of RA.

Morning Stiffness Morning stiffness is a common and significant symptom of RA (Orange et al. 2020). In people with RA, morning stiffness usually lasts for more than an hour, and frequently for several hours. It can be distinguished from other types of arthritis, including osteoarthritis, by its duration, as stiffness typically moves away within 30 min. Because stiffness is frequently severe, patients find it challenging to move and perform everyday tasks as soon as they wake up. It usually affects the hands, wrists, feet, and other peripheral joints, and usually affects the joints symmetrically. Stiffness tends to be worst in the morning after a period of inactivity (such as sleep) and can improve with movement and activity throughout the day.

Chronic Pain and Swelling A person's quality of life can be greatly affected by chronic pain and swelling, which are typical symptoms of RA. Both joint inflammation and injury can cause chronic pain, which frequently persists in RA. Pain can result from inflammation, joint damage, and central sensitization, where the nervous system becomes more sensitive to pain signals. Treatment includes medications such as DMARDs, biologics, NSAIDs, and pain management strategies such as physical therapy and psychological support (Ben Mrid et al. 2022). The buildup of synovial fluid in the joints is the result of inflammation of the synovial membrane, which causes swelling. Swollen joints are often tender and warm, and can lead to stiffness and reduced range of motion. Antiinflammatory medications, DMARDs, and lifestyle modifications can help in the management of swelling. Improving the quality of life of people with RA requires an understanding of and commitment to managing chronic pain and swelling. Understanding the pathophysiology of RA is essential to create tailored treatments and enhance patient outcomes.

12.3 Current Treatment Modalities of RA

Over time, RA treatment has changed dramatically, emphasizing early intervention and the utilization of a range of therapeutic approaches. Remission or minimal disease activity, minimization of joint degeneration, and enhancement of the quality of life are the objectives of RA treatment. Individualized treatment programs were created for each patient according to their unique requirements and therapeutic responses. These are the currently available RA therapy options.

12.3.1 Disease-Modifying Antirheumatic Drugs (DMARDs)

Methotrexate (MTX) Methotrexate is one of the most commonly prescribed and effective medications for treating RA. It prevents folic acid metabolism, which is essential for cell growth. This reduces the inflammation in autoimmune diseases. The most widely used DMARD is frequently regarded as the initial course of treatment for RA (Tanaka 2021). Methotrexate relieves symptoms such as joint pain, exhaustion, redness, and swelling by lowering inflammation. In addition, it can delay the development of joint injuries. As methotrexate effectively controls symptoms and slows the course of the disease, it is frequently used as the first-line treatment for RA. Many patients use methotrexate for long-term management of RA. Methotrexate has been a cornerstone of RA treatment for many years, and its benefits often outweigh the potential side effects when monitored properly.

Hydroxychloroquine Hydroxychloroquine is commonly used to treat RA (Nazir et al. 2021). Hydroxychloroquine is an antimalarial drug with antiinflammatory properties. Modification of the immune system lowers the synthesis of molecules that cause inflammation. Hydroxychloroquine has been studied for its immunomodulatory effects as well as its role in treating adverse conditions associated with RA (Nazir et al. 2021). It can prevent long-term joint degeneration and help reduce joint pain, stiffness, and edema. Hydroxychloroquine is a valuable option for managing RA, especially in patients who may not tolerate strong medications.

Sulfasalazine A DMARD called sulfasalazine is used to treat RA (Plosker and Croom 2005). 5-Aminosalicylic acid (5-ASA) and sulfapyridine are the two constituents of sulfasalazine. The precise mode of action is unclear; however, it appears to be related to immune system modulation and inflammation reduction. It can decrease the progression of joint deterioration and assist in reducing joint pain, stiffness, and edema. For many patients with RA, sulfasalazine is useful in reducing symptoms and avoiding joint deterioration. It is useful for managing RA in the long term, particularly when paired with other DMARDs. Sulfasalazine has been a valuable treatment option for RA for many years, and its benefits often outweigh its potential side effects when monitored properly.

Leflunomide RA is treated with leflunomide, a DMARD (Alldred and Emery 2001). Leflunomide functions by blocking the pyrimidine-synthesizing enzyme dihydroorotate dehydrogenase (DHODH). This suppression reduces the growth of activated lymphocytes, which contributes to RA's inflammatory process of RA. This helps reduce inflammation and slow the progression of RA. It can be used in individuals who are unable to tolerate or do not respond to MTX (Tanaka 2021). Leflunomide is a valuable option for managing RA, and its benefits often outweigh its potential side effects when monitored properly.

12.3.2 Biologic DMARDs

Tumor Necrosis Factor (TNF) Inhibitors Tumor necrosis factor (TNF) is considered a major proinflammatory cytokine, affecting various aspects of the immune reaction (Lis et al. 2014). Tumor necrosis factor (TNF) inhibitors are a class of biologic medications used to treat autoimmune diseases like RA. All five TNF inhibitors currently available on the market (i.e., etanercept, infliximab, adalimumab, certolizumab, and golimumab) are top sellers, although indicated only in autoimmune diseases, including rheumatoid arthritis, Crohn's disease, and psoriasis (Lis et al. 2014). TNF inhibitors are a family of medications used to treat RA and other inflammatory diseases. TNF is a cytokine and a signaling molecule that is vital for inflammation and is produced by the immune system. In RA, excessive TNF production results in joint injury and inflammation. TNF inhibitors, such as etanercept, infliximab, and adalimumab, are generally used to treat RA (Tanaka 2021).

Interleukin-6 (IL-6) Inhibitors Interleukin-6 (IL-6) is a cytokine that plays a significant role in the immune response and inflammation (Pandolfi et al. 2020). It promotes inflammation within the joints, leading to pain, swelling, and stiffness. A class of medications known as IL-6 inhibitors is used to treat RA. IL-6 Inhibitors can target IL-6 or its receptor, such as tocilizumab and sarilumab, which have been developed to treat RA (Favalli 2020). These drugs can help reduce inflammation and improve symptoms. Clinical studies have shown that IL-6 inhibitors can be effective in reducing disease activity and improving the quality of life for RA patients (Nishimoto 2006).

T-Cell Costimulation Modulators T-cell costimulation modulators are a class of medications used to treat RA by inhibiting the activation of T cells, which play a key role in the autoimmune response (Östör 2008). Abatacept is a fusion protein that unites the Fc region of human immunoglobulin G1 (IgG1) with the extracellular domain of cytotoxic T-lymphocyte–associated antigen 4 (CTLA-4) (Iannone and Lapadula 2012). It functions by blocking the costimulatory signal required for T-cell activation. It specifically attaches itself to antigen-presenting cells, CD80 and CD86, preventing them from interacting with T cells' CD28. Abatacept decreases the activation of T cells, which are essential for the inflammatory process in RA, by blocking this costimulatory signal. Because it targets the underlying immunological system that causes RA, abatacept offers a significant leap in the treatment of the illness.

B-Cell Depletion Agents Rituximab is a biological DMARD that targets B cells to treat RA (Dass et al. 2006). A monoclonal antibody called rituximab targets the CD20 antigen of B cells. Rituximab depletes B lymphocytes, which are essential for the inflammatory process in RA, by binding to CD20. B-cell depletion aids in lowering RA-related joint degeneration, discomfort, and inflammation. There is currently evidence that B-cell depletion with rituximab is safe and effective in treating RA in a range of patient populations. Because of its safety profile and therapeutic results, rituximab has been approved for the treatment of RA patients who have not responded well to anti-TNF-alpha medications (Dass et al. 2006). For patients with RA who have not responded to previous treatments, rituximab has been a useful addition to the available therapeutic alternatives.

Janus Kinase (JAK) Inhibitors Janus kinases (JAKs) are a family of enzymes that play a crucial role in the signaling pathways of various cytokines and growth factors (Shen et al. 2024). In the context of RA, JAK inhibitors (JAKis) have emerged as an important class of medications (Harrington et al. 2023). RA is treated by a class of drugs called Janus Kinase (JAK) inhibitors, which include tofacitinib and baricitinib. They function by preventing the action of Janus kinases, which play important roles in the inflammatory process. JAK medications can lower inflammation and decrease the course of RA by blocking these enzymes (Tanaka 2021).

12.3.3 Targeted Synthetic DMARDs

JAK Inhibitors A Janus Kinase (JAK) inhibitors, a more recent class of drugs, are used to treat RA. They prevent the action of Janus kinases, which are inflammatory process-related enzymes. JAK inhibitors aid in reducing inflammation and delaying the course of RA by blocking these enzymes. They are a new class of DMARDs that target specific pathways involved in the inflammatory process (Tanaka 2021). JAK inhibitors are useful supplements to the available treatment alternatives for patients with RA who might not respond to conventional therapy.

Glucocorticoids

Glucocorticoids, often referred to as steroids, are a class of medications widely used in the treatment of RA (Hua et al. 2020). Glucocorticoids help reduce inflammation and alleviate symptoms such as pain and swelling. They suppress the immune system, which can help control the autoimmune response in RA. Often used as a bridge while waiting for slower-acting DMARDs to take effect. RA is frequently treated with a class of drugs called glucocorticoids, sometimes known as steroids. Glucocorticoids exhibit strong immunosuppressive and antiinflammatory properties

(Hua et al. 2020). They function by lowering the activity of immune cells involved in the inflammatory process and by preventing the synthesis of inflammatory cytokines. They help reduce joint pain, swelling, and stiffness, providing rapid relief from RA symptoms. Prednisone, for example, is used as a bridge therapy, whereas DMARDs temporarily relieve acute symptoms (Fraenkel et al. 2021). Glucocorticoids are an essential part of the RA treatment arsenal, especially for managing acute symptoms, and as a bridge therapy while waiting for other DMARDs to take effect.

12.3.4 Nonpharmacological Interventions

Physical Therapy Physical therapy is a crucial component of managing RA. It helps improve joint function, reduce pain, and maintain overall physical fitness. Maintenance of Joint Function and Muscle Strength. Enhances joint flexibility and range of motion. Alleviation of joint pain using various therapeutic techniques. Build muscle strength to support and stabilize the affected joints. Boost the capacity to perform everyday tasks. Keeping joints healthy can lower the chances of developing long-term disability. A vital component of the all-encompassing treatment of RA, physical therapy aids patients in preserving their physical well-being and enhancing their standard of living.

Occupational Therapy Occupational therapy (OT) is an effective strategy for controlling RA and enhancing day-to-day functioning. The goals of occupational therapy for RA are to reduce pain during daily activities, conserve energy and avoid fatigue, protect joints from further damage, learn new ways to perform tasks to reduce strain on joints, and use tools and devices to make daily tasks easier. Occupational therapy can be a game-changer for individuals with RA, helping them maintain their independence and improve their quality of life.

Lifestyle Modifications Lifestyle modifications can significantly improve RA management. Engage in activities such as walking, swimming, or cycling to maintain joint flexibility and muscle strength without causing additional stress to the joints. Incorporate gentle stretching exercises to improve flexibility and reduce stiffness. Light weights or resistance bands are used to build muscle strength, which supports and stabilizes the joints. The strain on weight-bearing joints such as the knees and hips can be reduced by losing excess weight. Consumption of a diet high in lean proteins, whole grains, fruits, and vegetables promotes general health and lower inflammation. To aid in your body's healing and better pain management, ensure you get enough sleep every night. Smoking increases the risk of complications and exacerbates RA symptoms. Smoking cessation can reduce the symptoms of RA and enhance general health. Excessive alcohol consumption may worsen RA symptoms and interact with medicines. Reducing alcohol consumption can improve condition management. You can enhance your general quality of life and manage RA symptoms more effectively by adopting these lifestyle modifications.

12.3.5 Surgical Interventions

Synovectomy Synovectomy is a surgical procedure used to treat RA by removing the inflamed synovium and the tissue lining the inside of the joints. The synovium becomes inflamed and thickened in RA, producing excess fluid, which leads to joint swelling and pain. Removing inflamed synovium can help reduce pain and improve joint function. Although it does not cure RA, it can slow down cartilage damage and provide relief from symptoms. Synovectomy can be a valuable option for managing RA symptoms, particularly when other treatments are ineffective.

Joint Replacement When conventional therapies fail to adequately relieve severe RA, joint replacement surgery, commonly referred to as arthroplasty, is a popular and successful treatment. A prosthesis, an artificial joint composed of ceramic, metal, or plastic, is used to replace the injured joint. This enhances mobility, reduces pain, and restores joint function. Reducing inflammation, easing symptoms, protecting joints and organs, and improving the overall quality of life of RA patients are the objectives of various therapeutic modalities. For patients with severe RA, joint replacement surgery can be a life-changing choice because it can enhance joint function and provide pain relief.

12.4 Nanomedicine in RA Treatment

By providing creative ways to optimize medication distribution, lessen side effects, and boost treatment success, nanotechnology has demonstrated tremendous potential for the treatment of RA (Ren et al. 2024). Some key applications of nanotechnology in RA are listed in Table 12.1.

12.4.1 Targeted Drug Delivery

Nanoparticles (NPs) To guarantee that the therapeutic medicines reach the location of inflammation, NPs can be engineered to target particular cells or tissues, such as inflammatory synovial joints (Zheng et al. 2021) (Table 12.1). NPs represent a promising breakthrough in the area of tailored medication delivery for RA. NPs can minimize systemic side effects by delivering medications directly to inflamed joints. They can provide the medication with a regulated release, guaranteeing a consistent, long-term therapeutic impact. Owing to their small size, NPs can enter tissues more deeply and reach places that traditional medications might miss. NPs lessen the possibility of systemic drug delivery side effects by focusing on the inflammatory joints. Enhanced drug delivery to the affected area can improve the overall effectiveness of the treatment. Targeted delivery allows lower drug dosages

Table 12.1 Nanoparticles for the treatment of RA in preclinical findings

Nanoparticles	Advantage	Treatment of RA	References
Gelatin	Biocompatible, biodegradable, targeted delivery, enhanced drug stability	Drug delivery system, combinational therapy Gelatin nanoparticles can modulate macrophage polarization, reducing inflammation and cartilage destruction	Ahmad et al. (2021), Ansari et al. (2021), Jori et al. (2024) and Sun et al. (2024)
Chitosan	Biocompatible, biodegradable, controlled release, targeted delivery, enhanced drug stability, antiinflammatory properties	Drug delivery, combinational therapy It can alleviate RA by modulating oxidative stress and inflammatory parameters	Ansari et al. (2019), Gravandi et al. (2024)
Polymeric	Biocompatibility, targeted delivery, controlled release, enhanced drug stability, antiinflammatory properties	Drug delivery, combination therapy, stimuli-responsive systems	Xie et al. (2020)
Solid-lipid NPs	Biocompatibility, targeted delivery, controlled release, enhanced drug stability, antiinflammatory properties	Drug delivery, combination therapy, theranostic systems	Zewail et al. (2024)
Dendrimer	Controlled structure, targeted delivery, high drug loading capacity, biocompatibility, versatility	Drug delivery systems, combination therapy, inhibition of inflammatory processes, macrophage modulation	Leah (2011), Li et al. (2021)
Metallic	Therapeutic agents, drug delivery vectors, diagnostic capabilities, anti-angiogenesis effect	Au NPs to reduce inflammation and joint damage Au NPs combination with tocilizumab (TCZ), an anti-interleukin-6 (IL-6) receptor inhibitor Ag NPs have antimicrobial properties and can help reduce infection risks associated with RA Ag NPs ameliorate RA by macrophage apoptosis and repolarization Hydroxyapatite (HA) NPs can protect cartilage and provide lubrication, which is beneficial for joint health	Li et al. (2022), Venkatesan et al. (2022) and Yang et al. (2021)

and reduces the risk of toxicity. Researchers are continuously exploring new types of NPs and their applications for RA treatment. Future advancements may lead to personalized NP-based therapies that are tailored to individual patient needs.

Enhanced Bioavailability NPs can increase drug bioavailability, enabling lower dosages and fewer systemic adverse effects (Zheng et al. 2021). There are numerous approaches to increasing the bioavailability of medications using nanotechnology, which improves the effectiveness and efficiency of therapies. Drugs that are poorly soluble can become more soluble when NPs are used, improving the body's ability to absorb them. NPs can be encapsulated by other NPs to improve their stability and solubility. Drugs can be targeted to particular tissues or cells by binding particular ligands to the surface of NPs, which decreases off-target effects and improves the therapeutic efficacy. Drugs can be gradually released from NPs over time, thereby maintaining a steady therapeutic dose. Medicines can reach their target without being broken down by altering the surface of the NPs to help them dodge the immune system (Egwu et al. 2024). Drug stability and shelf life can be improved by encapsulating them in NPs to prevent degradation (Patra et al. 2018). The rapidly developing field of nanotechnology has great promise for transforming medication delivery and enhancing patient outcomes.

12.4.2 Controlled Release

Sustained Release NPs can enable regulated and prolonged drug release, thereby lowering the frequency of delivery and preserving the therapeutic levels over time (Ren et al. 2024). This can help improve the treatment efficacy and patient compliance. Minimizes peak-trough fluctuations in drug levels, thereby reducing the risk of side effects. It simplifies treatment regimens, making it easier for patients to adhere to their medication schedules. Sustained-release drug delivery systems are significant advancements in biomedical applications, offering improved treatment outcomes and patient convenience.

Responsive Systems To provide more efficient and individualized care, responsive systems are designed to adjust to the unique requirements of the patient and the stage of the disease. To improve the accuracy of drug administration, nanosystems have been designed to release medications in response to particular stimuli, including variations in pH or temperature (Ren et al. 2024). A major development in the treatment of RA is responsive systems that provide more individualized and efficient care.

12.4.3 Combination Therapy

Multidrug Delivery The codelivery of several medications with distinct modes of action is made possible by nanotechnology, which enhances therapeutic results and has a synergistic impact (Ren et al. 2024). Targeting different RA symptoms, such as pain, inflammation, and joint degeneration, with a variety of medications has distinct modes of action. This enables combination therapy by supplying medications with improved controlled-release profiles, targeted delivery, and improved physicochemical characteristics. Creating medicinal nanoparticles capable of delivering various medications and therapeutic approaches, including sonodynamic therapy and phototherapy. The treatment results may improve if several RA pathways are targeted. The risk of adverse effects is decreased and systemic exposure is minimized with targeted delivery. Combining several medications into a single delivery system simplifies treatment plans. By providing more efficient and patient-friendly choices, multidrug delivery systems show promise for enhancing the treatment of RA.

12.4.4 Diagnostic and Monitoring Tools

Imaging Agents As NPs can be used to track the course of RA and evaluate the effectiveness of therapies (Qamar et al. 2019). NPs are increasingly being used as imaging agents for RA because of their unique properties and their ability to enhance imaging quality. Because gold, silver, and iron oxide NPs have superior optical and magnetic properties, they are frequently used in imaging (Li et al. 2022). These semiconductor NPs exhibit high brightness and stability, making them suitable for imaging applications. Biodegradable polymers can be used to encapsulate imaging agents, providing a controlled release and improved stability. Iron oxide NPs can enhance MRI contrast, providing clearer images of the inflamed joints. Metal NPs, such as gold and silver, can improve CT imaging by increasing contrast (Li et al. 2022). Quantum dots and dye-loaded NPs can be used for fluorescence imaging, allowing real-time visualization of RA progression. NPs improve the resolution and clarity of imaging, aiding the accurate diagnosis and monitoring of RA. Functionalized NPs can target specific tissues or cells, reduce background noise, and improve imaging specificity. NPs can enhance imaging efficiency, allow for lower doses of imaging agents, and reduce potential toxicity. NPs present a viable strategy to improve RA imaging methods, provide better diagnostic instruments, and support the creation of more potent treatments.

Biomarker Detection NPs are revolutionizing the detection of biomarkers for RA, offering more sensitive and accurate diagnostic tools. Gold, silver, and iron oxide NPs are commonly used because of their excellent optical and magnetic properties.

These can encapsulate biomarkers or imaging agents and enhance detection sensitivity. Semiconductor NPs provide high brightness and stability for imaging applications. NPs can be used to detect anti-CCP antibodies, which are key biomarkers for RA (Tai et al. 2024). NPs can enhance the detection of C-reactive protein (CRP), an indicator of inflammation. Improved detection of RF is another important biomarker for RA. NPs can be used to detect specific miRNAs that are associated with RA (Imas et al. 2020). NPs improve the sensitivity of biomarker detection, allowing for an earlier diagnosis and better monitoring of disease progression. It requires smaller sample volumes, making the tests less invasive. Provides quicker diagnostic results, aiding timely treatment decisions. NPs offer a promising approach for improving the detection of RA biomarkers, leading to better diagnostic accuracy and patient outcomes.

12.4.5 Prodrug-Based Nanomedicines

Prodrug Strategy Nanotechnology and prodrug techniques are used in prodrug-based nanomedicines to increase drug stability, target activity, and bioavailability while lowering adverse effects (Li et al. 2024a). Creating inert substances (prodrugs) that transform the body into active medications, frequently at the site of inflammation, is the goal of the prodrug strategy for RA. To minimize systemic negative effects, prodrugs can be preferentially activated in inflamed joints. Prodrugs also offer improved drug stability and solubility, enhancing absorption and effectiveness. By limiting exposure of healthy tissues to the active medication, the risk of adverse effects is reduced. DMARDs can be more efficiently administered to irritated joints using prodrugs. Improving the effectiveness and distribution of biological medicines used to treat RA. Prodrugs can be used in conjunction with other therapeutic agents to offer a more thorough treatment. The prodrug method is a viable way to improve RA treatment by optimizing drug distribution, lowering side effects, and improving patient outcomes.

Higher drug-loading capacities are possible with these devices, which increases their efficacy in treating RA (Li et al. 2024a). In the context of treating RA, higher drug loading means that more therapeutic agents may be delivered to the target location, thereby increasing the effectiveness of treatment. Better therapeutic results may be due to the presence of more drug molecules at the target site. Increased drug loading can decrease dosing frequency by extending the duration of drug release, thereby simplifying treatment plans, which helps patients remember taking their medications on time. The delivery and effectiveness of biological medicines and DMARDs can be enhanced by higher drug loading. The coadministration of many medications with various modes of action may offer a more thorough treatment course. A promising strategy for RA treatment is a higher medication loading, which improves patient outcomes and drug delivery.

12.4.6 Safety and Biocompatibility

Biodegradable Materials Many nanomaterials are made from biodegradable and biocompatible materials, minimizing their long-term accumulation and potential toxicity. The capacity of biodegradable materials to safely break down within the body after delivering therapeutic ingredients has led to an increase in their use for the treatment of RA. The breakdown of these substances produces nontoxic metabolites that the body can readily remove. Steady therapeutic impact can be achieved by engineering biodegradable polymers to gradually release medications over time. By focusing on particular tissues, such as irritated joints, they can be used to improve therapeutic results and reduce adverse effects. Biodegradable materials can be used to create nanoparticles (NPs), microparticles, and hydrogels for targeted and sustained drug delivery. Encapsulating genetic materials in biodegradable materials can enhance the delivery and expression of therapeutic genes. Biodegradable systems can deliver multiple drugs or therapeutic agents and provide a comprehensive treatment approach. Researchers at UC San Diego have developed a biodegradable polymer system that works in concert with the immune system to treat RA (McBride et al. 2023). This approach can treat or prevent diseases by reprogramming immune cells into regulatory T cells using encapsulated all-trans retinoic acid (ATRA) (McBride et al. 2023). Biodegradable materials are a promising strategy for improving the treatment of RA by improving the medication distribution, lowering side effects, and producing long-lasting therapeutic effects. By overcoming the drawbacks of traditional medicines and offering more efficient, individualized, and targeted therapeutic choices, nanotechnology presents a possible way to transform the treatment of RA.

12.5 Routes of Drug Delivery

Numerous novel medication delivery methods have been made possible by nanotechnology, each with unique benefits and uses. Drugs can be delivered directly to the gastrointestinal tract via ingestion of NPs. Although this method is easy and noninvasive, drug stability and absorption issues can arise. Drugs can be delivered across the skin barrier by applying NPs. This method can increase patient compliance and is helpful for continuous drug delivery. Drugs can be delivered via the nasal route, which is advantageous for targeting the central nervous system because it permits quick absorption and avoids the blood-brain barrier. To treat ocular illnesses, NPs can be used to transport drugs to the eyes. Inhaling medications into the lungs can help to treat respiratory conditions and promote rapid absorption. This includes injections administered subcutaneously (SC), intramuscularly (IM), and

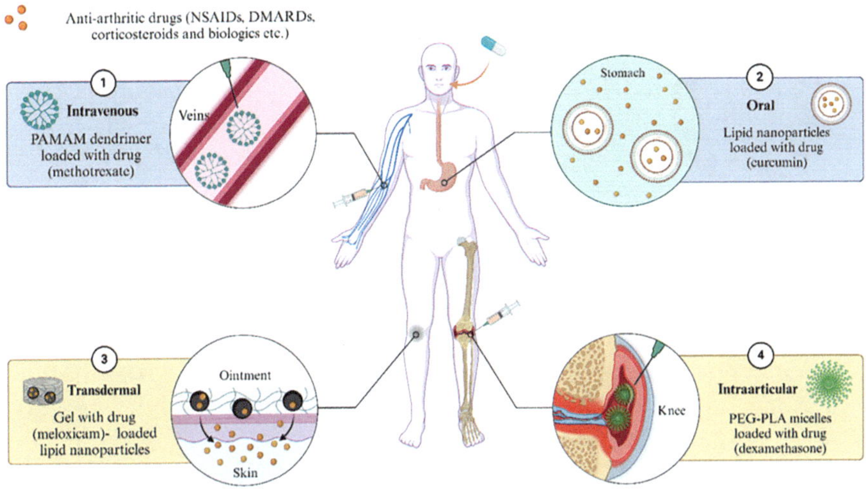

Fig. 12.2 Schematic representation of drug delivery routes in RA. (Image created with Biorender.com)

intravenously (IV). NPs can increase the bioavailability and improve drug delivery to specific tissues. To increase the effectiveness and reduce negative effects, NPs may be engineered to target particular cells or regions. Over time, NPs can maintain the best therapeutic dose by delivering a regulated and prolonged release of medications. NPs can increase the stability and solubility of drugs, thereby increasing their bioavailability. Treatment and monitoring can be performed simultaneously owing to the ability of NPs to combine therapeutic and diagnostic roles. Drug delivery has been revolutionized by nanotechnology, which provides safer, more targeted, and more effective treatment alternatives. A schematic representation of some common routes of drug delivery is shown in Fig. 12.2.

12.6 Nanoparticles

NPs' distinct physical and chemical characteristics have made them promising for the treatment of RA. Because nanotechnology can overcome the drawbacks of traditional medicines and provide more individualized, focused, and effective therapeutic alternatives, it presents a viable way to transform the treatment of RA. NPs' special qualities and adaptability of NPs make them popular for biomedical applications (Fig. 12.3). Here, we review some of the most popular NP types used to treat RA.

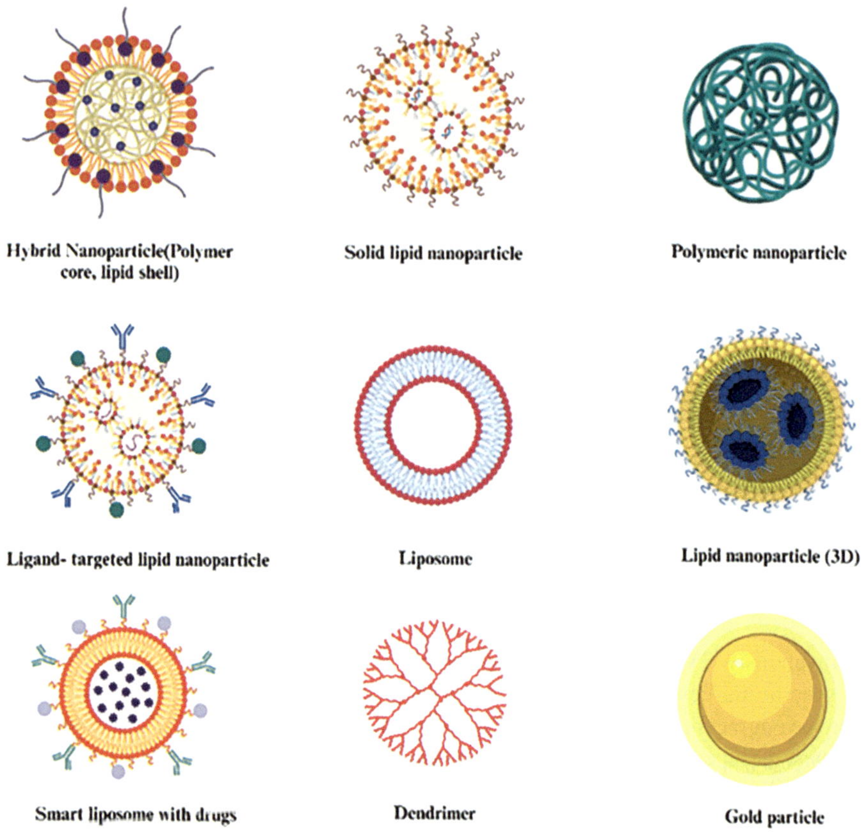

Fig. 12.3 Schematic representation of next-generation nanoparticles for the management of RA. (The images were created using Biorender.com)

12.6.1 Lipid-Based NPs

RA can be treated using lipid-based NPs (LNPs) owing to their unique qualities and benefits (Zewail et al. 2024). LNPs can increase the effectiveness of treatment by delivering therapeutic medicines to inflamed joints. LNPs minimize systemic exposure and reduce the risk of side effects by focusing on certain tissues (Zewail et al. 2024). LNPs provide better control over drug release, guaranteeing consistent therapeutic impact. NPs can be safely broken down in the body and are generally biocompatible. Lipid-based NPs are a viable strategy for improving RA treatment by boosting drug delivery, lowering side effects, and producing long-lasting therapeutic effects (Zewail et al. 2024).

Liposomes Drug delivery and gene therapy are two applications of spherical vesicles with phospholipid bilayers. One type of lipid-based NP that has demonstrated significant promise for the treatment of RA is liposomes. To increase the effectiveness of therapeutic drugs, liposomes can be engineered to target particular tissues, including inflamed joints. Liposomes minimize systemic exposure and lower the likelihood of adverse effects by delivering medications directly to the site of inflammation. Liposomes can prolong the stability and shelf life of medications by preventing disintegration (Kim et al. 2024). Glucocorticoids and nonsteroidal antiinflammatory medications (NSAIDs) can be delivered directly to inflamed joints using liposomes (Hu et al. 2019). DMARDs' ability of DMARDs to reduce inflammation and prevent joint deterioration can be improved by targeted administration.

Liposomes can enhance the distribution and effectiveness of biological medicines used to treat RA. Studies have demonstrated that neutrophils may absorb liposomes adorned with particular chemicals, such as sialic acid, and transfer them to inflamed joints, intensifying the antiinflammatory effect (Hu et al. 2019). Tofacitinib citrate, a DMARD, can be delivered directly to inflamed joints using a liposomal method, increasing its effectiveness and lowering its negative effects (Shen et al. 2020). As liposomes improve medication distribution, minimize side effects, and produce long-lasting therapeutic effects, they offer a viable way to improve the treatment of RA. Functionalized liposomes can improve medication transport and targeting, and represent an interesting new development in the treatment of RA (Wang et al. 2020) (Fig. 12.4). To increase the effectiveness of therapeutic drugs, functionalized liposomes can be used to target particular tissues, including inflamed joints. Liposomes can prolong the stability and shelf life of medications by preventing disintegration. Over time, they can maintain the best therapeutic levels by delivering regulated and prolonged release of medications. Functionalized liposomes reduce systemic exposure and the chance of negative effects by delivering medications directly to the site of inflammation (Nasra et al. 2022).

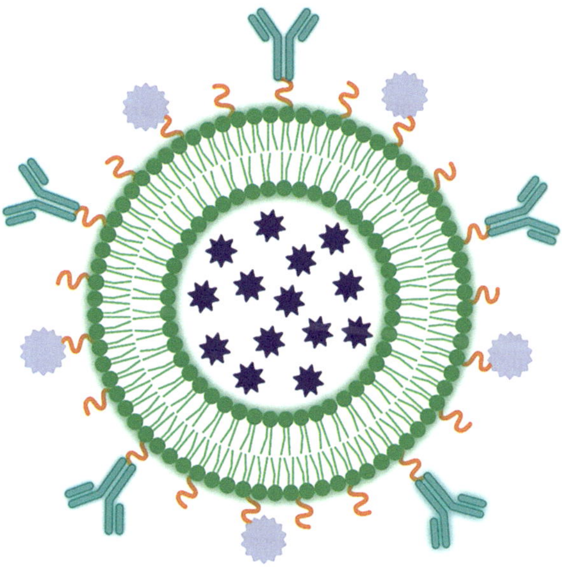

Fig. 12.4 Schematic representation of liposomes functionalized with drugs for targeted drug delivery at the site of inflammation. (Image created with Biorender.com)

Solid Lipid NPs (SLNs) NPs composed of solid lipids provide better stability and regulated medication release. RA therapy may benefit from the use of SLNs because of their special qualities and benefits. SLNs can increase the effectiveness of treatment by delivering therapeutic drugs to inflamed joints. SLNs minimize systemic exposure and reduce the chances of adverse effects by focusing on certain organs. SLNs can increase their stability and shelf life by shielding medications from deterioration. The body can safely break down NPs and is typically biocompatible. Glucocorticoids and nonsteroidal antiinflammatory medications (NSAIDs) can be administered directly to inflamed joints by SLNs. DMARDs' ability of DMARDs to reduce inflammation and prevent joint deterioration can be improved by targeted administration. SLNs can enhance the distribution and effectiveness of biological medicines used to treat RA. To precisely target macrophages in RA, SLNs can be functionalized with targeting molecules such as antibodies (Albuquerque et al. 2015). This makes it possible for diagnostic and therapeutic purposes. For combination therapeutic and imaging applications, SLNs can encapsulate several medicinal drugs including methotrexate (MTX) and superparamagnetic iron oxide nanoparticles (SPIONs) (Albuquerque et al. 2015). SLNs are a promising strategy for improving RA treatment by boosting drug delivery, reducing side effects, and producing long-lasting therapeutic effects.

12.6.2 Polymeric NPs

The special qualities and benefits of polymeric NPs make them a potential treatment option for RA. To increase the effectiveness of therapeutic medicines, polymeric nanoparticles (NPs) can be engineered to target particular tissues, including inflamed joints. With the help of these NPs, medications can be released under control and maintained at a constant therapeutic dose. Polymeric nanoparticles can be safely broken down in the body and are generally biocompatible. By preventing deterioration, they help prolong the stability and shelf life of the medications. Glucocorticoids and nonsteroidal antiinflammatory medications (NSAIDs) can be administered directly to inflamed joints using polymeric nanoparticles. If delivered precisely, DMARDs can be more effective at lowering inflammation and avoiding joint damage. The distribution and effectiveness of biological medicines used to treat RA can be enhanced by using polymeric nanoparticles. Targeted drug release at the site of inflammation is possible with polymeric nanoparticles (NPs) that are engineered to react to particular stimuli, such as pH, temperature, or enzymes (Xie et al. 2020). NPs can encapsulate multiple therapeutic substances and offer a thorough therapeutic strategy.

PLGA (Poly(Lactic-co-Glycolic Acid)) NPs PLGA NPs' special qualities and benefits make them an attractive therapeutic option for RA. The body naturally breaks down into lactic acid and glycolic acid, which are biodegradable and biocompatible. Over time, PLGA NPs can sustain a consistent therapeutic effect by delivering regulated and prolonged release of therapeutic ingredients (Ma et al. 2023). The effectiveness of medicinal medicines can be increased by engineering NPs to target particular tissues, including inflamed joints (Higaki et al. 2005). PLGA can increase its stability and shelf life by shielding medications from deterioration (Ma et al. 2023). Glucocorticoids and nonsteroidal antiinflammatory medications (NSAIDs) can be administered directly to inflamed joints using PLGA NPs (Gandhi and Shende 2024). DMARDs' ability of DMARDs to reduce inflammation and prevent joint deterioration can be improved by targeted administration. PLGA NPs can enhance the effectiveness and distribution of biological medicines used to treat RA (Gandhi and Shende 2024). Gold and methotrexate were combined to create anti-CD64 antibody-conjugated PLGA NPs, which were intended to target inflamed joints (Gandhi and Shende 2024). Compared to nonconjugated NPs and free medicines, in vivo investigations have demonstrated improved clinical indices and arthritis ratings (Gandhi and Shende 2024). Drugs such as rapamycin have been investigated for intra-articular administration of PLGA NPs, which have demonstrated encouraging outcomes in lowering synovial inflammation and joint cartilage degradation (Ma et al. 2023). PLGA NPs are a viable strategy for improving RA treatment by boosting drug delivery, lowering side effects, and producing long-lasting therapeutic effects.

Chitosan NPs Chitosan NPs' special qualities and benefits make them a possible therapeutic option for RA. As chitosan is biocompatible and biodegradable, it can be safely used. The effectiveness of medicinal medicines can be increased by designing chitosan nanoparticles to target particular tissues, such as inflamed joints. Chitosan can prolong the stability and shelf life of medications by preventing degradation. Chitosan has inherent antiinflammatory properties that can be beneficial for treating RA (Gravandi et al. 2024). Chitosan NPs can encapsulate and deliver antiinflammatory drugs, DMARDs, and biological agents directly to inflamed joints (Saleem et al. 2022). In combination therapy, these NPs can be used to deliver multiple drugs with different mechanisms of action. It has been demonstrated that rutin-loaded chitosan nanoparticles can reduce RA symptoms by adjusting inflammatory and oxidative stress markers (Gravandi et al. 2024). In animal models, it has been demonstrated that methotrexate-loaded chitosan NPs reduce proinflammatory cytokines to produce anti-arthritic effects (Saleem et al. 2022). Zinc gluconate-loaded chitosan nanoparticles reduced joint swelling, erythema, and edema (Ansari et al. 2019). By boosting medication distribution, reducing side effects, and producing long-lasting therapeutic effects, chitosan nanoparticles are a viable strategy for improving RA therapy.

12.6.3 Dendrimers

Because of their distinct structures and characteristics, dendrimers are a type of nanotechnology that has shown promise in the treatment of RA. Over time, a constant therapeutic level can be maintained via the controlled and prolonged release of therapeutic substances that dendrimers can provide. The effectiveness of therapeutic drugs can be increased by designing them to target particular tissues, such as inflamed joints. Owing to their high drug-loading capacity, dendrimers can deliver more therapeutic agents. Dendrimers are biocompatible and can be safely degraded in the body. Dendrimers can encapsulate and deliver antiinflammatory drugs, DMARDs, and biological agents directly to the inflamed joints. These NPs can be used for combination therapy to deliver multiple drugs with different mechanisms of action. Studies have shown that peptide dendrimers grafted with drugs such as methotrexate (MTX) can enhance drug delivery and efficacy, reduce inflammation, and protect cartilage (Li et al. 2024b). To treat RA, researchers have created hierarchical microcarriers containing methotrexate grafted from peptide dendrimers. These devices can improve therapeutic results by releasing medication in the acidic environment of inflamed joints (Li et al. 2024b). Dendrimers can reduce inflammation and cartilage degradation by modifying macrophage polarization. Through improved drug distribution, fewer side effects, and longer-lasting therapeutic effects, dendrimers represent a viable strategy for improving RA treatment.

Polyamidoamine (PAMAM) Dendrimers: Used in drug delivery and gene therapy because of their highly branched structures. Polyamidoamine (PAMAM) dendrimers are dendrimers with unique properties that make them suitable for the treatment of RA. Therapeutic substances can be released gradually and under control using PAMAM dendrimers, sustaining a constant therapeutic level over time. To increase the effectiveness of therapeutic drugs, they might be used to target particular tissues, including inflamed joints. Owing to their high drug-loading capacity, PAMAM dendrimers can deliver more therapeutic agents. The PAMAM dendrimers are biocompatible and can be safely degraded in the body. PAMAM dendrimers can encapsulate and deliver antiinflammatory drugs, DMARDs, and biological agents directly to the inflamed joints. These dendrimers can be used for combination therapy to deliver multiple drugs with different mechanisms of action. Studies have shown that peptide dendrimers grafted with drugs such as methotrexate (MTX) can enhance drug delivery and efficacy, reduce inflammation, and protect the cartilage. Cathepsin-B-cleavable oligopeptide-modified PAMAM dendrimers have shown promise in improving medication administration and effectiveness (Lee et al. 2017). By improving drug distribution, lowering side effects, and producing long-lasting therapeutic effects, PAMAM dendrimers represent a viable strategy for improving RA therapy.

12.6.4 Gelatin NPs

The special qualities and benefits of gelatin NPs make them a potential therapeutic option for RA. Gelatin is safe for the body to use because it is both biocompatible and biodegradable. The effectiveness of medicinal drugs can be increased by tailoring gelatin nanoparticles to target particular tissues, such as inflamed joints. Gelatin can prolong the stability and shelf life of medications by preventing their breakdown. Because of its natural antiinflammatory properties, gelatin may help cure RA. Gelatin nanoparticles have the ability to encapsulate and distribute biological medicines, DMARDs, and antiinflammatory medications to inflamed joints. By delivering several medications with various modes of action, NPs can be utilized in combination therapy. Research has demonstrated that tragacanth gum-gelatin composite nanocapsules modified with chondroitin sulfate and loaded with curcumin nanocrystals can dramatically reduce oxidative stress and inflammatory factor expression (Sun et al. 2024). In experimental inflammatory arthritis, it has been demonstrated that aescin-comprised zein-coated gelatin nanoparticles reduce synovial inflammation (Jori et al. 2024). These NPs showed excellent cytocompatibility and markedly decreased edema, erythema, and joint swelling. Gelatin nanoparticles can alter macrophage polarization, which lowers inflammation and cartilage degradation. By boosting medication delivery, reducing side effects, and producing long-lasting therapeutic effects, gelatin nanoparticles (NPs) are a viable strategy for improving RA therapy.

12.6.5 Metallic NPs

Because metallic NPs have special qualities and benefits, they are a prospective treatment option for RA. Metallic NPs can increase the effectiveness of medicinal drugs by targeting particular tissues, such as inflamed joints. The effectiveness of treatment can be increased by delivering therapeutic drugs to inflamed joints. RA can be treated and monitored concurrently, owing to the dual-use capabilities of metallic NPs (Li et al. 2022). Certain metallic nanoparticles, including gold NPs, are naturally antiinflammatory. Owing to their antiinflammatory and anti-angiogenic properties, gold nanoparticles (AuNPs) can disrupt the function of inflammatory cytokines such as IL-6 in the pathophysiology of RA (Zheng et al. 2021). Silver NPs can reduce inflammation and exhibit antibacterial properties. Owing to their magnetic properties, iron oxide (Fe3O4) NPs can be directed to particular areas by applying external magnetic fields. Metallic NPs are a viable strategy for improving RA therapy by boosting medication delivery, lowering side effects, and producing long-lasting therapeutic effects.

12.7 Conclusion

Nanotechnology provides novel approaches for the treatment of RA by taking advantage of the unique qualities of nanomaterials. The effectiveness of medicinal drugs can be increased, and negative effects can be decreased by engineering NPs to target particular tissues, such as inflamed joints. Over time, NPs can maintain the best therapeutic dose by delivering a regulated and prolonged release of medications. Drug solubility and bioavailability can be enhanced using nanomaterials, resulting in more efficient treatments. A holistic therapeutic approach is made possible by nanotechnology, which enables the codelivery of several medications with various modes of action. Certain NPs have dual therapeutic and diagnostic uses, enabling RA to be treatment and monitoring concurrently. Liposomes and SLNs are utilized because of their biocompatibility and capacity to encapsulate a variety of medications. Gelatin, chitosan, and PLGA NPs provide biodegradability and controlled release. Dendrimer-branched molecules can target particular cells or regions and can carry several therapeutic molecules. Targeted drug delivery is possible using gold and silver nanoparticles, which also have antiinflammatory properties. Advanced combination therapies that codeliver numerous medications with improved physicochemical qualities and targeted delivery capabilities have been made possible by emerging nanotechnology. To provide a comprehensive therapeutic strategy, NPs can be structured to incorporate various modalities, including imaging, sonodynamic therapy, and phototherapy. The goal of next-generation NPs for RA treatment is to improve medication administration, reduce adverse effects, and produce better therapeutic results. The codelivery of several medications with improved physicochemical characteristics, targeted delivery, and controlled release profiles is made possible by the emerging nanotechnology. This strategy can overcome the drawbacks of traditional combination therapy, namely its poor selectivity for arthritic joints and the different pharmacokinetics of the coupled medications. Advanced nanomedicines offer a holistic treatment approach by combining several modalities, including imaging, phototherapy, and sonodynamic therapy. These multipurpose nanoplatforms can simultaneously treat RA and monitor their effectiveness at the same time. To treat RA, researchers have created hierarchical microcarriers containing methotrexate grafted from peptide dendrimers. These systems can improve treatment results by releasing the medication into the acidic environment of inflamed joints. For targeted drug release at the site of inflammation, NPs can be engineered to react to specific stimuli such as pH, temperature, or enzymes. With their more potent, precise, and secure therapeutic alternatives, these next-generation NPs represent important breakthroughs in the treatment of RA. An important development in the treatment of RA is nanotechnology, which provides safer, more targeted, and effective therapeutic alternatives.

Conflicts of Interest The authors declare that they have no competing interests.

References

Ahmad A, Ansari MM, AlAsmari AF, Ali N, Maqbool MT, Raza SS, Khan R (2021) Dose dependent safety implications and acute intravenous toxicity of aminocellulose-grafted-polycaprolactone coated gelatin nanoparticles in mice. Int J Biol Macromol 192:1150–1159

Albuquerque J, Moura CC, Sarmento B, Reis S (2015) Solid lipid nanoparticles: a potential multifunctional approach towards rheumatoid arthritis Theranostics. Molecules 20:11103–11118

Aletaha D, Smolen JS (2018) Diagnosis and Management of Rheumatoid Arthritis: a review. JAMA 320:1360–1372

Alldred A, Emery P (2001) Leflunomide: a novel DMARD for the treatment of rheumatoid arthritis. Expert Opin Pharmacother 2:125–137

Ansari MM, Ahmad A, Mishra RK, Raza SS, Khan R (2019) Zinc gluconate-loaded chitosan nanoparticles reduce severity of collagen-induced arthritis in Wistar rats. ACS Biomater Sci Eng 5:3380–3397

Ansari MM, Ahmad A, Kumar A, Alam P, Khan TH, Jayamurugan G, Raza SS, Khan R (2021) Aminocellulose-grafted-polycaprolactone coated gelatin nanoparticles alleviate inflammation in rheumatoid arthritis: a combinational therapeutic approach. Carbohydr Polym 258:117600

Ben Mrid R, Bouchmaa N, Ainani H, El Fatimy R, Malka G, Mazini L (2022) Anti-rheumatoid drugs advancements: new insights into the molecular treatment of rheumatoid arthritis. Biomed Pharmacother 151:113126

Burmester GR, Pope JE (2017) Novel treatment strategies in rheumatoid arthritis. Lancet 389:2338–2348

Chen D, Liu H, Wang Y, Chen S, Liu J, Li W, Dou H, Hou W, Meng M (2019) Study of the adoptive immunotherapy on rheumatoid arthritis with Thymus-derived invariant natural killer T cells. Int Immunopharmacol 67:427–440

Cross M, Smith E, Hoy D, Carmona L, Wolfe F, Vos T, Williams B, Gabriel S, Lassere M, Johns N, Buchbinder R, Woolf A, March L (2014) The global burden of rheumatoid arthritis: estimates from the global burden of disease 2010 study. Ann Rheum Dis 73:1316–1322

Dass S, Vital EM, Emery P (2006) Rituximab: novel B-cell depletion therapy for the treatment of rheumatoid arthritis. Expert Opin Pharmacother 7:2559–2570

Egwu CO, Aloke C, Onwe KT, Umoke CI, Nwafor J, Eyo RA, Chukwu JA, Ufebe GO, Ladokun J, Audu DT, Agwu AO, Obasi DC, Okoro CO (2024) Nanomaterials in drug delivery: strengths and opportunities in medicine. Molecules 29:2584

Elshabrawy HA, Chen Z, Volin MV, Ravella S, Virupannavar S, Shahrara S (2015) The pathogenic role of angiogenesis in rheumatoid arthritis. Angiogenesis 18:433–448

Favalli EG (2020) Understanding the role of Interleukin-6 (IL-6) in the joint and beyond: a comprehensive review of IL-6 inhibition for the Management of Rheumatoid Arthritis. Rheumatol Ther 7:473–516

Fraenkel L, Bathon JM, England BR, St Clair EW, Arayssi T, Carandang K, Deane KD, Genovese M, Huston KK, Kerr G, Kremer J, Nakamura MC, Russell LA, Singh JA, Smith BJ, Sparks JA, Venkatachalam S, Weinblatt ME, Al-Gibbawi M, Baker JF, Barbour KE, Barton JL, Cappelli L, Chamseddine F, George M, Johnson SR, Kahale L, Karam BS, Khamis AM, Navarro-Millán I, Mirza R, Schwab P, Singh N, Turgunbaev M, Turner AS, Yaacoub S, Akl EA (2021) 2021 American College of Rheumatology Guideline for the Treatment of Rheumatoid Arthritis. Arthritis Care Res (Hoboken) 73:924–939

Gandhi S, Shende P (2024) Anti-CD64 antibody-conjugated PLGA nanoparticles containing methotrexate and gold for Theranostics application in rheumatoid arthritis. AAPS PharmSciTech 25:22

Gibofsky A (2014) Epidemiology, pathophysiology, and diagnosis of rheumatoid arthritis: a synopsis. Am J Manag Care 20:S128–S135

Gravandi MM, Pourmanouchehri Z, Behbood L, Fakhri S, Mohammadi-Noori E, Zhaleh M, Shirvani S, Kiani A, Farzaei MH (2024) Rutin-loaded chitosan nanoparticles alleviated

Freund's adjuvant induced rheumatoid arthritis via modulating oxidative stress and inflammatory parameters in Wistar rats. Naunyn Schmiedeberg's Arch Pharmacol 397:4771–4790

Guo Q, Wang Y, Xu D, Nossent J, Pavlos NJ, Xu J (2018) Rheumatoid arthritis: pathological mechanisms and modern pharmacologic therapies. Bone Res 6:15

Harel-Meir M, Sherer Y, Shoenfeld Y (2007) Tobacco smoking and autoimmune rheumatic diseases. Nat Clin Pract Rheumatol 3:707–715

Harrington R, Harkins P, Conway R (2023) Janus kinase inhibitors in rheumatoid arthritis: an update on the efficacy and safety of tofacitinib, baricitinib and upadacitinib. J Clin Med 12:6690

Hauw WWS, Chia JSJ, Nandurkar HH, Sashindranath M (2022) The potential role of protease systems in hemophilic arthropathy. Blood Adv 6:5505–5515

Higaki M, Ishihara T, Izumo N, Takatsu M, Mizushima Y (2005) Treatment of experimental arthritis with poly(D, L-lactic/glycolic acid) nanoparticles encapsulating betamethasone sodium phosphate. Ann Rheum Dis 64:1132–1136

Hu L, Luo X, Zhou S, Zhu J, Xiao M, Li C, Zheng H, Qiu Q, Lai C, Liu X, Deng Y, Song Y (2019) Neutrophil-mediated delivery of dexamethasone palmitate-loaded liposomes decorated with a sialic acid conjugate for rheumatoid arthritis treatment. Pharm Res 36:97

Hua C, Buttgereit F, Combe B (2020) Glucocorticoids in rheumatoid arthritis: current status and future studies. RMD Open 6:e000536

Iannone F, Lapadula G (2012) The inhibitor of costimulation of T cells: abatacept. J Rheumatol Suppl 89:100–102

Imas JJ, Ruiz Zamarreño C, Zubiate P, Sanchez-Martín L, Campión J, Matías IR (2020) Optical biosensors for the detection of rheumatoid arthritis (RA) biomarkers: a comprehensive review. Sensors (Basel) 20:6289

Jori C, Ansari MM, Ahmad A, Ali N, Raza SS, Khan R (2024) Biomaterial-based combinatorial approach of aescin-comprised zein-coated gelatin nanoparticles alleviates synovial inflammation in experimental inflammatory arthritis. Nanoscale 16:7965–7975

Kim M, Shin M, Zhao Y, Ghosh M, Son Y-O (2024) Transformative impact of Nanocarrier-mediated drug delivery: overcoming biological barriers and expanding therapeutic horizons. Small Sci 4:2400280

Kumar LD, Karthik R, Gayathri N, Sivasudha T (2016) Advancement in contemporary diagnostic and therapeutic approaches for rheumatoid arthritis. Biomed Pharmacother 79:52–61

Leah E (2011) Experimental arthritis: dendrimer drug mends monocytes. Nat Rev Rheumatol 7:376

Lee S, Son SJ, Song SJ, Ha TH, Choi JS (2017) Polyamidoamine (PAMAM) dendrimers modified with Cathepsin-B cleavable oligopeptides for enhanced gene delivery. Polymers (Basel) 9:224

Li S, Su J, Cai W, Liu JX (2021) Nanomaterials manipulate macrophages for rheumatoid arthritis treatment. Front Pharmacol 12:699245

Li Y, Liang Q, Zhou L, Liu J, Liu Y (2022) Metal nanoparticles: a platform integrating diagnosis and therapy for rheumatoid arthritis. J Nanopart Res 24:84

Li P, Wang C, Huo H, Xu C, Sun H, Wang X, Wang L, Li L (2024a) Prodrug-based nanomedicines for rheumatoid arthritis. Discov Nano 19:9

Li Y, Zhu H, Liu R, Zhao Y, Sun L (2024b) Hierarchical microcarriers loaded with peptide dendrimer-grafted methotrexate for rheumatoid arthritis treatment. Small Sci 4:2300097

Lindell DM, Lukacs NW (2010) Cytokines and chemokines in inflammation. In: Serhan CN, Ward PA, Gilroy DW (eds) Fundamentals of inflammation. Cambridge University Press, Cambridge, pp 175–185

Lis K, Kuzawińska O, Bałkowiec-Iskra E (2014) Tumor necrosis factor inhibitors – state of knowledge. Arch Med Sci 10:1175–1185

Liu S, Maeyama K (2016) Gene therapy for rheumatoid arthritis. Crit Rev Immunol 36:149–161

Ma JC, Luo T, Feng B, Huang Z, Zhang Y, Huang H, Yang X, Wen J, Bai X, Cui ZK (2023) Exploring the translational potential of PLGA nanoparticles for intra-articular rapamycin delivery in osteoarthritis therapy. J Nanobiotechnol 21:361

McBride DA, Kerr MD, Johnson WT, Nguyen A, Zoccheddu M, Yao M, Prideaux EB, Dorn NC, Wang W, Svensson MND, Bottini N, Shah NJ (2023) Immunomodulatory microparticles

epigenetically modulate T cells and systemically ameliorate autoimmune arthritis. Adv Sci (Weinh) 10:e2202720
Mueller AL, Payandeh Z, Mohammadkhani N, Mubarak SMH, Zakeri A, Alagheband Bahrami A, Brockmueller A, Shakibaei M (2021) Recent advances in understanding the pathogenesis of rheumatoid arthritis: new treatment strategies. Cells 101:3017
Nasra S, Bhatia D, Kumar A (2022) Recent advances in nanoparticle-based drug delivery systems for rheumatoid arthritis treatment. Nanoscale Adv 4:3479–3494
Nazir AM, Koganti B, Gupta K, Memon MS, Aslam Zahid MB, Shantha Kumar V, Tappiti M, Mostafa JA (2021) Evaluating the use of hydroxychloroquine in treating patients with rheumatoid arthritis. Cureus 13:e19308
Nishimoto N (2006) Interleukin-6 in rheumatoid arthritis. Curr Opin Rheumatol 18:277–281
Orange DE, Blachere NE, DiCarlo EF, Mirza S, Pannellini T, Jiang CS, Frank MO, Parveen S, Figgie MP, Gravallese EM, Bykerk VP, Orbai AM, Mackie SL, Goodman SM (2020) Rheumatoid arthritis morning stiffness is associated with synovial fibrin and neutrophils. Arthritis Rheumatol 72:557–564
Östör AJK (2008) Abatacept: a T-cell co-stimulation modulator for the treatment of rheumatoid arthritis. Clin Rheumatol 27:1343–1353
Paleolog EM (2002) Angiogenesis in rheumatoid arthritis. Arthritis Res Ther 4:S81
Pandolfi F, Franza L, Carusi V, Altamura S, Andriollo G, Nucera E (2020) Interleukin-6 in rheumatoid arthritis. Int J Mol Sci 21:5238
Patra JK, Das G, Fraceto LF, Campos EVR, Rodriguez-Torres MDP, Acosta-Torres LS, Diaz-Torres LA, Grillo R, Swamy MK, Sharma S, Habtemariam S, Shin HS (2018) Nano based drug delivery systems: recent developments and future prospects. J Nanobiotechnol 16:71
Plosker GL, Croom KF (2005) Sulfasalazine: a review of its use in the management of rheumatoid arthritis. Drugs 65:1825–1849
Qamar N, Arif A, Bhatti A, John P (2019) Nanomedicine: an emerging era of theranostics and therapeutics for rheumatoid arthritis. Rheumatology (Oxford) 58:1715–1721
Ren S, Xu Y, Dong X, Mu Q, Chen X, Yu Y, Su G (2024) Nanotechnology-empowered combination therapy for rheumatoid arthritis: principles, strategies, and challenges. J Nanobiotechnol 22:431
Rengel Y, Ospelt C, Gay S (2007) Proteinases in the joint: clinical relevance of proteinases in joint destruction. Arthritis Res Ther 9:221
Saleem MU, Muhammad F, Sharif A, Arshad MI, Akhtar K, Javed Y, Akhtar B (2022) Methotrexate-loaded biodegradable nanoparticles exert anti-arthritic effect by downregulating pro-inflammatory cytokines in Freund's complete adjuvant-induced arthritic rats. Inflammopharmacology 30:1079–1091
Sanchez-Castillo JA, Reinoso-Cobo A, Gijon-Nogueron G, Caliz-Caliz R, Exposito-Ruiz M, Ramos-Petersen L, Ortega-Avila AB (2021) Symmetry criterion for patients with rheumatoid arthritis of the foot: a Cross-sectional study. Int J Environ Res Public Health 18:3619
Sharif M, John P, Bhatti A, Paracha RZ, Majeed A (2024) Evaluation of the inhibitory mechanism of Pennisetum glaucum (pearl millet) bioactive compounds for rheumatoid arthritis: an in vitro and computational approach. Front Pharmacol 15:1488790
Shen Q, Shu H, Xu X, Shu G, Du Y, Ying X (2020) Tofacitinib citrate-based liposomes for effective treatment of rheumatoid arthritis. Pharmazie 75:131–135
Shen X, Liu X, Guo X, Hou X, Huang H, Feng Z (2024) Systematic review of Janus kinases inhibitors for rheumatoid arthritis: methodology, reporting, and quality of evidence evaluation. Front Pharmacol 15:1459511
Sparks JA (2019) Rheumatoid arthritis. Ann Intern Med 170:Itc1-itc16
Steffen U, Schett G, Bozec A (2019) How autoantibodies regulate osteoclast induced bone loss in rheumatoid arthritis. Front Immunol 10:1483
Sun J, Du J, Liu X, An J, Hu Y, Wang J, Zhu F, Feng H, Cheng S, Tian H, Mei X, Wu C (2024) Chondroitin sulfate-modified tragacanth gum-gelatin composite nanocapsules loaded with curcumin nanocrystals for the treatment of arthritis. J Nanobiotechnol 22:270

Tai WY, Chen TL, Wang HM, Fu LM (2024) Rapid Microfluidic biosensor for point-of-care determination of rheumatoid arthritis via anti-cyclic citrullinated peptide antibody detection. Biosensors (Basel) 14:545

Takeuchi Y, Hirota K, Sakaguchi S (2019) Synovial tissue inflammation mediated by autoimmune T cells. Front Immunol 10:1989

Tanaka Y (2021) Recent progress in treatments of rheumatoid arthritis: an overview of developments in biologics and small molecules, and remaining unmet needs. Rheumatology (Oxford) 60:vi12-vi20

Turner MD, Nedjai B, Hurst T, Pennington DJ (2014) Cytokines and chemokines: at the crossroads of cell signalling and inflammatory disease. Biochim Biophys Acta 1843:2563–2582

Venkatesan R, Xiong H, Yao Y, Reddy Nakkala J, Zhou T, Li S, Fan C, Gao C (2022) Immunomodulating theranostic gold nanocages for the treatment of rheumatoid arthritis in vivo. Chem Eng J 446:136868

Vivier E, Malissen B (2005) Innate and adaptive immunity: specificities and signaling hierarchies revisited. Nat Immunol 6:17–21

Vojdani A, Vojdani E, Vojdani C (2020) The immune system: our body's homeland security against disease. In: Noland D, Drisko JA, Wagner L (eds) Integrative and functional medical nutrition therapy: principles and practices. Springer International Publishing, Cham, pp 285–302

Wang Q, He L, Fan D, Liang W, Fang J (2020) Improving the anti-inflammatory efficacy of dexamethasone in the treatment of rheumatoid arthritis with polymerized stealth liposomes as a delivery vehicle. J Mater Chem B 8:1841–1851

Xie Y, Tuguntaev RG, Mao C, Chen H, Tao Y, Wang S, Yang B, Guo W (2020) Stimuli-responsive polymeric nanomaterials for rheumatoid arthritis therapy. Biophys Rep 6:193–210

Yamamoto K, Okada Y (2019) Shared genetic factors and their causality in autoimmune diseases. Ann Rheum Dis 78:1449–1451

Yang Y, Guo L, Wang Z, Liu P, Liu X, Ding J, Zhou W (2021) Targeted silver nanoparticles for rheumatoid arthritis therapy via macrophage apoptosis and Re-polarization. Biomaterials 264:120390

Zewail MB, Doghish AS, El-Husseiny HM, Mady EA, Mohammed OA, Elbadry AMM, Elbokhomy AS, Bhnsawy A, El-Dakroury WA (2024) Lipid-based nanocarriers: an attractive approach for rheumatoid arthritis management. Biomater Sci 12:6163–6195

Zhao J, Chen X, Ho K-H, Cai C, Li C-W, Yang M, Yi C (2021) Nanotechnology for diagnosis and therapy of rheumatoid arthritis: evolution towards theranostic approaches. Chin Chem Lett 32:66–86

Zheng M, Jia H, Wang H, Liu L, He Z, Zhang Z, Yang W, Gao L, Gao X, Gao F (2021) Application of nanomaterials in the treatment of rheumatoid arthritis. RSC Adv 11:7129–7137

Zheng H, Liu Y, Deng Y, Li Y, Liu S, Yang Y, Qiu Y, Li B, Sheng W, Liu J, Peng C, Wang W, Yu H (2024) Recent advances of NFATc1 in rheumatoid arthritis-related bone destruction: mechanisms and potential therapeutic targets. Mol Med 30:20

GPSR Compliance

The European Union's (EU) General Product Safety Regulation (GPSR) is a set of rules that requires consumer products to be safe and our obligations to ensure this.

If you have any concerns about our products, you can contact us on ProductSafety@springernature.com

In case Publisher is established outside the EU, the EU authorized representative is:

Springer Nature Customer Service Center GmbH
Europaplatz 3
69115 Heidelberg, Germany

Batch number: 08977999

Printed by Printforce, the Netherlands